Biomechanics V-B

International Series on Biomechanics

This series publishes the major biennial symposia on biomechanics, beginning with the Fifth International Congress of Biomechanics held at the University of Jyväskylä, and related seminars and congresses sponsored by the International Society on Biomechanics. Future volumes may also include monographs on special topics in the field of biomechanics, as determined by the International Society on Biomechanics.

Published:

Volume 1A: Komi-BIOMECHANICS V-A (Fifth International Congress of Biomechanics: sections on neuromuscular control, electromyography, rehabilitation, and ergonomics)

Volume 1B: Komi-BIOMECHANICS V-B (Fifth International Congress of Biomechanics: sections on fundamental movements, biomechanics of sport, and instrumentation and methodology)

International Series
on Biomechanics, Volume 1B

BIOMECHANICS V-B

Proceedings of the Fifth International Congress
of Biomechanics, Jyvaskyla, Finland

Edited by **Paavo V. Komi, Ph.D.**
Kinesiology Laboratory
Department of Biology of Physical Activity
University of Jyväskylä

UNIVERSITY PARK PRESS
Baltimore • London • Tokyo

UNIVERSITY PARK PRESS
International Publishers in Science and Medicine
Chamber of Commerce Building
Baltimore, Maryland 21202

Typset by Service Composition Co.

Manufactured in the United States of America by
Universal Lithographers, Inc., and The Maple Press Co.

Library of Congress Cataloging in Publication Data

International Congress of Biomechanics, 5th, Jyväskylä,
Finland, 1975.
Biomechanics V.

(International series on biomechanics; v. 1A-1B)
1. Human mechanics—Congresses. I. Komi, Paavo V.
II. Title. III. Series.
QP303.I5 1975 612'.76 76-15635
ISBN 0-8391-0946-6 (v. 1A)

ISBN 0-8391-0947-4 (v. 1B)

Contents

Track and Field

Water Sports

Winter Sports

INSTRUMENTATION AND METHODOLOGY

Introductory Paper

Instrumentation

Methods

The Fifth International Congress of Biomechanics

Contributors

Contributors to both volumes, Biomechanics V-A *and* Biomechanics V-B, *are listed. The volume and page number of each contribution are given in parentheses after the contributor's name.*

Ačimović, Ruža (A:444), Department for Hemiplegics, Rehabilitation Institute Ljubljana, Linhartova 51, 61000 Ljubljana, Yugoslavia

Adachi, Nagahiko (B:135), Department of Physical Education, College of General Education, University of Tokyo, 3-8-1, Komaba Meguroku, Tokyo, Japan

Adrian, Marlene (B:161), Department of Physical Education for Women, Washington State University, Pullman, Washington 99163, USA

Albisser, A. M. (A:474), Division of Biomedical Research, The Hospital for Sick Children, 555 University Avenue, Toronto M5G 1X8, Canada

Aleshinsky, S. (B:387), Department of Biomechanics, Central Institute of Physical Culture, Sirenevij Bulvar 4, Moscow, USSR

Amstutz, Harlan C. (A:394), Division of Orthopedic Surgery, UCLA Medical School, Le Conte Avenue, Los Angeles, California 90024, USA

Andersson, B. J. G. (A:520), Department of Orthopaedic Surgery I, University of Göteborg, Sahlgren Hospital, S-413 45 Göteborg, Sweden

Antti, Carl-Johan (A:515), Unit of Work Physiology, Department of Occupational Medicine, Umeå Hospital, S-901 85 Umeå, Sweden

Arcan, Mircea (B:415), School of Engineering, Tel-Aviv University, Ramat Aviv, Tel-Aviv 69978, Israel

Ariel, Gideon B. (B:361), Biomechanics Laboratory, Center of System Neurosciences, University of Massachusetts, C.B.A., Inc., 316 College Street, Amherst, Massachusetts 01002, USA

Arlien-Søborg, Peter (A:171, 177), Department of Neurology, Rigshospitalet, Blegdamsvej 9, DK-2100-Kobenhavn Ø, Denmark

Asami, Toshio (B:94, 135), Department of Physical Education, College of General Education, University of Tokyo, 3-8-1, Komaba Meguroku, Tokyo, Japan

Asmussen, Erling (A:23), University of Copenhagen, August Krogh Institute, Laboratory for the Theory of Gymnastics, 13 Universitetsparken, 2100 Copenhagen Ø, Denmark

Atomi, Yoriko (A:240), Laboratory for Exercise Physiology and Biomechanics, Faculty of Education, University of Tokyo, Hongo 7-3-1, Bunkyo-ku, Tokyo, Japan

Bajd, Tadej (A:444), Byokibernetic Depertement, Jožef Stefan Institute, University of Ljubljana, Jamova 39, 61000 Ljubljana, Yugoslavia

Ballreich, R. (B:208), Institut für Sport und Sportwissenschaft, der Johann Wolfgang Goethe-Universität, 6000 Frankfurt/Main, Ginnheimer Landstr. 39, West Germany

Balsevich, V. (B:141), Scientific Research Laboratory of Age Sports Pedagogics, Omsk State Institute of Physical Culture, Maslennikova 144, Omsk 644063, USSR

Barker, Phillip (B:314), Department of Physical Education, Walsall Technical College, Walsall, England

Basmajian, John V. (A:297), Department of Anatomy, Emory University School of Medicine, Atlanta, Georgia 30322, USA

Baumann, Wolfgang (A:328; B:194), Fachrichtung Sportwissenschaft, Westf. Wilhelms-Universität, D-4400 Münster, Horstmarer Landweg 62 b, West Germany

Becker, Theodore (A:452), Biomechanics Laboratory, School of Health, Physical Education and Recreation, Indiana University, Bloomington, Indiana 47401, USA

Belokovsky, V. V. (B:235), Department of Swimming, The Central Institute of Physical Culture, Skatertnyi Pereulok 4, Moscow 69, USSR

Bishop Patrick J. (B:299), Faculty of Human Kinetics and Leisure Studies, Department of Kinesiology, University of Waterloo, Waterloo, Ontario, Canada

Björksten, Marianne (A:515), Unit of Work Physiology, Department of Occupational Medicine, Umeå Hospital, S-901 85 Umeå, Sweden

Bober, Tadeusz (B:52), Biomechanics Laboratory, Academy of Physical Education, ul. Banacha 11, 51-617 Wrocław, Poland

Bodine-Reese, P. (A:96), 13104 Valleywood Drive, Wheaton, Maryland 20906, USA

Bone, John P. (A:96), Physical Education Department, University of Maryland, College Park, Maryland 20742, USA

Borms, Jan (B:309), Vrije Universiteit Brussel, HILO, Laboratory of Human Biometry and Biomechanics, A. Buyllaan, 105, B-1050 Brussel, Belgien

Bosco, Carmelo (B:174), Kinesiology Laboratory, Department of Biology of Physical Activity, University of Jyväskylä, SF 40100 Jyväskylä 10, Finland

Bouisset, S. (A:273), Laboratoire de Physiologie du Mouvement, Bâtiment 470, Université de Paris-Sud, 91405 Orsay, France

Boykin, William H. (A:233), Department of Engineering Science, Mechanics and Aerospace Engineering, University of Florida, Gainesville, Florida 32611, USA

Brandell, Bruce R. (A:319), Department of Anatomy, University of Saskatchewan, Saskatoon, Saskatchewan, Canada S7N OWO

Broman, Holger (A:165), Department of Applied Electronics, Chalmers University of Technology, Fack, 402 20 Göteborg, Sweden

Brudermann, Uwe (A:412), Abteilung für Biomedizinische Technik und Krankenhaustechnik, Medizinische Hochschule Hannover, 3000 Hannover 61, Karl-Wiechert-Allee 9, West Germany

Brull, Maurice (B:415), School of Engineering, Tel-Aviv University, Ramat Aviv, Tel-Aviv 69978, Israel

Brussatis, F. (B:34), Orthopädische Univ.-Klinik, Klinikum der Johannes Gutenberg-Universität, D-6550 Mainz, Langenbeckstrasse 1, West Germany

Butler, David L. (A:139), Department of Metallurgy, Mechanics and Materials Science, Michigan State University, East Lansing, Michigan 48824, USA

Cappozzo, A. (A:366), Instituto di Fisiologia Umana — Universitá degli Studi, P. le delle Scienze 5, 00185 Roma, Italy

Cavanagh, Peter R. (B:26, 399), Biomechanics Laboratory, The Pennsylvania State University, University Park, Pennsylvania 16802, USA

Clarys, J. P. (B:243), Instituut voor Morfologie, Vrije Universiteit Brussel, Eversstraat 2, 1000 Brussel, Belgium

Cooper, John M. (A:452), Indiana University, School of Health, Physical Education and Recreation, Graduate Division — Room 121, Bloomington, Indiana 47401, USA

Cordey, Jacques (A:373, 379), Laboratory for Experimental Surgery, Swiss Research Institute, Obere Strasse 22, CH-7270 Davos-Platz, Switzerland

Czeglédi, Peter (A:253), Institute for Co-ordination of Computer Techniques, Budapest V., Akadémia u. 17, Hungary

Dainis, Andrew (B:127), Biomechanics Laboratory, Department of Physical Education, University of Maryland, College Park, Maryland 20742, USA

Dal Monte, A. (B:258), C.O.N.I. Sports Medicine Institute, Via dei Campi Sportivi 46, 00197 Roma, Italy

Daniel, Klaus (B:250), Fachgebiet Schwimmsport, Deutsche Sporthochschule Köln, D-5000 Köln 41, Carl-Diem-Weg 5, West Germany

Dapena, Jesus (B:13), Biomechanics Laboratory, Department of Physical Education, University of Iowa, Iowa City, Iowa 52242, USA

Davies, C. T. M. (A:484), MRC Environmental Physiology Unit, London School of Hygiene and Tropical Medicine, Keppel Street, London WC1E 7HT, England

Desiprés, M. (B:73), Research Laboratory, Department of Physical Education, University of the Orange Free State, P.O. Box 339, Bloemfontein 9300, Republic of South Africa

Donskoi, D. (B:347), Department of Biomechanics, Central Institute of Physical Culture, 105483 Moscow, Sirenevyi Boulevar 4, USSR

Doré, Roland (B:277, 287), Département de Génie Mécanique, Ecole Polytechnique de Montréal, C.P. 6079, Succursale "A", Montreal H3C 3A7, Canada

Drager, Jörg (A:537), Opthalmic Hospital, St. Jurgens-Strasse, 28 Bremen, West Germany

Dupuis, Heinrich (A:537; B:34), Anthropotechnics Working Group, Max-Planck-Institute for Farm Work and Agricultural Engineering, 655 Bad Kreuznach, Am Kauzenberg, West Germany

East, David J. (B:65), School of Physical and Health Education, Hutchinson Hall DX-10, University of Washington, Seattle, Washington 98195, USA

Eickelberg, Warren (A:530), Adelphi University, INA MEND Laboratories, Human Resources Center, Albertson, New York 11507, USA

Einer-Jensen, Niels (A:152), Institute of Physiology, Odense University, DK-5000 Odense, Denmark

Eng, P. (A:394), Visiting Associate Professor of Surgery, University of California, Los Angeles, California 90024, USA

Erdmann, Włodzimierz (B:52), Biomechanics Laboratory, Higher School of Physical Education, ul. Wiejska 1, 80-336 Gdańsk, Poland

Ericson, Bengt-Eric (A:515), Unit of Work Physiology, Department of Occupational Medicine, Umeå Hospital, S-901 85 Umeå, Sweden

Fedina, T. I. (A:124), Department of Biological Problems of Sport, All-Union Research Institute of Physical Culture, 103064 Kazakova st. 18, Moscow, USSR

Fetchero, P. (A:280), Restorative Dentistry, School of Dentistry, UMAB, University of Maryland, Baltimore, Maryland 21201, USA

Fidelus, Kazimierz (B:351), Department of Biomechanics, Institute of Biological Sciences, Academy of Physical Education, 01-813 Warsaw, Marymoncka Street 34, Poland

Figura, F. (A:366), Instituto di Fisiologia Umana — Università degli Studi, P. le delle Scienze 5, 00185 Roma, Italy

Flatt, A. E. (A:355), Biomechanical Laboratory, Division of Hand Surgery, Department of Orthopaedics, University Hospitals, Iowa City, Iowa 52242, USA

Florin, P. (A:379), Laboratory for Experimental Surgery, Swiss Research Institute, CH-7270 Davos-Platz, Switzerland

Forsberg, Artur (A:112), Department of Physiology, Gymnastik-Koch Idrottshögskolan, Lidingövägen 1, 114 33 Stockholm, Sweden

Francis, Peter R. (B:456), Physical Education Department, Iowa State University, Ames, Iowa 50011, USA

Gabel, Ronald H. (A:437; B:456), Biomechanics Laboratory, Department of Orthopedics, University of Iowa, Iowa City, Iowa 52240, USA

Garrett, Gladys E. (B:371), Department of Physical Education for Women, Purdue University, West Lafayette, Indiana 47907, USA

Geret, V. (A:379), Laboratory for Experimental Surgery, Swiss Research Institute, CH-7270 Davos-Platz, Switzerland

Gibson, D. A. (A:474), Orthopaedic Surgery, The Hospital for Sick Children, 555 University Avenue, Toronto M5G 1X8, Canada

Goig, Juan-Roman (A:347), Centro de Rehabilitatión, Giudad Sanitaria de la Seguridad Social, Pº Valle Hebrón, s/n, Barcelona (16), Spain

Gomez-Pellico, L. (A:289), Departamento de Morfología, Iª Cátedra de Anatomía, Facultad de Medicina, Universidad Complutense, Madrid, Spain

Goto, Sayoko (A:246), Physical Activity Laboratory, Aichi Prefectural University, Takada-cho, Mizuho-ku, Nagoya, Japan

Greenisen, M. C. (A:199), Physical Education Department, McGill University, 475 Pine Avenue, Montreal, P.Q., Canada

Groh, Herbert (A:328), Institut für Biomechanik, Deutsche Sporthochschule Köln, D-5000 Köln 41, Carl-Diem-Weg 5, West Germany

Gros, Nuša (A:444), Department for Hemiplegics, Rehabilitation Institute Ljubljana, Linhartova 51, 61000 Ljubljana, Yugoslavia

Gross, D. (A:199), Meakins-Christie Laboratories, Experimental Medicine Department, McGill University, 3775 University Street, Montreal, P.Q., Canada

Gross, H. H. (B:347), Biomechanics Laboratory, Institute of Pedagogics, 200101 Tallin, Narva Road 41, USSR

Gutewort, Wolfgang (B:441), Laboratory of Biomechanics and Motorics, Department of Sportscience, Friedrich-Schiller-University of Jena, 69 Jena, Seidelstrasse 20, German Democratic Republic

Gydikov, Alexander A. (A:45, 65), Laboratory of Motor Control, Institute of Physiology, Bulgarian Academy of Sciences, Acad. G. Bonchev Str., Bl. 1, 1113 Sofia, Bulgaria

Häkkinen, Veikko K. (B:492), Laboratory of Clinical Neurophysiology, Central Hospital of Tampere, SF 33520 Tampere 52, Finland

Hamabuchi, M. (A.468), Address unavailable.

Hamanishi, C. (A:468), Address unavailable.

Hartung, Emil (A:537; B:34), Antropotechnics Working Group, Max-Planck-Institute for Farm Work and Agricultural Engineering, 655 Bad Kreuznach, Am Kauzenberg, West Germany

Hatze, Herbert (B:5), National Research Institute for Mathematical Sciences, CSIR, P.O. Box 395, Pretoria 0001, Republic of South Africa

Havu, Matti (A:118), Kinesiology Laboratory, Department of Biology of Physical Activity, University of Jyväskylä, SF 40100 Jyväskylä 10, Finland

Hay, James G. (B:13, 467), Biomechanics Laboratory, Department of Physical Education, University of Iowa, Iowa City, Iowa 52242, USA

Hayes, Keith C. (A:194), Department of Kinesiology, University of Waterloo, Waterloo, Ontario N2L 3G1, Canada

Hebbelinck, Marcel (B:309), Vrije Universiteit Brussel, HILO, Laboratory of Human Biometry Biomechanics, A. Buyllaan, 105, B-1050 Brussel, Belgien

Heinilä, Kalevi (A:13), Department of Planning for Physical Culture, University of Jyväskylä, SF 40100 Jyväskylä 10, Finland

Hennig, Ewald M. (B:433, 449), Institut für Sport und Sportwissenschaften, Universität Frankfurt, D-6000 Frankfurt, Ginnheimer Landstrasse 39, West Germany

Herberts, P. (A:520), Department of Orthopaedic Surgery I, University of Göteborg, Sahlgren Hospital, S-413 45 Göteborg, Sweden

Hiki, Takeo (A:88), Department of Orthopaedics, Nippon Medical School, 1-1-5, Sendagicho, Bunkyo-ku, Tokyo 113, Japan

Hoshikawa, Tamotsu (A:247; B:109, 200), Physical Activity Laboratory, Aichi Prefectural University, Takada-cho, Mizuho-ku, Nagoya, Japan

Ikegami, Yasuo (B:200), Research Center of Health, Physical Fitness and Sports, University of Nagoya, Furo-cho, Chikusa-ku, Nagoya-city, Japan

Ireland, D. C. R. (A:355), "Stanhill" 34 Queens Road, Melbourne, 3004, Australia

Isherwood, P. A. (A:406), Biomechanical Research and Development Unit, Roehampton Lane, London SW 15 5PR, England

Ishida, H. (A:303), Department of Physical Anthropology, Faculty of Science, Kyoto University, Kyoto, Japan

Ito, Tadaatsu (A:88), Department of Orthopaedics, Nippon Medical School, 1-1-5, Sendagicho, Bunkyo-ku, Tokyo 113, Japan

Jaakkola, Pekka (B:492), Research Institute for Bioengineering, Satamakatu 17 B, SF 33200 Tampere 20, Finland

Jensen, Robert K. (B:380), School of Physical and Health Education, Laurentian University, Sudbury, Ontario P3E 2C6, Canada

Jiskoot, J. (B:243), Akademie voor Lichamelijke Opvoeding, Willinklaan 5, Amsterdam, The Netherlands

Johnston, Richard C. (A:437), Biomechanics Laboratory, Department of Orthopedics, College of Medicine, University of Iowa, Iowa City, Iowa 52242, USA

Jokl, Ernst (A:15), University of Kentucky, Lexington, Kentucky 40506, USA

Jonsson, Bengt (A:261, 509, 515), Unit of Work Physiology, Department of Occupational Medicine, Umeå Hospital, S-901 85 Umeå, Sweden

Jonsson, Solveig (A:261), Unit of Work Physiology, Department of Occupational Medicine, Umeå Hospital, S-901 85 Umeå, Sweden

Jørgensen, Finn (A:152, 159), Institute of Physiology, Odense University, DK-5000 Odense, Denmark

Jørgensen, Kjell Ola (A:544), National Institute of Technology, Rogaland dep., Strandkaien 2, N-4000 Stavanger, Norway

Jørgensen, Kurt (A:145), Laboratory for the Theory of Gymnastics, August Krogh Institute, University of Copenhagen, Universitetsparken 13, DK-2100 Copenhagen Ø, Denmark

Kadefors, Roland (A:165), Laboratory of Clinical Neurophysiology, Sahlgren Hospital, Fack, 413 45 Göteborg, Sweden

Karlsson, Jan (A:112, 118), Department of Physiology, Gymnastik-Koch Idrottshögskolan, Lidingövägen 1, 114 33 Stockholm, Sweden

Kasahara, Yoshitaka (A:468), Department of Orthopedic Surgery, Kyoto University, School of Medicine, 53 Shogoin Kawara-cho, Sakyo-ku, Kyoto 606, Japan

Kaufmann, David A. (B:181), 19 Florida Gym, College of Physical Education, Health and Recreation, University of Florida, Gainesville, Florida 32611, USA

Kazai, Nobuyuki (A:311), Physical Education, Department of Literature, Bukkyo University, Murasakino, Kita-ku, Kyoto 603, Japan

Kelley, David L. (B:127), Biomechanics Laboratory, Department of Physical Education, University of Maryland, College Park, Maryland 20742, USA

Kemper, H. C. G. (B:243), Coronel Laboratorium en Laboratorium voor Psychofysiologie, Universiteit van Amsterdam, 1ste Constantijn Huygensstraat 20, Amsterdam, The Netherlands

Khalil, Tarek M. (A:233), Department of Industrial Engineering, School of Engineering and Environmental Design, University of Miami, P.O. Box 248294, Coral Gables, Florida 33124, USA

Kikuchi, Takemichi (B:135), Department of Health and Physical Education, College of Arts and Science, Chiba University, 1-33, Yayoi-cho, Chiba-shi, Japan

Kimura, T. (A:303), Department of Forensic Medicine, Faculty of Medicine, Teikyo University, Teiko, Japan

Kitagawa, Kaoru (B:135), Department of Physical Education, Faculty of Education, University of Tokyo, 7-3-1, Hongo Bunko-ku, Tokyo, Japan

Kitamura, Kiyokazu (B:271), Research Center of Health, Physical Fitness and Sports, University of Nagoya, Furo-cho, Chikusa-ku, Nagoya 464, Japan

Klauck, Jürgen (B:250), Institut für Biomechanik, Deutsche Sporthochschule Köln, D-500 Köln 41, Carl-Diem-Weg 5, West Germany

Kobayashi, Kando (B:121), Research Center of Health, Physical Fitness and Sports, University of Nagoya, Furo-cho, Chikusa-ku, Nagoya, Japan

Kobayashi, Shigeo (A:240), Laboratory for Exercise Physiology and Biomechanics, Faculty of Education, University of Tokyo, Hongo 7-3-1, Bunkyo-ku, Tokyo, Japan

Komi, Paavo V. (A:5, 118; B:174), Kinesiology Laboratory, Department of Biology of Physical Activity, University of Jyväskylä, SF 40100 Jyväskylä 10, Finland

Koreska, J. (A:474), Research Division, Department of Medical Engineering, Hospital for Sick Children, 555 University Avenue, Toronto M5G 1X8, Ontario, Canada

Kuhlow, A. (B:291), Institut für Sport und Sportwissenschaft, der Johann Wolfgang Goethe-Universität, 6000 Frankfurt/Main, Ginnheimer Landstrasse 39, West Germany

Kumamoto, Minayori, (A:311; B:41, 215, 222, 326), Laboratory of General Education, Kyoto University, Nihonmatsu-cho, Yoshida, Sakyo-ku, Kyoto 606, Japan

Kuznetsov, V. V. (B:235), Department of Swimming, The Central Institute of Physical Culture, Skatertnyi Pereulok 4, Moscow 69, USSR

Lagassé, Pierre P. (A:208), Département d'Education Physique, Université Laval, Quebec G1K 7P4, Canada

Lammert, Ole R. (A:152, 159), Institute for Physical Education, Odense University, DK-5000 Odense, Denmark

Laurig Wolfgang (A:219), Institut für Arbeitwissenschaft, Fachbereich Maschinenbau, Technische Hochschule Darmstadt, Petersenstrasse 18, 6100 Darmstadt, BRD

Leo, T. (A:366), CSSCCA-CNR — Instituto di Automatica, Università degli Studi, Via Eudossiana, 18, 00184 Roma, Italy

Leonardi, L. M. (B:258), C.O.N.I. Sports Medicine Instiute, Via dei Campi Sportivi 46, 00197 Roma, Italy

Less, Menahem (A:530), Adelphi University, INA MEND Laboratories, Human Resources Center, Albertson, New York, 11507, USA

Lestienne, F. (A:273), Laboratoire de Physiologie du Mouvement, Bâtiment 470, Université de Paris-Sud, France

Lewillie, Léon (B:230), Unité de Recherche de Biomécanique du Mouvement, Laboratoire de l'Effort — I.S.E.P.K., Université Libre de Bruxelles, Avenue Paul Héger 28 — C.P. 168, B-1050 Bruxelles, Belgium

Liemohn, Wendell P. (A:452), Indiana University, Developmental Training Center, 2853 East Tenth Street, Bloomington, Indiana 47401, USA

Little, Robert W. (A:139), Department of Mechanical Engineering, Michigan State University, East Lansing, Michigan 48824, USA

Llanos-Alcazar, L. F. (A:289), Departamento de Morfología, Iª Cátedra de Anatomía, Facultad de Medicina, Universidad Complutense, Madrid, Spain

Lohberger, Volker (A:412), Abteilung Informatik, Universität Dortmund, 4600 Dortmund 50, August-Schmidt-Strasse, West Germany

Louhivaara, Ilppo-Simo (A:11), Department of Mathematics, University of Jyväskylä, SF 40100 Jyväskylä 10, Finland

Luhtanen, Pekka (B:174), Kinesiology Laboratory, Department of Biology of Physical Activity, University of Jyväskylä, SF 40100 Jyväskylä 10, Finland

Mai, Jesper (A:171, 177), Neurological Laboratory, Department of Neurology, Århus Kommunehospital, DK-8000 Århus C, Denmark

Maldonado, Fernando (A:347), Centro de Rehabilitación, Ciudad Sanitaria de la Seguridad Social, Pº Valle Hebrón, s/n, Barcelona, Spain

Mann, Ralph (B:161), Department of Physical Education for Men, Washington State University, Pullman, Washington 99163, USA

Marchetti, M. (A:366), Instituto di Fisiologia Umana — Universitá degli Studi, P. le delle Scienze, 5, 00185 Roma, Italy

Martinez, Arnaldo G. (A:233), International Paper Company, 3600 Michael Boulevard No. 114, Mobile, Alabama 36609, USA

Matake, Tomokazu (B:426), Department of Mechanical Engineering, Nagasaki University, 1-14, Bunkyo-machi, Nagasaki 852, Japan

Maton, B. (A:273), Laboratorie de Physiologie du Travail du C.N.R.S., 91, Boulevard de l'Hospital, 75013 Paris, France

Matsui, Hideji (B:121, 200, 271), Research Center of Health, Physical Fitness and Sports, University of Nagoya, Furo-cho, Chikusa-ku, Nagoya 464, Japan

McNeice, Gregory M. (A:394), Solid Mechanics Division, University of Waterloo, Waterloo, Ontario, Canada N2L 3G1

Michels, A. (A:267), Department of Physical Education, Section: Physiotherapy, University of Leuven, Tervuurse vest 191, 3030 Heverlee, Belgium

Miller, Doris I. (B:65), School of Physical and Health Education, Hutchinson Hall DX-10, University of Washington, Seattle, Washington 98195, USA

Miura, Mochiyoshi (B:271), Research Center of Health, Physical Fitness and Sports, University of Nagoya, Furo-cho, Chikusa-ku, Nagoya 464, Japan

Miyashita, Mitsumasa (A:240; B:94, 151), Laboratory for Exercise Physiology and Biomechanics, Faculty of Education, University of Tokyo, Hongo 7-3-1, Bunkyo-ku, Tokyo, Japan

Moers, R. (B:309), Dambruggestraat 79, B-2000 Antwerpen, Belgien

Murase, Yutaka (B:200), Nagoya Gakuin University, 1350, Kamishinano-cho, Seto-city, Aichi-prefecture, Japan

Nagai, Jun (A:468), Department of Orthopedic Surgery, Kyoto University, School of Medicine, 53 Shogoin Kawahara-cho, Sakyo-ku, Kyoto 606, Japan

Nagypál, Tibor (A:253), Institute of Neurosurgery, 1145 Budapest XIV, Amerikai ut 57, Hungary

Neukomm, Peter A. (B:476), Biomechanics Laboratory, Swiss Federal Institute of Technology, Gloriastrasse 35, 8006 Zürich, Switzerland

Nicol, Klaus (B:433, 449), Institut für Sport und Sportwissenschaft, der Johann Wolfgang Goethe-Universität, 6000 Frankfurt/Main, Ginnheimer Landstrasse 39, West Germany

Nigg, Benno M. (B:476), Biomechanics Laboratory, Swiss Federal Institute of Technology, Gloriastrasse 35, 9006 Zürich, Switzerland

Noble, Earl G. (B:87), Department of Kinesiology, University of Waterloo, Waterloo, Ontario, Canada

Nordeen, Katherine S. (B:26), Biomechanics Laboratory, The Pennsylvania State University, University Park, Pennsylvania 16802, USA

Norman, Robert W. (B:87), Department of Kinesiology, University of Waterloo, Waterloo, Ontario, Canada

Oka, Hideo (B:326), Kinesiology Laboratory, Attached Senior High School of Osaka Kyoiku University, Midorigaoka, Ikeda-shi, Osaka 563, Japan

Okada, Morihiko (A:303), Primate Research Institute, Kanrin, Inuyama-shi, Aichi 484, Japan

Okamoto, Tsutomu (A:311; B:215, 222, 326), Kinesiology Laboratory, Department of Liberal Arts, Kansai Medical School, Uyama Hirakata-shi, Osaka 573, Japan

Ohtsuki, Tatsuyuki (A:240), Laboratory of Physical Activity, Faculty of Literature, Nara Women's University, Nara City, Nara, Japan

Olsson, Tord H. (B:502), Department of Diagnostic Radiology, University Hospital, S-221 85 Lund, Sweden

Ono, Mitsutsugu (B:94), Department of Physiology and Kinesiology, School of Physical Education, Tokyo Gakugei University, Koganei City, Tokyo code 184, Japan

Örtengren, Roland (A:520), Department of Clinical Neurophysiology, University of Göteborg, Sahlgren Hospital, S-413 45 Göteborg, Sweden

Payne, A. H. (B:314), Department of Physical Education, University of Birmingham, Edgbaston, Birmingham B15 2TT, England

Pedersen, Ejner (A:171, 177), Department of Neurology, Århus Kommunehospital, DK-8000 Århus C, Denmark

Perren, Stephan M. (A:373, 379), Laboratory for Experimental Surgery, Swiss Research Institute, Obere Strasse 22, CH-7270 Davos-Platz, Switzerland

Person, Raisa S. (A:61), Laboratory of Physiology of Human Movement, Institute for Problems of Information Transfer, Academy of Sciences of USSR, Moscow, E-24, Aviamotornajia ul., 8-A, USSR

Petersén, Ingemar (A:165), Laboratory of Clinical Neurophysiology, Sahlgren Hospital, Fack, 413 45 Göteborg, Sweden

Pezzack, John C. (B:87), Department of Kinesiology, University of Waterloo, Waterloo, Ontario, Canada

Piotrowski, George (B:181), 128 Mechanical Engineering, Department of Mechanical Engineering, University of Florida, Gainesville, Florida 32611, USA

Plaja, Juan (A:347), Centro de Rehabilitación, Ciudad Sanitaria de la Seguridad Social, Pº Valle Hebrón, s/n, Barcelona (16), Spain

Povetkin, J. (B:118), Committee for Physical Education and Sport of the USSR Council of Ministers, Moscow, USSR

Quanbury, Arthur O. (A:334), Biomedical Engineering Research, Shriners Hospital for Crippled Children, 633 Wellington Crescent, Winnipeg, Manitoba R3M 0A8, Canada

Radford, P. F. (A:82; B:188), Director, Department of Physical Education and Recreation, University of Glasgow, Glasgow, Scotland

Rahn, Berton A. (A:373), Laboratory for Experimental Surgery, Swiss Research Institute, Obere Strasse 22, CH-7270 Davos-Platz, Switzerland

Ramey, Melvin R. (B:167), Department of Civil Engineering, University of California, Davis, California 95616, USA

Ranta, Matti A. (B:337), Department of General Sciences, Helsinki University of Technology, Otakaari 1, SF 02150 Espoo 15, Finland

Ratov, I. P. (B:357), Biomechanics Laboratory, All-Union Scientific Research Institute of Physical Culture, ul. Kazakova 18, Moscow, USSR

Rau, Günter (B:485), Research Institute for Human Engineering, Lüftelberger Strasse 1123, D-5309 Meckenheim, West Germany

Reid, J. G. (A:461), School of Physical and Health Education Queen's University, Kingston, Ontario, Canada

Reimer, Gary D. (A:334), Health Sciences Computer Centre, University of Manitoba, Winnipeg, Manitoba, Canada

Reiss, Robert (A:437), Materials Engineering, College of Engineering, University of Iowa, Iowa City, Iowa 52242, USA

Rijken, H. (B:243), Nederlands Scheepsbouwkundig Proefstation, Haagsteeg 2, Wageningen, The Netherlands

Riley, Donald R. (B:371), Computer Aided Design and Graphics Laboratory, School of Mechanical Engineering, Purdue University, West Lafayette, Indiana 47907, USA

Roberts, Elizabeth M. (B:20), Department of Physical Education, Lathrop Hall, 1050 University Avenue, Madison, Wisconsin 53706, USA

Roselle, N. (A:267), Department of Physical Medicine and Clinical Neurophysiology, University of Leuven, Kapucijnenvoer 33, 3000 Leuven, Belgium

Roy, Benoit (B:277, 287, 332), Laboratoire des Sciences de l'Activité Physique, Faculté des Sciences de l'Education, Université Laval, Québec G1K 7P4, Canada

Samson, J. (B:332), Department d'Education Physique, Université Laval, Quebec G1K 1P4, Canada

Sano, Yuji (B:135), Department of Physical Education, College of General Education, University of Tokyo, 3-8-1, Komaba Meguroku, Tokyo, Japan

Sargeant, A. J. (A:484), MRC Environmental Physiology Unit, London School of Hygiene and Tropical Medicine, Keppel Street, London WC1E 7HT, England

Schaaf, E. (A:185), Address unavailable.

Schaldach, M. (A:385), Zentralinstitut für Biomedizinische Technik, Turnstrasse 5, D-852 Erlangen, Germany

Schreiner, Kristian E. (A:131), Norwegian College of Physical Education and Sport, P.O. Box 40, Kringsjå, Oslo 8, Norway

Selvik, Göran (B:502), Department of Anatomy, University of Lund, Biskopsgatan 7, S-223 62 Lund, Sweden

Sharratt, Michael T. (B:87), Department of Kinesiology, University of Waterloo, Waterloo, Ontario, Canada

Shirai, Yasumasa (A:88), Department of Orthopaedics, Nippon Medical School, 1-1-5, Sendagicho, Bunkyo-ku, Tokyo 113, Japan

Simkin, Ariel (B:415), Department of Orthopaedics, Hadassah University Hospital, Jerusalem, Israel

Sjödin, Bertil (A:112), Department of Physiology, Gymnastik- och Idrottshögskolan, Lidingövägen 1, 114 33 Stockholm, Sweden

Skorupski, Lechoslaw (B:351), Department of Biomechanics, Institute of Biological Sciences, Academy of Physical Education, 01-813 Warsaw, Marymoncka Street 34, Poland

Soderberg, Gary L. (A:437), Programs in Physical Therapy, College of Medicine, University of Iowa, Iowa City, Iowa 52242, USA

Sodeyama, Hiroshi (B:271), Physical Education Laboratory, Kinjo Gakuin University, 2282 Omori, Moriyama-ku, Nagoya 463, Japan

Sorensen, Harold (B:161), Department of Civil Engineering, Washington State University, Pullman, Washington 99163, USA

Sprague, B. L. (A:355), Biomechanical Laboratory, Division of Hand Surgery, Department of Orthopaedics, University Hospitals, Iowa City, Iowa 52242, USA

Staling, L. M. (A:280), Physiology Department, UMAB Dental School, University of Maryland School of Dentistry, Baltimore, Maryland 21201, USA

Stanič, Uroš (A:444), Byokibernetic Depertement, Jožef Stefan Institute, University of Ljubljana, Jamova 39, 61000 Ljubljana, Yugoslavia

Steeger, D. (B:34), Orthopädische Univ.-Klinik, Klinikum der Johannes Gutenberg-Universität, D-6500 Mainz, Langenbeckstrasse 1, West Germany

Stevens, A. (A:267), Department of Physical Medicine and Clinical Neurophysiology, University of Leuven, Kapucijnenvoer 33, 3000 Leuven, Belgium

Stoboy, H. (A:423), Applied Physiology, Department of Orthopaedics, Freie Universität Berlin, Clayallee 229, D 1000 Berlin 33, Germany

Sust, Martin (B:441), Laboratory of Biomechanics and Motorics, Department of Sportscience, Friedrich-Schiller-University of Jena, 69 Jena, Seidelstrasse 20, German Democratic Republic

Suzuki, Ryohei (A:341), Department of Orthopaedic Surgery, Nagasaki University, School of Medicine, 7-1 Sakamoto-machi, 852 Nagasaki, Japan

Talishev, F. M. (A:124), Department of Biological Problems of Sport, All-Union Research Institute of Physical Culture, 103064 Kazakova st. 18, Moscow, USSR

Tankov, Nicolas T. (A:65), Laboratory of Motor Control, Institute of Physiology, Bulgarian Academy of Sciences, Acad. G. Bonchev Str., Bl. 1, 1113 Sofia, Bulgaria

Taubenheim, H. (A:185), Address unavailable.

Terauds, Juris (B:497), Biomechanics Laboratories, Department of Physical Education, University of Alberta, Edmonton, Alberta, Canada

Tesch, Per (A:112), Department of Physiology, Gymnastik- och Idrottshögskolan, Lidingövägen 1, 114 33 Stockholm, Sweden

Thorstensson, Alf (A:112, 118), Department of Physiology, Gymnastik- och Idrottshögskolan, Lidingövägen 1, 114 33 Stockholm, Sweden

Thull, R. (A:385), Zentralinstitut für Biomedizinische Technik, Turnstrasse 5, D-852 Erlangen, Germany

Tichauer, Erwin R. (A:493), Division of Biomechanics, Institute of Rehabilitation Medicine (RR-311), New York University Medical Center, 400 East 34th Street, New York, New York 10016, USA

Tichonov, V. N. (B:103), Moscow District, Station of Perlovskaia, Moscow Institute of Cooperation, Physical Education Department, Moscow, USSR

Togari, Haruhiko (B:135), Department of Physical Education, College of General Education, University of Tokyo, 3-8-1, Komaba Meguroku, Tokyo, Japan

Tokuyama, Hiroshi (B:215, 222), Kinesiology Laboratory, Osaka Kyoiku University, Jyonancho, Ikeda-shi, Osaka 563, Japan

Torm, Rein Y. (B:322), Department of Gimnastic, University of Tartu, 202400 Tartu, Ulikoli 18, USSR

Tóth, Szabolcs (A:253), Institute of Neurosurgery, 1145 Budapest XIV, Amerikai ut 57, Hungary

Toyoshima, Shintaro (A:247; B:109), Physical Activity Laboratory, Aichi Prefectural University, Takada-cho, Mizuho-ku, Nagoya, Japan

Tveit, Per (B:81), Department of Kinesiology, The Norwegian College of Physical Education and Sport, Sognsveien 220, Oslo 8, Norway

Umeno, Katsumi (A:105), Department of Anatomy, Kyorin University, School of Medicine, Shinkawa, Mitaka, Tokyo 181, Japan

Upton, Adrian R. M. (A:82; B:188), McMaster University Medical Centre, McMaster University, 1200 Main Street West, L8S 4J9, Hamilton, Ontario, Canada

Vain, Arved A. (B:58, 322), Department of Sport Physiology, University of Tartu, 18 Ülikooli Street, 202400 Tartu, Estonian S.S.R., USSR

Veihelmann, D. (A:379), Laboratory for Experimental Surgery, Swiss Research Institute, CH-7270 Davos-Platz, Switzerland

Verschuur, R. (B:243), Coronel Laboratorium en Laboratorium voor Psychofysiologie, Universiteit van Amsterdam, 1ste Constantijn Huygensstraat 20, Amsterdam, The Netherlands

Viitasalo, Jukka T. (A:118), Kinesiology Laboratory, Department of Biology of Physical Activity, University of Jyväskylä, SF 40100 Jyväskylä 10, Finland

Vorro, Joseph (A:280), Department of Anatomy, Michigan State University, East Lansing Michigan 48824, USA

Ward, Terry (B:46), Biomechanics Laboratory, 211 Montgomery, Florida State University, Tallahassee, Florida 32306, USA

Wartenweiler, Jurg (A:16), Biomechanics Laboratory, Swiss Federal Institute of Technology, Plattenstrasse 26, 8032 Zürich, Switzerland

Willner, Stig (B:502), Department of Orthopaedics, General Hospital, S-214 01 Malmö, Sweden

Wilson, Barry D. (B:13, 467), Biomechanics Laboratory, Department of Physical Education, University of Iowa, Iowa City, Iowa 52242, USA

Winter, David A. (A:334), Department of Kinesiology, University of Waterloo, Waterloo, Ontario N2L 3G1, Canada

Wit, Andrzej (B:351), Department of Biomechanics, Institute of Biological Sciences, Academy of Physical Education, 01-813 Warsaw, Marymoncka Street 34, Poland

Wittekopf, G. (A:185), Deutsche Hochschule für Körperkultur, Friedrich-Ludwig-Jahn-Allee 59, 701 Leipzig, Germany

Wood, George K. (B:127), Department of Physical Education, College of Charleston, Charleston, South Carolina 29400, USA

Yabe, Kyonosuke (A:75), Laboratory of Ergonomics, Department of Therapeutics, Institute for Developmental Research, Aichi Prefectural Colony, Kamiya-cho, Kasugai City, Aichi 480-03, Japan

Yamamoto, Keizo (B:135), Department of Physical Education, College of General Education, University of Tokyo, 3-8-1, Komaba Meguroku, Tokyo, Japan

Yamashita, Noriyoshi (B:41), Laboratory of Physiology, College of General Education, Kyoto University, Nihonmatsu-cho, Yoshida, Sakyo-ku, Kyoto 606, Japan

Yasuda, Noriaki (B:200), Department of Physical Education, Chukyo University, Kaizu-cho, Toyota-city, Japan

Yoshizawa, Masatada (B:222), Kinesiology Laboratory, Department of Education, Fukui University, Bunkyo Fukui-shi, Fukui 910, Japan

Youm, Y. (A:355), Biomechanical Laboratory, Division of Hand Surgery, Department of Orthopaedics, University Hospitals, Iowa City, Iowa 52242, USA

Zatziorsky, V. (B:387), Department of Biomechanics, Central Institute of Physical Culture, Sirenevij Bulvar 4, Moscow, USSR

Zernicke, Ronald F. (B:20), Biomechanics Laboratory, Department of Kinesiology, University of California, Los Angeles, California 90024, USA

Ziegler, Walter J. (A:373), Laboratory for Experimental Surgery, Swiss Research Institute, Obere Strasse 22, CH-7270 Davos-Platz, Switzerland

Zinkovsky, Anatoly V. (B:322), Science Research Laboratory, Department of Physical Education, Polytechnical Institute 195251, Leningrad, Polytechnickeskaya st., 29, USSR

Preface

Biomechanics V (Parts A and B) includes most of the papers presented at the Fifth International Congress of Biomechanics, which was held at The University of Jyväskylä, Jyväskylä, Finland, June 29 to July 3, 1975. The proceedings of the first three international seminars held in Zürich (Switzerland), Eindhoven (The Netherlands), and Rome (Italy), respectively, were published in the *Medicine and Sport Series* by S. Karger, Basel (Switzerland). The Fourth International Seminar on Biomechanics was held at The Pennsylvania State University, University Park, Pennsylvania (USA), and its proceedings were published by University Park Press, Baltimore, Maryland (USA), as Volume 1 of the *International Series on Sport Sciences.*

Biomechanics V launches the new *International Series on Biomechanics,* published by University Park Press, under the direction of the International Society of Biomechanics, which was founded in 1973. The Fifth Congress at Jyväskylä was the first to be held under the sponsorship of the newly formed Society and was the largest of any of the meetings held to date. This trend may reflect an increased development of the field of biomechanics; it may have been due to the prestige and publicity provided by the new International Society, or it could have been the lure of the wonderful Finnish lakes and forests surrounding the city of Jyväskylä. Most likely, it was a combination of these three factors.

The response to the Congress was so great that it was necessary to publish the proceedings in two bound volumes. The two volumes, Parts A and B, include a total of 140 contributions by researchers from more than 30 countries. This is approximately 75 percent of the papers presented during the Congress. In this volume, Part A, are papers dealing with the general topics of neuromuscular control, electromyography, orthopaedics, rehabilitation, and ergonomics, in addition to the presentations in the Opening Session of the Congress. Part B, the companion volume, includes various studies of fundamental movements, biomechanics of sport, and new developments in instrumentation and methodology for biomechanical research. Part B concludes with information concerning the organization of the Congress.

These two volumes contain a wealth of information from original research —a tribute to the excellent research in biomechanics being conducted throughout the world. The Editor thanks the contributors to these Proceedings for following the printed guidelines quite carefully in preparing their manuscripts and for responding to the submission deadline dates so faithfully. A special thanks is extended to colleagues Richard C. Nelson and Chauncey A. Morehouse for their technical help during the editorial work. Their experienced support and the cooperation of the authors helped considerably in reducing the time required to edit these two volumes and consequently resulted in a much earlier publication date.

Fundamental movements

Basic kinetics and kinematics

Biomechanical aspects of a successful motion optimization

H. Hatze
National Research Institute for Mathematical Sciences, CSIR, Pretoria

One of the major objectives of biomechanical research is the development of methods for the optimization of human motions. For a given individual and performance criterion (such as time, total energy expended, etc.) it is often desired to find that motion which will minimize (maximize) the performance criterion under consideration. For instance, one may wish to predict that walking motion which, for a given individual, renders the total energy expended a minimum for a certain distance covered. This particular problem has attracted much attention in recent years (see, e.g., Beckett and Chang, 1968; Morrison, 1970; and Zarrugh, Todd, and Ralston, 1974). It has also provoked the impressive study of Chow and Jacobson (1971), which probably constitutes the first attempt to employ optimal control for the optimization of a human motion. However, the controls used by Chow and Jacobson are not related to the actual physiological controls, motor unit recruitment and change of stimulation frequency, and the theoretical predictions of their model have not been verified experimentally.

A different kind of motion, in which the execution time was to be minimized, has been selected by the writer for the purpose of performing an optimization. Consider the situation depicted in Figure 1. A male subject (23 yr old, 1.82 m tall, and weighing 87.0 kg) is standing on his left leg with his pelvis fixed to a steel frame by means of a belt. On his right foot he wears a special boot having a mass of 10.0 kg. This additional 10.0-kg mass serves to slow down the otherwise too rapid motion. The ankle is fixed at 90° and the right leg is free to move in the sagittal plane.

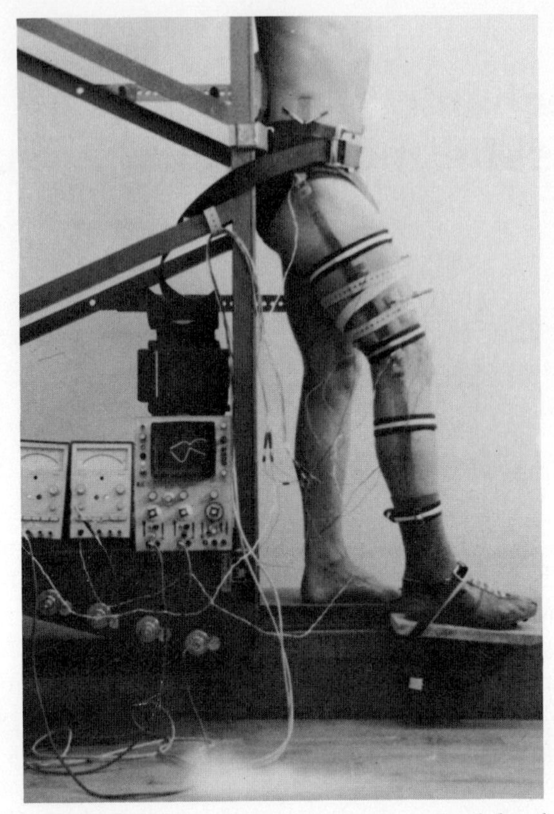

Figure 1. Photograph of the experimental arrangement used for the observations made on the living system.

The subject is required to kick at an appropriately positioned target (Figure 2) with the mass W (attached to his boot) in such a way that the execution of the motion takes the *shortest possible time.* The motion is started with the leg "hanging" down, i.e., all the leg muscles relaxed. The target is so placed that when contact occurs, the leg must be stretched (knee angle 0.05 rad, angular knee velocity zero) while the hip is at an angle of 0.8 rad from the vertical.

The aim of the present research project was to predict the time-optimal motion, in all detail, by means of an optimization performed on the mathematical model of the musculo-skeletal link system under consideration, and then to verify the result by measurements taken from the living system. The link-mechanical part of the mathematical model is based on previous work of the author (Hatze, 1973) while the musculo-mechanical part is derived from a modified version of a recently published control model of skeletal muscle (Hatze, 1974a). It turns out

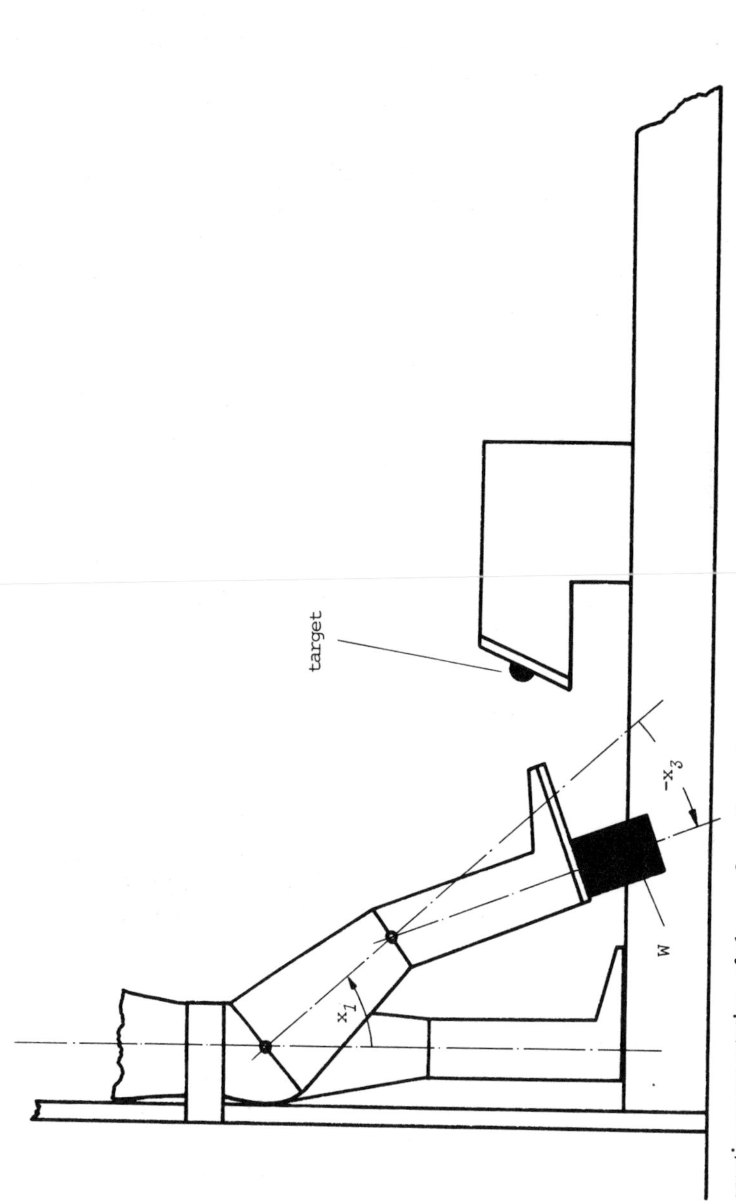

Figure 2. Schematic representation of the configuration of the target and the subject's right leg. The hip angle is denoted by x_1 and the knee angle by x_3. The symbol W denotes the 10.0-kg mass that is attached to the subject's boot.

that the musculo-skeletal link system under consideration can be described by a system of 19 nonlinear first-order differential equations that contain the control parameters of the five muscle groups involved. The numerical values of the subject-specific constants appearing in these equations were determined by means of special methods, some of which are new and have been developed in the course of this project. For instance, a new method for the simultaneous measurement of the moment of inertia, the location of the center of mass, and the damping coefficient of a body segment in situ was devised. This method yielded accurate and highly reproducible results (Hatze, 1975). All the methods used as well as a detailed description of the model and the optimization procedure (which was carried out on an IBM 360/50 digital computer) can be found in Hatze (1974b) and Hatze (1976b).

The optimization was successful beyond expectations. Not only could the optimality of the predicted motion be confirmed unequivocally by measurements made on the living system (i.e., on the subject's leg), but the simulation also produced other biomechanically highly significant results. Moreover, it was possible in a relatively short period of time to train the subject to perform nearly optimally. It is this latter aspect of the optimization on which the present paper focuses.

METHODS

The observable motion of the subject's leg is defined by the histories $x_1(t)$ and $x_3(t)$ of the hip angle x_1 and the knee angle x_3 (these angles are shown in Figure 2). On the living system, the functions $x_1(t)$ and $x_3(t)$ were measured by means of electrogoniometers positioned over the respective average joint centers (see Figure 1). The output signal was displayed on the screen of a two-beam storage oscilloscope and then photographed with an oscilloscope camera. The motion as predicted by the mathematical optimization was available in the form of a computer print-plot. Figure 3 depicts a motion as observed on the living system as well as the predicted optimal motion.

A new method that permits the concise description of the motion of any complex biosystem (Hatze, 1976a) was used to compare the motion as observed on the biosystem with the predicted optimal motion. Basically, the set of functions $\{x_1(t), x_3(t) \mid 0 \leq t \leq \tau\}$, where τ is the execution time of the motion, is converted into a Fourier representation. The motion Ω of the whole system can then be written as the matrix product:

$$\Omega = [c_{ij}]\,[\cos \frac{2\pi jl}{(2l+1)\,\tau} t] \quad , \qquad i = 1,2, \quad j = 0, \ldots, l. \quad (1)$$

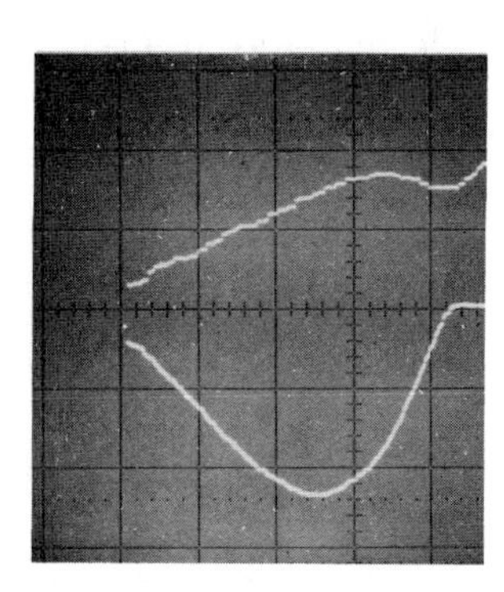

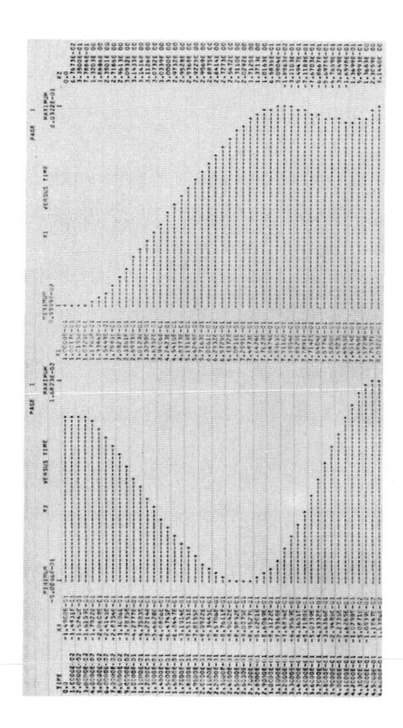

Figure 3. *Left*, photograph of the oscilloscope-screen display of the hip angle trajectory $x_1(t)$ (*top*) and the knee angle trajectory $x_3(t)$ (*bottom curve*) of a near-optimal motion as observed on the living system. Horizontal scale: 0.1 sec/div. Vertical scale: 0:5 rad/div. *Right*, photograph of the computer print-plot displaying the predicted optimal trajectories $x_1(t)$ (*top curve*) and $x_3(t)$ (*bottom curve*).

For the particular problem at hand, a value of $l = 23$ was considered appropriate giving a sampling time interval of 0.02 sec for the optimal time of $\tau_{\text{opt}} = 0.46$ sec.

If two motions Ω_r and Ω_{opt} are to be compared it is advantageous to define a *normalized difference function* Δ by:

$$\Delta(r) = \| \Omega_r - \Omega_{\text{opt}} \| =$$

$$\frac{1}{l+1} \sum_{i=1}^{n} \sum_{j=0}^{l} \frac{| c_{r,ij} - c_{\text{opt},ij} |}{|c_{\text{opt},i0}|} + \frac{|\tau_r - \tau_{opt}|}{\tau_{\text{opt}}} , \qquad (2)$$

where the execution time τ_r of the rth motion is determined by the terminal conditions $x_1(\tau_r) = 0.8$ rad, $x_3(\tau_r) = 0.05$ rad, and $\dot{x}_3(\tau_r) = 0$ rad/sec. Obviously, $\Delta(r)$ converges to zero as Ω_r approaches Ω_{opt}, i.e., when the two motions become congruent.

RESULTS

Only one of the most significant results of the trainability aspect of the motion optimization is discussed here. When the experiments with the living system began, the computer optimization had already been carried out but the result was not known to the subject. For an initial period of 2 wk, during which he performed a total of 120 kicks (20 kicks/day, three sessions per week), the subject was only told the execution time of each kick. After this initial period of natural adaptation, the training towards the optimal performance commenced. He was repeatedly shown a film displaying the optimal motion as well as superimpositions of the optimal motion on his own performances. The degree of congruency between the observed biomotion and the predicted optimal motion can be measured by the function $\Delta(r)$, as defined by equation (2), and is shown in Figure 4.

DISCUSSION AND CONCLUSION

As can be seen from Figure 4, the natural adaptation is at first fast but soon becomes quite slow. On the other hand, it can be seen that after the commencement of training the approach towards the optimal motion is rapid. This was also evident from the execution times τ_r, which for $r > 140$ were very near the predicted optimal time of $\tau_{opt} = 0.46$ sec.

Other biomechanically significant results of the optimization include the prediction, by the model, of the well-known stretch reflex, the theoretical proof for the incompatibility of speed and exact coordination in a motion, and the indication of the occurrence of a change from time optimality to energy optimality of the motion when the subject becomes fatigued. In addition, the successful optimization is, of course, also proof for the adequacy of the muscle model employed.

However, the predictive power of the model does not end here. The optimum found for the investigated motion is actually a *relative* one since it corresponds to the subject's present state of muscular development. It is possible to perform a so-called parameter optimization that yields the *absolute* optimum and, at the same time, provides us with a complete training program for all the muscle groups involved. This program is, however, specific for the motion under consideration.

It is believed that this is the first time that a complete optimization of a human motion, which could be verified on the living system, has been performed successfully. It would thus appear that the problem of finding an optimal motion for a given individual and a given task to be performed, by means of a mathematical procedure, has been solved in principle.

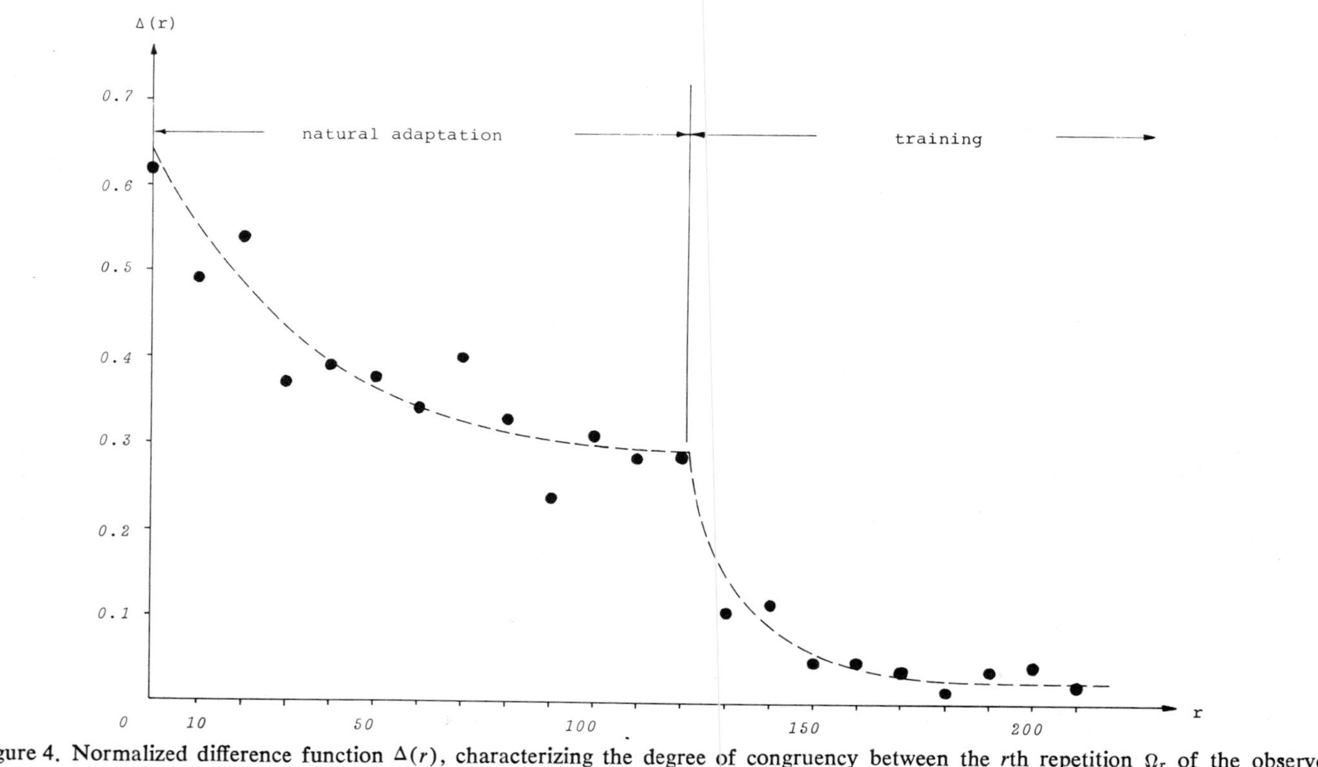

Figure 4. Normalized difference function $\Delta(r)$, characterizing the degree of congruency between the rth repetition Ω_r of the observed biomotion and the predicted optimal motion Ω_{opt}. A small value of $\Delta(r)$ indicates that the observed biomotion is near the optimal motion. Only the values corresponding to each tenth repetition are displayed.

REFERENCES

Beckett, R., and K. Chang. 1968. An evaluation of the kinematics of gait by minimum energy. J. Biomech. 1: 147–159.

Chow, C. K., and D. H. Jacobson. 1971. Studies of human locomotion via optimal programming. Math. Biosci. 10: 239–306.

Hatze, H. 1973. Optimization of human motions. *In:* S. Cerquiglini, A. Venerando, and J. Wartenweiler (eds.), Biomechanics III, pp. 138–142. S. Karger, Basel.

Hatze, H. 1974a. A model of skeletal muscle suitable for optimal motion problems. *In:* Richard C. Nelson and Chauncey A. Morehouse (eds.), Biomechanics IV, pp. 417–422. University Park Press, Baltimore.

Hatze, H. 1974b. A control model of skeletal muscle and its application to a time-optimal biomotion. Ph.D. thesis, University of South Africa, Pretoria, Rep. of South Africa.

Hatze, H. 1975. A new method for the simultaneous measurement of the moment of inertia, the location of the centre of mass, and the damping coefficient of a body segment in situ. Eur. J. Appl. Physiol. 34: 217–226.

Hatze, H. 1976a. A method for describing the motions of biological systems. J. Biomech. 9: 101–104.

Hatze, H. 1976b. The complete optimization of a human motion. Math. Biosci. 28: 99–135.

Morrison, J. B. 1970. The mechanics of muscle function in locomotion. J. Biomech. 3: 431–451.

Zarrugh, M. Y., F. N. Todd, and H. J. Ralston. 1974. Optimization of energy expenditure during level walking. Eur. J. Appl. Physiol. 33: 293–306.

Identification of the limiting factors in the performance of a basic human movement

J. G. Hay, B. D. Wilson, and J. Dapena
University of Iowa, Iowa City

According to Zatziorsky (1974) relationships between biomechanical parameters and the criterion for physical performance generally take one of three forms, linear, curvilinear, and tapering (Figure 1). These differing types of relationship would seem to have different implications in the practical situation. For instance, if a biomechanical parameter exhibited a strong linear relationship with the criterion, and if a logical cause–effect relationship existed, it would seem reasonable to suggest that improvements in performance could be sought via improvements in this parameter, *regardless of the performer's present level of achievement*. Further, in those cases where a strong curvilinear relationship

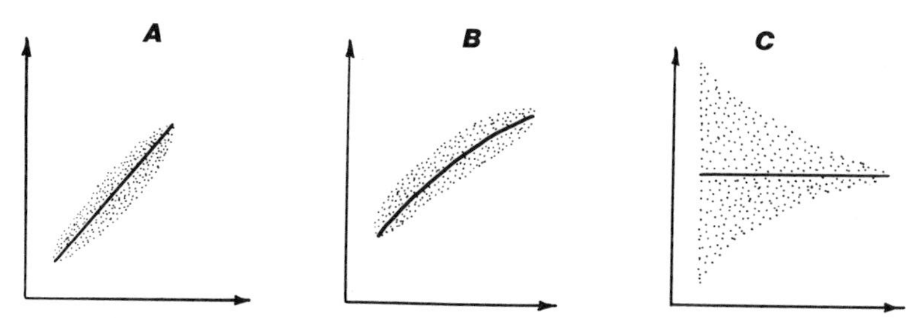

Figure 1. The most frequently encountered types of relationships between biomechanical characteristics (*ordinate*) and performances (*abscissa*), after Zatziorsky (1974).

existed, and where, again, a logical cause–effect relationship could be identified, it would seem reasonable to place instructional emphasis on the parameter in question *at those levels of performance at which such emphasis was indicated.*

PURPOSE

This study is the first in a series of studies whose broad purposes are to develop a research strategy (1) to determine the nature and strength of the causative relationships that exist between biomechanical parameters and the quality of performance of a motor skill and (2) to identify those biomechanical parameters that are instrumental in limiting the quality of performance at successive levels of performance.

PROCEDURES

The research strategy employed in this first attempt at a solution to the problem consisted of three steps: (1) the development of a theoretical model of the relationships between the dependent variable and the contributing independent variables; (2) the collection, reduction, and analysis of data on the performances of a large number of subjects whose abilities extended over a wide range; (3) the evaluation of the recorded performances in terms of the theoretical model.

A standing vertical jump and reach with two hands was chosen as the skill to be investigated because of its inherent simplicity, its essentially two-dimensional nature, and its importance in a wide variety of activities.

THEORETICAL MODEL

The height to which a person can jump and reach with both hands (H) can be regarded as the sum of three lesser heights, the height of his center of gravity at take-off (H_1), the height which his center of gravity is elevated from take-off to the peak of flight (H_2), and the height to which he can reach beyond the peak height attained by his center of gravity (H_3).

The magnitudes of H_1 and H_3 are determined by the lengths of the segments, the orientations of the segments, and in the case of H_3, by the performer's ability to time his movements so that his body is fully extended at the instant his center of gravity reaches its peak height.

The magnitude of H_2 is determined by the vertical velocity at take-off, which, in turn, is determined by the vertical impulse and the mass of the performer. While the precise nature of the factors and relationships that determine the magnitude of the vertical impulse was not known, it was considered likely that segment positions, linear and angular displacements, and linear and angular velocities were among the factors of importance.

The various relationships among the dependent and independent variables are summarized in the theoretical model in Figure 2.

DATA COLLECTION AND REDUCTION

Two hundred and thirteen male students in the Physical Education Skills Program at the University of Iowa were used as subjects. The performances of each subject were recorded with a Locam 16-mm motion-picture camera and the resulting films analyzed with the aid of a Vanguard Motion Analyzer linked on-line to a digitizer and paper-tape punch. The data thus obtained were processed using a series of computer programs especially written for the purpose.

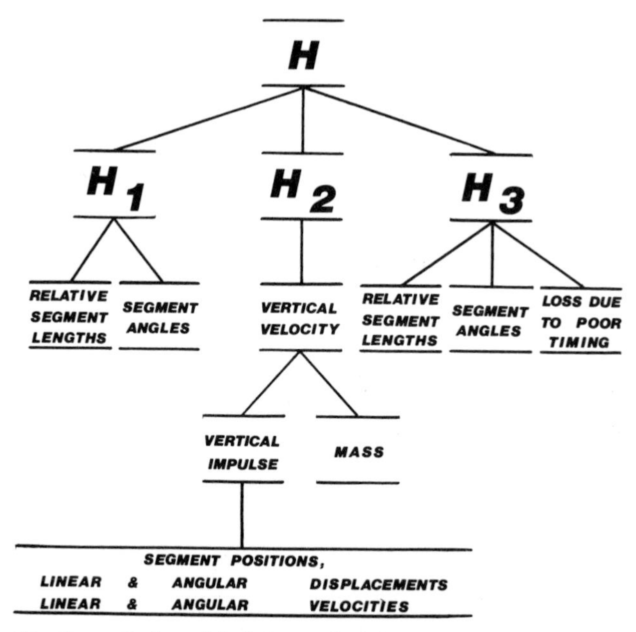

Figure 2. The theoretical model of the vertical jump.

DATA ANALYSIS

Since the study was concerned with the effect of variations in the techniques employed by the subjects rather than with effects attributable to variations in physique, each biomechanical parameter was first correlated with stature and weight. If the zero-order correlation coefficient was significantly greater than zero ($p \leq .001$, $r = .221$), the physique measure concerned was partialled-out in the subsequent statistical analysis.

Each of the component H-values (H_1, H_2, and H_3) was correlated with H and each of the other biomechanical parameters was correlated with the component H-value to which they logically contributed, according to the theoretical model. In addition, scattergrams were plotted in order to establish the nature of the relationship (i.e., linear, curvilinear, or tapering) between each of the biomechanical parameters and the component H-value with which it was correlated.

Where indicated by the significance of the zero-order or partial correlations, multivariate analyses—specifically, multiple regression and factor analyses—were undertaken.

RESULTS AND DISCUSSION

H

With stature partialled out, significant relationships were found between H and each of the component H-values, with H_2 clearly the dominant influence. This result contrasted with that obtained when zero-order correlations were computed for the same variables. In this latter case, the dominant roles of H_1 and H_3 were clearly evident.

H_1

Significant relationships were found between H_1 and four of the 14 individual segment lengths (stature partialled out) and between H_1 and five segmental angles of inclination at the instant of take-off. Multiple regression analysis revealed that approximately 37 percent of the variability observed in H_1 could be attributed to variability in the relative lengths of the segments and approximately 49 percent to variability in the orientation of the segments. It was further revealed that the bulk of this latter variability was attributable to the feet and trunk segments.

H_2

A vertical jump consists essentially of two phases, a *preparatory phase* during which the performer's hips, knees, and ankles are flexed and his arms positioned in preparation for the explosive effort to follow, and a *propulsive phase* in which the arms are swung forward and upward and the legs vigorously extended to drive the performer upward and into the air. Since the first of these phases is characterized by a lowering and the second by an elevating of the performer's center of gravity, the moment at which his center of gravity is at its lowest point was used to indicate the end of the first phase and the beginning of the second.

Joint Angles When the smallest angles recorded at each of the hip, knee, and ankle joints were correlated with H_2, significant correlations were found indicating that the more the ankles were flexed, the greater was the magnitude of H_2. This finding, of course, might well have been expected. What is surprising, in view of the widespread interest that exists in the optimum angles of joint flexion for maximum jumping performance, is that neither of the other minimum joint angles was similarly correlated with H_2.

Linear and Angular Displacements With one minor exception, none of the correlations between H_2 and the linear and angular displacements undergone by the segments during the preparatory and propulsive phases was statistically significant.

Linear Velocities The average horizontal and vertical velocities of the individual segments during the propulsive phase of the jump were more highly correlated with H_2 than any of the other biomechanical parameters considered in this study. The average horizontal velocity of four upper limb segments and the head segment was significantly correlated with H_2. In addition, the correlations involving the remaining upper limb segments closely approached the level needed for statistical significance. Still stronger relationships were in evidence when the average vertical velocities of the segments were correlated with H_2: all six of the upper limb segments were significantly correlated with H_2, as too were the head and trunk segments. These various correlations indicated that the faster the arms were swung forward and upward and the faster the head and trunk were brought up and back towards the vertical, the greater was the magnitude of H_2.

The use of factor analysis revealed that the observed variability in the average vertical velocities could be accounted for in terms of three orthogonal factors, an upper limb factor, a lower limb factor, and a head-and-trunk factor. These factors accounted respectively for 65.5, 20.1, and 14.5 percent of the common variance. Subsequent multiple regression analysis revealed further that the head-and-trunk factor accounted for 30 percent and the upper limbs factor for an additional 7 percent of the observed variability in H_2.

In view of the generally held belief that the actions of the legs are primarily responsible for determining the outcome in jumping activities, the modest contribution of the leg factor in accounting for the variability observed in H_2 was somewhat surprising. One possible explanation of this is that, although the legs are the dominant influence, this dominance is not reflected in the average vertical velocities of the leg segments during the propulsive phase. A second possibility is that there is relatively little variation in the leg action from subject to subject and that variations observed in H_2 are primarily due to differences in the actions of head, trunk, and upper limb segments. A further explanation (which actually encompasses the previous two) is that the ability of the legs to transmit the forces generated in the upper body to the ground, while they themselves are moving upward, may place natural limits on the actions which the head, trunk, and arms can efficiently utilize. Such an explanation is consistent with the first suggestion made here in that some factor, other than those reflected in the average vertical velocities of the leg segments, permits the legs to contribute significantly to variations in H_2. It is also consistent with the second suggestion in that variations in head, trunk, and arm actions (themselves conditioned by the available leg strength) are primarily responsible for the observed variations in H_2. It is anticipated that further work on this question will reveal which, if any, of these possible explanations is correct.

Angular Velocities Significant positive correlations were found between H_2 and the magnitudes of the average angular velocities of six segments—the head, three upper limb, and two lower limb segments—during the propulsive phase.

H_3

No significant relationships were found between H_3 and either the individual segmental angles of inclination or the loss in H_3 due to poor timing. Significant relationships were found between H_3 and two of the 14 individual segment lengths (stature partialled out).

DISTRIBUTIONS

Visual inspection of the scattergrams for variables found to be significantly correlated revealed evidence of a linear relationship in every case. No evidence of curvilinear relationships was found among the scattergrams for variables whose zero-order correlation was nonsignificant. Several examples of the "tapering" relationships reported by Zatziorsky (1974) were found when angular measures were correlated with H_2.

These seemingly odd distributions were due to a ceiling effect that limited the maximum value of the joint angles (to the approximate limit of the joint range) and thereby truncated otherwise normal bivariate distributions.

CONCLUSIONS

On the basis of the results obtained to date it was concluded that the research strategy employed in this study has shown some promise of ultimately being effective in determining the factors that govern the quality of performance of a motor skill.

ACKNOWLEDGMENTS

The authors wish to express their sincere appreciation to Dr. George G. Woodworth, Statistical Consulting Center, University of Iowa, for his considerable assistance with the statistical analysis employed in this study.

REFERENCE

Zatziorsky, V. M. 1974. Studies of motion and motor abilities of sportsmen. *In:* Richard C. Nelson and Chauncey A. Morehouse (eds.), Biomechanics IV, pp. 273–276. University Park Press, Baltimore.

Human lower extremity kinetic relationships during systematic variations in resultant limb velocity

R. F. Zernicke
University of California, Los Angeles

E. M. Roberts
University of Wisconsin, Madison

The accurate description and prediction of coordinated movement kinetics are indeed some of the primary objectives of researchers in biomechanics. Relatively recently the study of human kinetics has begun to take on a quantitative nature, and vague explanations have given way to significant analytical investigations. Mathematical modeling of the human body in conjunction with high-speed cinematography has provided a fruitful approach to the quantitative assessment of movement patterns. The techniques of mathematical modeling have been employed in this study for a twofold purpose: first, to quantify kinetic parameters of the kicking lower limb during a soccer toe kick; and second, to analyze the relative contribution of selected kicking limb segments to systematic increments in resultant limb velocity.

METHODS

Five skilled male soccer players volunteered to be filmed performing a soccer toe kick. The subjects ranged from 20 to 27 yr of age, 66.4–80.6 kg in weight, and 169–182 cm in height. Using an official size and weight soccer ball, each subject was trained to perform a soccer toe kick under three conditions: resulting horizontal ball velocity was slow (15.24 ± 1.52 m/sec), medium (21.34 ± 1.52 m/sec), or fast (>27.42 m/sec). A Dark Field Velocimeter (Roberts, 1972) provided immediate feedback, to the nearest 0.3 m/sec. Velocity ranges were based on preliminary experiments and results reported in previous studies (Becker, 1963; Roberts and Metcalfe, 1968; Macmillan, 1975). A Milliken 16-mm motor-driven pin registered camera operating at a speed of 300 fps was used to photograph the side view of each performer through a 25-mm lens. Each subject performed at least three filmed kicks at each velocity level, in an object plane perpendicular to the optical axis of the camera. Subject-to-camera distance was 12.19 m. The shutter opening of 60° and film rate established an exposure time of 5.5×10^{-4} sec. A Recordak microfilm reader with the base removed (model P40) was used to project the film image (40× magnification) onto the digitizing platen of a Hewlett-Packard (HP) 9864A digitizer system. Rounding error of the system was 127 μ. Order of digitizing the trials was randomized to reduce systematic error resulting from day-to-day operator variations. Digitized planar rectangular coordinates of the subjects' right hip, knee, and ankle joints, and kicking foot centers of gravity were sequentially punched onto seven-track binary coded decimal paper tape by means of a HP 9810A programmable calculator. Three independent measurements of each trial were performed and averaged to increase reliability of coordinate estimation. Coordinate data for all trials were taken from the frame in which the subject's kicking foot left the ground to one frame prior to ball contact. After the data were transferred from paper tape to computer cards they were transformed and treated with four sequential FORTRAN programs to compute component and resultant joint forces and moments of force of the kicking limb (Zernicke, 1974).

RESULTS AND DISCUSSION

Results are based on a total of 45 soccer toe kicks, 15 executed at each of three different levels of kicking limb velocity. The component and resultant joint forces (proportion of body weight) and joint moments of force (kg-m) for each of the five subjects provided the primary interval data for analysis. The validity of the data presented in the study depended on the adequacy of the modeling technique (Youm et al., 1974) utilized for the analysis of the kicking limb. The results of a validation experiment provided evidence that the technique produced reasonable estimates of both magnitude and temporal sequencing of the kinetic variables of interest in this study. Mean percent agreement between vertical ground reaction forces computed from the modeling technique compared to those forces simultaneously recorded by a force platform was greater than 95 percent (Zernicke, Caldwell, and Roberts, 1976).

The data presented here provide a descriptive analysis of lower limb component and resultant joint forces and resultant moments that were derived from rigid-body, planar, three-segment equations of motion. Sign conventions adopted for interpretation of the data are provided in Figure 1. To facilitate comparisons between and within subjects, all force and moment of force data were normalized with respect to time. Time was set to unity for the period beginning at the frame the kicking foot left the ground and ending one frame before foot–ball contact. Figure 2 provides an example of the magnitude and direction of resultant joint forces for the lower limb during a typical fast kick.

Ankle Joint Forces

Ankle joint forces exhibited similar patterns of development across all velocities, although absolute magnitudes of the forces increased monotonically with increased limb velocity. Primary resultant forces in the ankle joint occurred between the time of deepest knee flexion and contact for all the kicks analyzed.

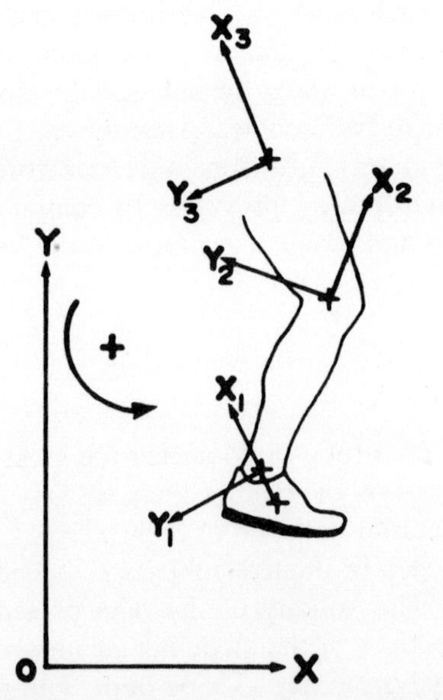

Figure 1. Sign conventions for segmental joint forces and moments of force: example of positive joint forces and moments of force. $X - Y$, inertial coordinate system. x_i, joint forces parallel to the particular limb segment ($i = 1,2,3$); generally a positive x force was directed proximally and a negative x force was directed distally (Youm et al., 1973). y_i, joint forces normal to the particular limb segment ($i = 1,2,3$); generally a positive y joint force was directed posteriorly and a negative y force was directed anteriorly (Youm et al., 1973). Counterclockwise motion is positive resultant moment of force.

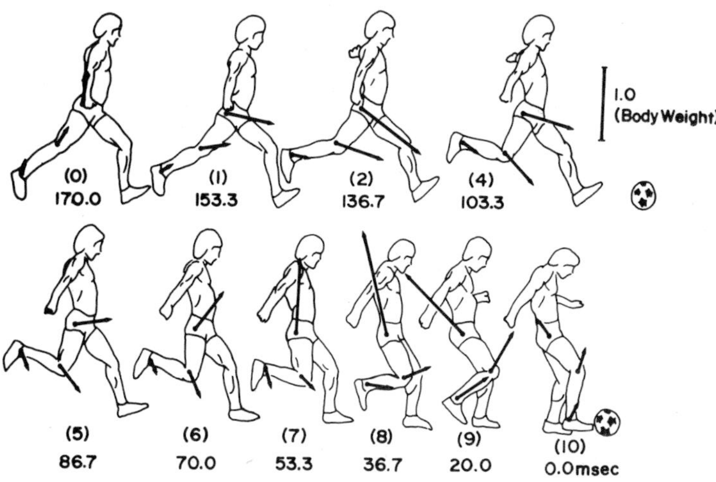

Figure 2. An example of resultant joint force vectors, fast kick (Trial 522). The resultant joint force vectors graphically depict the magnitude and direction of planar joint forces at the ankle, knee, and hip joints. During the 15 fast kicks, average maximum resultant joint forces were: ankle joint = 0.46 × body weight; knee joint = 0.68 × body weight; hip joint = 1.34 × body weight. Numbers in parentheses correspond to time unit divisions on the resultant moments of force graph presented in Figure 3.

Knee Joint Forces

Knee joint normal forces, in general, were anterior in direction before knee extension and posterior in direction during the latter part of knee extension for all kicks. Superior knee joint forces exhibited two discrete maxima. The first peak occurred before knee extension and corresponded to maximum positive thigh angular acceleration. The second occurred following initiation of knee extension and was directly related to maximum positive knee angular acceleration of the kicking limb.

Hip Joint Forces

Magnitudes of contributing joint force components at the hip joint were inversely related to the position of deepest knee flexion in the kicking limb. The resultant was composed of a predominant anterior component prior to initiation of knee extension, and predominant superior component after the knee began to extend. Kinematics of thigh angular acceleration indicated that the thigh segment was near maximum positive angular acceleration at the same time that anterior hip joint force was approaching maximum and superior component was approaching zero net force. When the superior component was rapidly increasing, following deepest knee flexion, the anterior component decreased and actually became a posterior component. The increase in superior joint force component was associated with an increase in thigh angular negative acceleration and maximum knee joint positive angular acceleration which followed initiation of knee extension.

Resultant Moments of Force

Moments of force at the ankle joint were consistently positive and increased monotonically with resultant foot velocity. Magnitudes of ankle torques were negligible with respect to knee and hip joint torques. During slow kicks extensor torques at the knee gradually increased and were maintained through initiation of knee extension. In contrast, knee joint extensor torques that occurred in medium and, particularly, in fast kicks dropped rapidly a few milliseconds before or during the start of knee extension (see Figure 3). Moments of force at the hip joint usually remained positive throughout the entire analyzed movement for slow and medium kicks. At the fast velocity, however, a characteristic retarding extensor torque was evident just before contact. The average

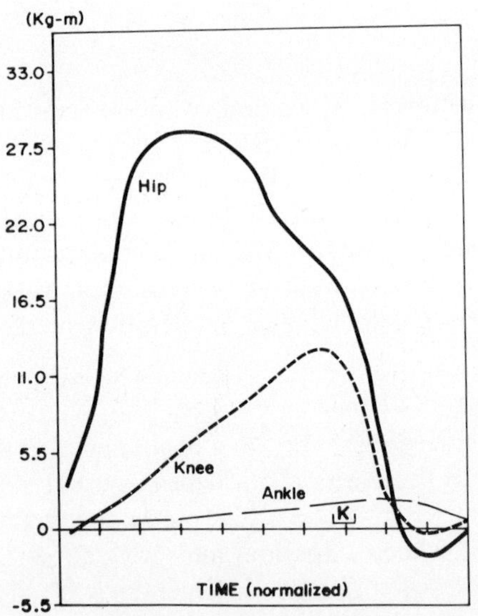

Figure 3. Resultant lower extremity moments of force; mean curve for 15 fast-velocity kicks. Each time division represents 10 percent of the elapsed time from the point where the kicking foot left the ground until foot-ball contact (see Figure 2). *K*, variability in the occurrence of deepest knee flexion across the 15 different trials. Solid line, hip joint moments of force; short-dash line, knee joint moments of force; long-dash line, ankle joint moments of force.

maximum positive hip torque, which exceeded 28 kg-m, usually developed 30–40 msec after the kicking foot left the ground.

Kinetic data that resulted from this investigation provided new information regarding magnitude, directionality, and temporal occurrence of human lower extremity forces. In particular, component and resultant joint forces derived in this study produced initial estimates of net joint forces in the lower limb during high velocity non-weight-supporting motion. Additional analyses will be necessary to begin to parcel out the effects of the individual mechanical elements in the equations of motion. Elaboration of these basic data should provide quantitative explanations and eventually lead to predictive kinetic information regarding human movement.

REFERENCES

Becker, J. W. 1963. The mechanical analysis of a football place kick. M. A. Thesis, University of Wisconsin, Madison.

Macmillan, M. B. 1975. The determinants of the flight of the kicked football. Res. Quart. 46: 48–57.

Roberts, T. W. 1972. Incident light velocimetry. Percept. Motor Skills 34: 263–268.

Roberts, E. M., and A. Metcalfe. 1968. Mechanical analysis of kicking. *In:* J. Wartenweiler, E. Jokl, and M. Hebbelinck (eds.), Biomechanics I, pp. 315–319. S. Karger, Basel.

Youm, Y., T. C. Huang, R. F. Zernicke, and E. M. Roberts. 1973. Mechanics of simulated kicking. *In:* J. L. Bleustein (ed.), Mechanics and Sport, pp. 181–195. A.S.M.E., New York.

Youm, Y., E. M. Roberts, T. C. Huang, and R. F. Zernicke. 1974. Kinematics and Kinetics of Human Motion. Kinesiology Res. Tech. Report No. 1: 02. Departments of Engineering Mechanics and Physical Education, University of Wisconsin, Madison.

Zernicke, R. F., G. Caldwell, and E. M. Roberts. 1976. Fitting biomechanical data with cubic spline functions. Res. Quart. (in press).

Zernicke, R. F. 1974. Human lower extremity kinetic parameter relationships during systematic variations in resultant limb velocity. Ph.D. dissertation, University of Wisconsin, Madison.

Simulation of lower limb kinematics during cycling

K. S. Nordeen and P. R. Cavanagh
The Pennsylvania State University, University Park

Cycling is a particularly suitable movement for simulation since at least one point on the body is constrained to move in a certain path, and this is a somewhat unusual occurrence in most human movements. The development of a satisfactory simulation would be useful for a number of purposes: first, under specified conditions the effect of such factors as pedaling speed upon rates of joint movement could be investigated without the need for performing a series of experiments. Second, the problem of fitting a bicycle to the dimensions of an individual rider and the consequences of a misfit could be investigated in a systematic manner.

Very few studies have examined the motion of the limbs during cycling. Sharpe (1896) presented various kinematic parameters in a comprehensive work that used examples from cycling to illustrate general concepts in mechanics. Fenn (1932) examined mechanical energy fluctuations in the lower limbs during cycling but gave no data on kinematic parameters during different experimental conditions. The determination of correct frame size and seat adjustment is largely carried out on a completely arbitrary basis by most bicycle riders. There is, therefore, a need to provide descriptive data on the movement of the lower limb segments during cycling at different speeds and with different bicycle adjustments. It is the purpose of this paper to develop a kinematic simulation of the lower limbs during cycling and to present initial results of its operation in the investigation of some of the problems discussed above.

METHODS

Input Data

A FORTRAN program called BLIMSIM (*B*icycling *Li*mb *M*ovement *Sim*ulation) was written to perform the necessary calculations. Input data in three categories are required to describe the dimensions of the subject (SDATA), the dimensions of the bicycle (BDATA), and the conditions of the simulated trial (TDATA).

SDATA Standard anthropometric dimensions are measured from the subject whose movements are to be simulated. These include thigh length, shank length, and lateral malleolus height together with a dimension called foot length. This last distance is obtained, with the foot in position on the pedal, by dropping a perpendicular from the lateral malleolus when the foot is horizontal and measuring the horizontal distance between this line and the center of the pedal spindle.

BDATA The bicycle dimensions (Figure 1) have been chosen so that they are relatively simple to measure. Seat height (BSH) is measured as the vertical distance between saddle surface and floor with the bicycle on level ground. This is an easier measurement to make than the usual measurement down the seat tube to the pedal. Bottom bracket height (BBBH) is measured in a similar manner and crank length is measured between the centers of the two spindles on each end. Pedal height (BPH) is taken as the distance between the pedal spindle and the surface of the pedal on which the foot rests. Finally the fore and aft movement of the saddle is taken into account by the dimension called seat adjustment (BSA), which is the horizontal distance of the front edge of the seat from the perpendicular line through the bottom bracket.

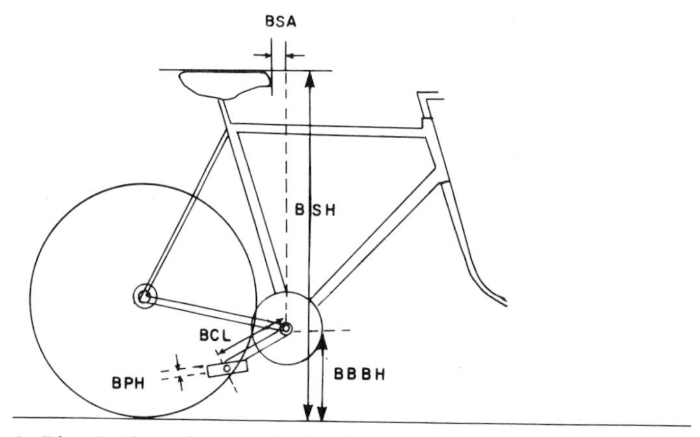

Figure 1. Bicycle dimensions required as input to the simulation.

TDATA The information regarding conditions of the experimental trial that is to be simulated include trial identification, pedal frequency, and film speed, the latter being the number of data points per second required on output. From these data an array of equally spaced crank angles (θ_1, in Figure 2*B*) is defined and a value for each output parameter is subsequently calculated for every crank angle between 0° and 360°. This assumes a constant angular velocity of the cranks throughout the pedal cycle.

Determination of Typical Foot and Hip Movements

Despite the fact that the pedal is constrained to move in a circle, the movements of the lower limb are still indeterminate for two reasons. Firstly, the most simple rigid body representation of the limb would be a three-bar linkage with thigh, shank, and foot as separate segments. An infinite number of movement patterns would thus be possible at the three joints while still satisfying constraints applied to the ends of this kinematic chain. Secondly, the hip joint does not remain fixed in space but moves as the cyclist redistributes his weight during the pedal cycle.

The approach taken in this simulation was to specify the pattern of movement of the foot and hip both as a function of crank angle. This procedure effectively reduces the problem to that of a two-bar linkage with known movement of the endpoints. There are then only two possible solutions, one of which is anatomically impossible. To obtain realistic values for hip position and foot angle, data were collected using high-speed cinematography from four racing cyclists pedal-

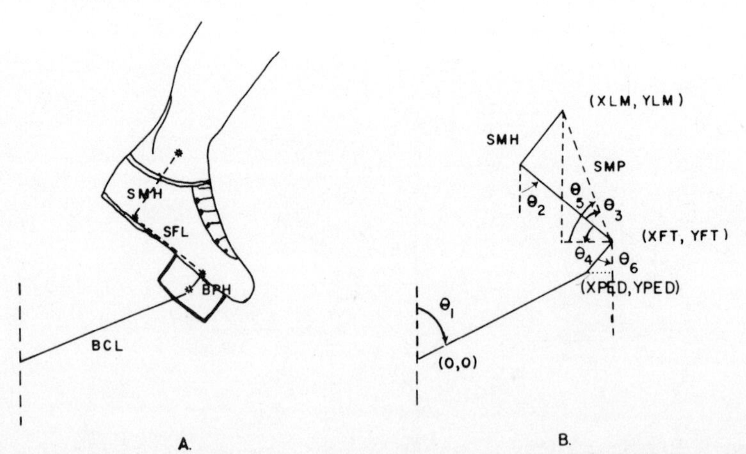

Figure 2. Linear dimensions (*A*) together with angles and geometrical construction needed (*B*), to determine the coordinates of the lateral malleolus (*X*LM,*Y*LM).

ling a racing bicycle on rollers at three workloads between 900 and 1600 kpm.

Foot Angle and Lateral Malleolus Coordinates It was found that foot angle (θ_2 in Figure 2*B*) as a function of crank angle could be approximated extremely well by a sine wave with suitable corrections for phase and amplitude. This is clear from Figure 3, where the simulated foot angle and the data from one of the experimental subjects are shown. The three values needed to generate the sine wave are also shown in Figure 3. These are the maximum and minimum foot angles and the foot angle at top dead center (TDC), (i.e., zero crank angle). With the nomenclature shown in Figures 2 and 3, the resulting approximation can be shown to be:

$$\theta_2\,(\theta_1) = \left(\frac{\text{FMAX-FMIN}}{2}\right)\ \sin\,(\theta_1 + \phi) + \left(\frac{\text{FMAX} + \text{FMIN}}{2}\right)$$

where the phase angle is $\phi = \text{ASIN}\ \left(\frac{2^*\text{FZERO}-(\text{FMAX}+\text{MIN})}{\text{FMAX-FMIN}}\right)$

Since no clear trend with workload was found from the experiments, mean values of 87.5, 42.0, and 61.9 degrees were calculated from all

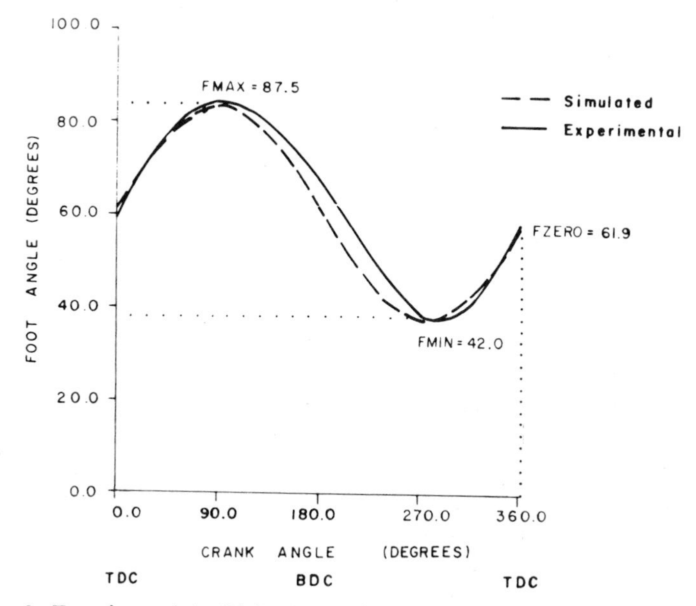

Figure 3. Experimental (*solid line*) and simulated (*dashed line*) foot angle to the vertical as a function of crank angle. The three values required to generate the sine wave approximation are shown in the figure.

subjects, all trials for FMAX, FMIN, and FZERO, respectively. This results in a phase angle of approximately —7 degrees.

Once the foot angle has been specified, the coordinates of the lateral malleolus can be calculated at each crank angle by simple trigonometry from the angles and lengths shown in Figure 2.

Hip Position The experimental data indicated that the movement of the hip joint center in the X direction was relatively small in all subjects (< 1 inch) and was thus neglected for present purposes. The migration of the hip joint in the Y direction as the cyclist rocks from side to side and shifts more of the body weight over the active leg was, however, considerable. This movement must be included in the simulation since an error in the hip–ankle distance of 1.5 inches at bottom dead center can cause an error of more than 20 degrees in knee angle. The Y coordinate of the hip was approximated by a sine wave in a similar manner to the foot angle.

Determination of Angles and Angular Velocities

With the coordinates of the hip and lateral malleolus defined, the remaining joint or segment angles could be calculated. The thigh angle to the vertical and the knee and ankle angles were found and their first derivatives were obtained by numerical differentiation (Lanczos, 1967). Full details of all the calculations and program coding are given in the BLIMSIM Users Manual (Cavanagh and Nordeen, 1975).

RESULTS

The simulation output from a trial at 90 rpm is shown in Figure 4, together with experimental data from the subject whose dimensions were input to the program. The thigh angle and knee angles are seen to be in fairly good agreement with maximum differences of the order of seven degrees. The ankle angle is the least satisfactory of the three simulated angles shown.

Within a reasonable range, preliminary experimental data suggest that seat height has minimal effect on the pattern of foot movement. The data presented in Figure 5 show thigh–knee diagrams similar to those described by Cavanagh and Grieve (1973) for two seat heights 1.4 inches apart, calculated with the assumption of similar foot movements. The effect of the change in seat height upon these angles is clearly seen, with maximum changes in the range of joint motion occurring in the region of bottom dead center.

Figure 6 shows the maximum angular velocities at the knee joint in both flexion and extension during the recovery and power phases of

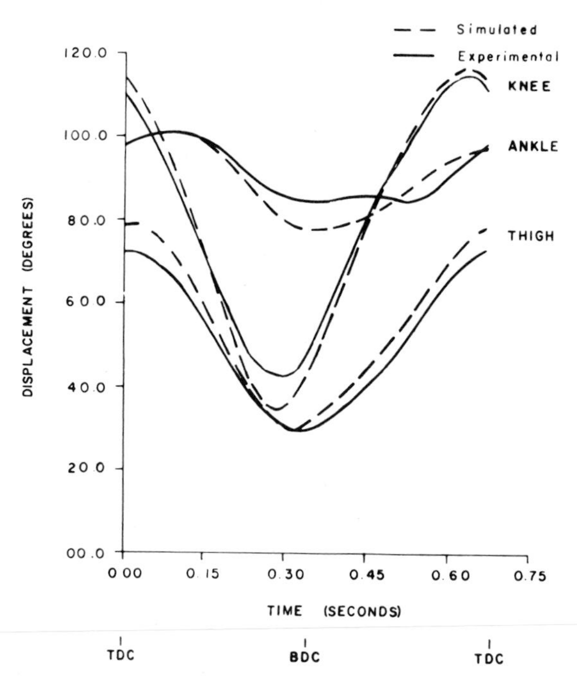

Figure 4. Lower limb joint angles for 90-rpm pedaling from experimental data (*solid lines*) and simulation output (*dashed lines*).

the pedal cycle, respectively, at simulated pedal frequencies between 40 and 120 rpm. The linear relationship between peak angular velocity in both flexion and extension and pedal frequency is apparent and it is interesting to note that at a given speed, the peak velocity in extension is greater than that during flexion. This is a consequence of dorsiflexion at the ankle during the downward power phase.

It is interesting to compare the power phase of the pedal cycle with the support phase of running. Data presented by Sinning and Forsyth (1970) indicate a peak angular velocity during extension of the knee of —2.6 rad/sec for running at a four-minute mile pace. Velocities of this magnitude would occur during cycling at pedaling rates below 40 rpm, indicating considerable differences between the two activities in this respect.

FUTURE WORK

This study has indicated the feasibility of simulating lower limb movements during cycling, and future work must attempt to put the simulation on a wider experimental base than the four subjects used in the

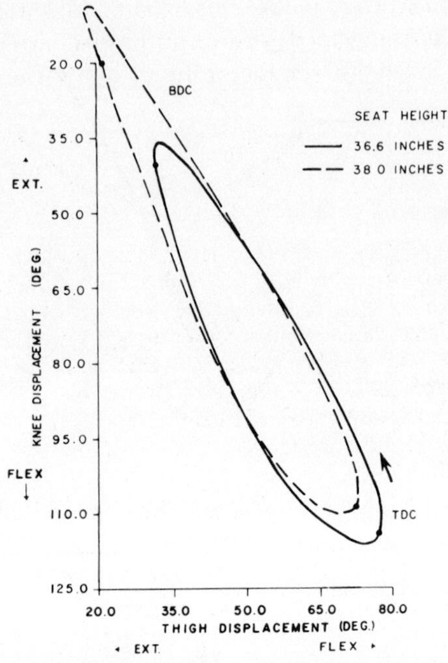

Figure 5. Thigh–knee angle diagrams from simulation output at two seat heights 1.4 inches apart.

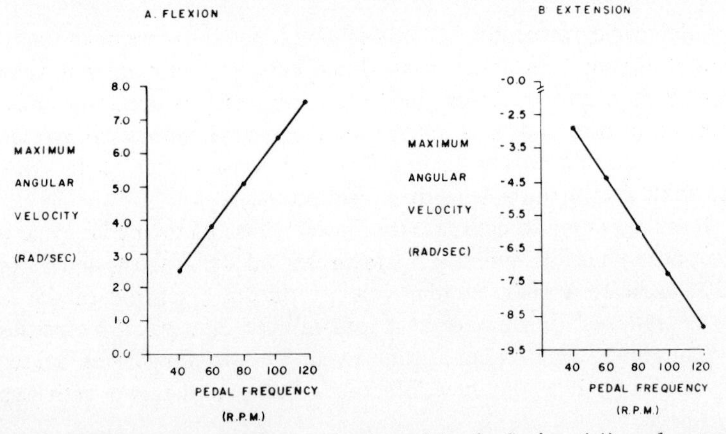

Figure 6. Peak angular velocities at the knee joint in flexion (A) and extension (B) from simulation output at five pedaling rates.

present paper. Further validation studies are planned, particularly with respect to foot angle patterns at various seat heights and on the effects of a wider range of workloads than was used in the present experiments. To eliminate the need for cinematography, a bicycle has been instrumented with potentiometers on both pedal and crank and electronic subtraction circuitry is being used to give an on-line value for foot angle. Extension of this kinematic simulation to include various kinetic factors is also underway.

REFERENCES

Cavanagh, P. R., and D. W. Grieve. 1973. The graphic representation of angular movements of the body. Br. J. Sports Med. 7, 1-2: 129–133.
Cavanagh, P. R., and K. S. Nordeen. 1975. BLIMSIM Users Manual. Unpublished document. Biomechanics Laboratory, Penn State University.
Fenn, W. O. 1932. Zur mechanik des radfahrens in vergleich zu der des laufens. Pflüg. Arch. 229: 354–366.
Lanczos, C. 1967. Applied Analysis. Pitmans, New York.
Sinning, W. E., and H. L. Forsyth. 1970. Lower limb actions while running at different velocities. Med. Sci. Sports. 2, No. 1: 28–34.
Sharp, A. Bicycles and Tricycles. 1896. Longmans, Green and Co., London.

Distribution of load forces in the hip joint under different parameters

F. Brussatis
Universität Mainz, Mainz

H. Dupuis
Max-Planck-Institut für Landarbeit und Landtechnik, Bad Kreuznach Am Kauzenberg

E. Hartung
Max-Planck-Institut für Landarbeit und Landtechnik, Bad Kreuznach Am Kauzenberg

D. Steeger
Universität Mainz, Mainz

Several authors (Rydell, 1966) attempted to record the forces in the hip joint during walking. Some representative examples include Koch (1917) who found these active forces to be 1.6 times greater than body weight. Grunewald (1918) estimated the muscle power acting on the femur to be about 400 kp. Storck (1931) also found the muscle power to be twice the value of the body weight. Further in his investigations with subjects standing on one leg or walking, Pauwels (1935) found the forces to be equivalent to 3 and 4.5 times body weight, respectively. Similar results were obtained by Inman (1947), Cabot and Peralba (1952), Knese (1955) and Müller (1957). Denham (1959) considered in his investigations various aspects of the problem such as shortening of the neck of the femur, protusio acetabuli, abductortenotomy, medial displacement of the femoral shaft and walking with a stick. Osborne and Fahrni (1950) noted that a change in the lever of the hip joint could produce changes in the forces acting on the joint. Hackenbroch (1961) found the active forces on the hip joint to be 4 times body weight minus the weight of the standing leg. Williams and Lissner (1962), Hochmann (1964), Rossi (1963), and Hauge (1965) came to similar conclusions with some slight alternations.

Rydell (1966) experimented on living subjects who had a total prosthesis of the hip joint. He made intravital measurements of the pressure on a prosthesis head during standing, walking on the flat, climbing stairs and running. He reported values between 2 and 3 times body weight. Pauwels (1973) summarized his investigations concerning biomechanics of the hip joint by stating that coxa valga and coxa vara are acting on the hip joint with different lever forces.

There has been until now, however, no experimental evidence published on the quantitative changes in the distribution of internal pressure caused by changes in biomechanical conditions in the hip joint. Experiments with a mechanical model were therefore designed specifically to find out what changes take place in the distribution of static forces at the point of contact between the acetabulum and the head of the femur under various conditions. In particular, we wanted to examine the effect of the adduction osteotomy (varus position) and abduction osteotomy (valgus position) with regard to achieving a reduction in the joint pressure. The following anthropometric parameters were determined:

1. Effective lever (h) of the abductors of the hip on the neck of femur (4 stages).
2. Effective lever (d_5) of the partial body weight between the vertical line of gravity and the point of rotation of the hip joint, with the subject standing on one leg.
3. Partial body weight (G_5) (3 stages).

It was necessary to develop a mechanical model which, apart from
All definitions and abbreviations are according to Pauwels (1973).
showing the three anthropometric variables mentioned, would make it possible to measure the force distribution by means of measuring the forces at five points in the joint, as well as the traction force of the abductors.

METHOD

A model (scale 1:1) was developed and constructed in the experimental workshop (Figures 1 and 2). It was designed to represent an adult of medium build, and permitted measurement of selected values of the anthropometric parameters. Thread spindles were used for the levers h and d_5. In this way, h could be adjusted between 33 and 81 mm, d_5 between 58 and 190 mm.

From the chosen total body weights G_6 (55, 75, and 95 kg) the partial body weights G_5 (45, 61, and 77 kg) were computed by subtracting the weight of the standing leg $G_B = 0.1864\ G_6$ (Schulte, 1951).

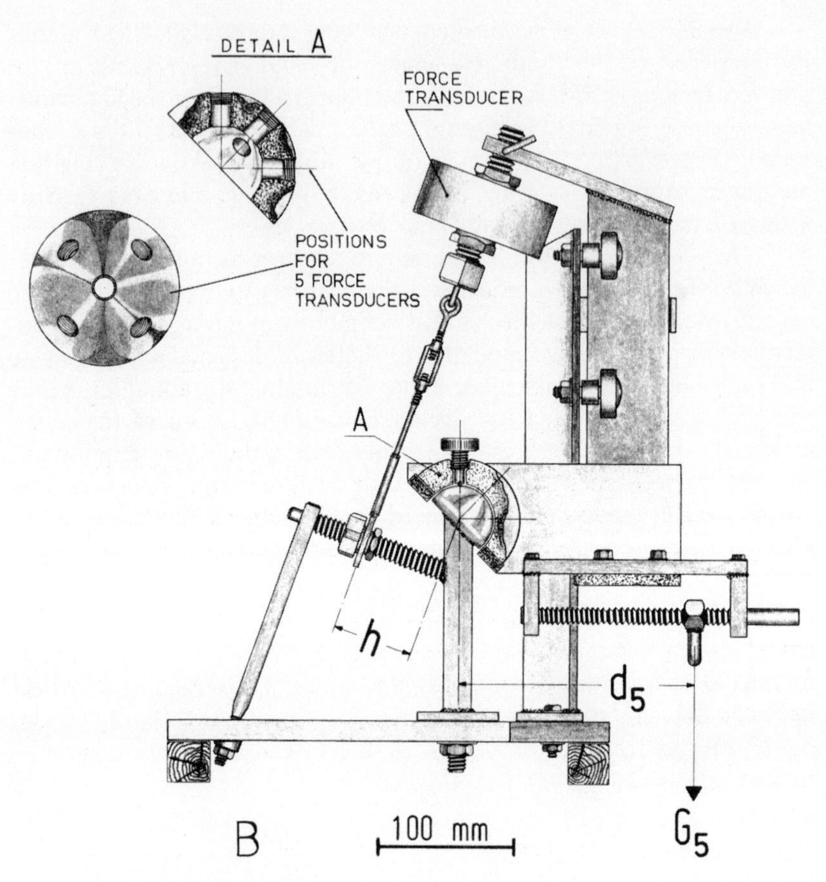

Figure 1. *A,* Model of hip joint.
B, Measuring model.

The traction force of the abductors (muscle force, M) was registered by means of an inductive force transducer, which was applied between the point of application of M on the variable lever h of the neck of the femur, and the simulated fixed constant pelvic attachment of the abductors of the hip.

The specially designed hip with its acetabulum and head of the femur was constructed so that the force transducers for measurement of force in the hip joint could be mounted into the socket walls. The force proceeding from the head of the femur is applied to the measuring cylinder through a 3-mm-diameter ball-bearing. The cylinder presses onto a circular metal sheet, to the back of which was stuck a strain gauge. The elastic flexion of the sheet measured in this way gives the value for the transmitted force. The electric voltage so measured shows

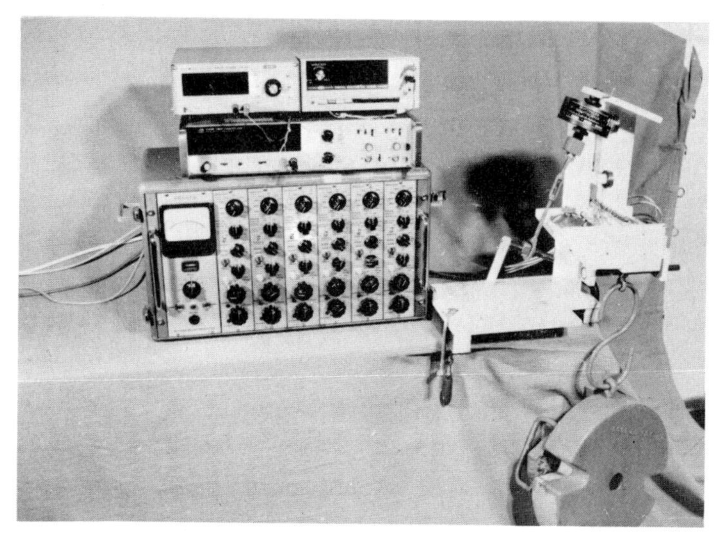

Figure 2. Experimental equipment with measuring model, amplifiers, and digital voltmeters.

a nearly linear relationship to the mechanical forces. In order to account for any temperature influence a second strain gauge was stuck to the back of a plate not under pressure.

Five identical force transducers were placed in the acetabulum as follows: the direction of the first transducer (no. 1) lay in the vertical anterior plane through the center of the joint, the second (no. 2) at an angle of 45°, the third (no. 3) horizontal, the fourth (no. 4) at 45° ventro-medial, the fifth (no. 5) at 45° dorso-medial.

Changes in the traction force M and pressure forces F_1 to F_5 (forces at no. 1 to 5) as electrical values were magnified by amplifiers. The voltage was displayed on a digital voltmeter (Figure 2). The static values were read off and noted directly.

Before and after each series of experiments, the six force transducers were calibrated by removing them from their fittings and applying known weights. In each of the three series (repetitions) of experiments, all values of h, d_5, and G_5 were adjusted sequentially to the following specific values:

$h_1 = 33$ mm, $h_2 = 49$ mm, $h_3 = 65$ mm, $h_4 = 81$ mm
$d_{5.1} = 58$ mm, $d_{5.2} = 124$ mm, $d_{5.3} = 190$ mm
$G_{5.1} = 45$ kp, $G_{5.2} = 61$ kp, $G_{5.3} = 77$ kp

The results of all six measurements (M and F_1 to F_5) were noted.

RESULTS

The values gained from the three experimental series are presented with respect to all registration points and experimental conditions in Figures 3–5. An example of the interdependence of all variables at point 1 is demonstrated in the three-dimensional diagram of Figure 6.

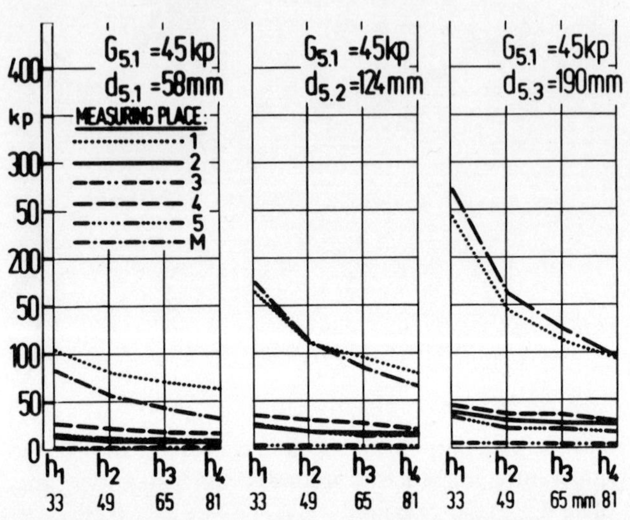

Figure 3. Partial forces F_1–F_5 and abductor forces M depending on h and d_5 for partial body weight G_5 = 45 kp.

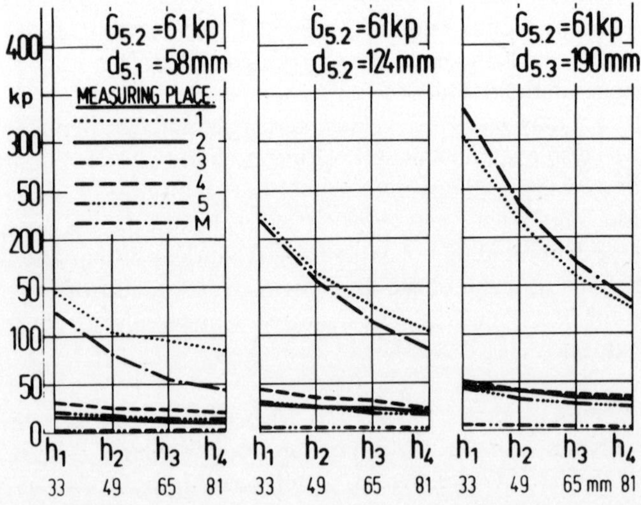

Figure 4. Partial forces F_1–F_5 and abductor forces M depending on h and d_5 for partial body weight G_5 = 61 kp.

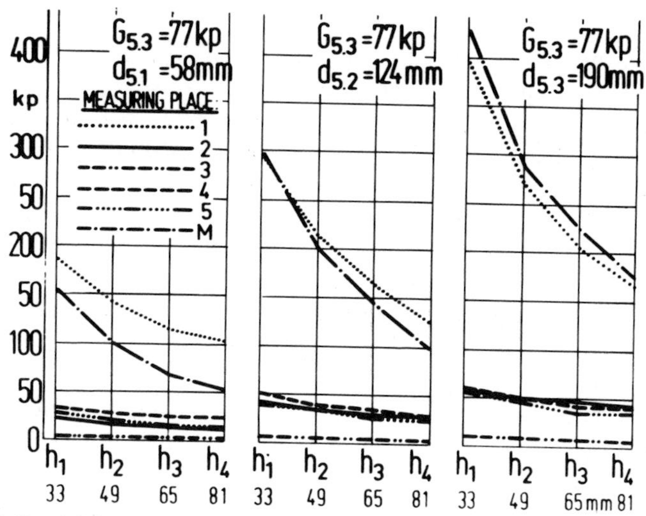

Figure 5. Partial forces F_1–F_5 and abductor forces M depending on h and d_5 for partial body weight $G_5 = 77$ kp.

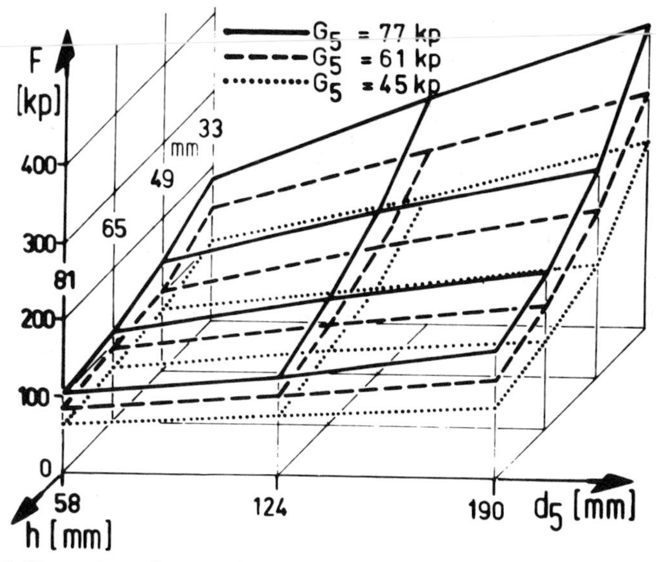

Figure 6. Force dependence on h, d_5, and G_5 at measuring place 1.

From the diagrams in Figures 3–5 it is clear that under the influence of the variants of the levers h and d_5 and of the partial body weight G_5, the force M of the abductors and the vertical force (F_1) on the hip joint reach the overall highest values (up to 400 kp) when compared with the other forces (F_2–F_5). There is a close relationship between forces M and F_1 under all conditions; as the lever h increases, they decrease. In absolute terms, they increase from $G_{5.1}$ over $G_{5.2}$ to $G_{5.3}$. When lever d is small, the force F_1 in the hip is greater than force M of the abductors. With a long lever ($d_{5.3}$) the force F_1 is somewhat lower than M while these forces are in a state of transition for the medium lever ($d_{5.2}$). The transducers in plane 45° (nos. 2, 4, and 5) show considerably lower forces (up to 50 kp) than the vertically measuring transducer (no. 1). These values also fall with an increase of h. The measuring horizontal force (no. 3) transducer shows the lowest absolute forces, which do not exceed 5 kp. The three-dimensional graph in Figure 6 depicts which combinations produce the highest values for the most important force, F_1.

Maximum forces for F_1 are found by a combination of $G_{5.3}$ (77 kp), $d_{5.3}$ (190 mm), and h_1 (33 mm), and minimum forces F_1 by $G_{5.1}$ (45 kp), $d_{5.1}$ (58 mm) and h_4 (81 mm).

DISCUSSION

Experimental work in biomechanical model demonstrates the distribution of load forces at the hip joint between the head of the femur and the acetabular roof under different conditions. The results show that very strong forces are acting on the upper lateral part of the acetabulum (100–400 kp) that are dependent on the muscular force of the abductors and the lever arm on which they act. These results are important in clinical cases of arthrosis in the upper lateral part of the hip joint, because lengthening of the abductor muscle lever reduces the load in the lateral part of the hip joint. This principle is realized in the operation of adduction osteotomy. In cases involving central arthrosis of the hip joint, such a surgical procedure of limited value since it does not reduce significantly the load forces in the center of the joint.

REFERENCES

Rydell, N. W. 1966. Forces acting on the femoral tread prosthesis. University of Göteborg, Sweden.

Pauwels, F. 1973. Atlas of biomechanics of normal and abnormal hip joints.

Force generation in leg extension

N. Yamashita and M. Kumamoto
Kyoto University, Kyoto

Based on the hypothesis that force exerted outward is limited by the muscle force at the weakest joint, the mechanism of generation and transmission of force was discussed by Takagi, Kumamoto, and Okamoto (1963) in their analysis of strokes in kayak. Yamashita (1976) supported experimentally the hypothesis of Takagi, Kumamoto, and Okamoto, that is, in maximal isometric leg extension the total force exerted outward as the resultant force of leg extension showed values close to the lesser of the two forces produced at the knee and at the hip joint from dynamic and electromyographical viewpoints. The present study was designed to investigate the generation and transmission of the resultant isometric force of leg extension when the force exerted was varied from minimum to maximum based upon the electrical activity of the muscles involved in the movement.

METHODS

Leg extension in which knee and hip extensions were combined was performed by one subject while in a supine position with both knee and hip joints kept at 90° angles. Analyses were carried out of the forces exerted along the line (g) from the ankle to the center of gravity of the body, and another line (k) from the ankle to a point near the knee (Figure 1*a*). Both knee extension (seated position, Figure 1*b*) and hip extension (supine position, Figure 1*c*) were performed independently. A dynamometer was pulled in a direction perpendicular to the longitudinal axis of the bone from the ankle joint for knee extension and from the distal end of the femur for hip extension. Direct and integrated EMGs were recorded simultaneously while the constant individual joint extension force or combined resultant force was exerted for a few seconds.

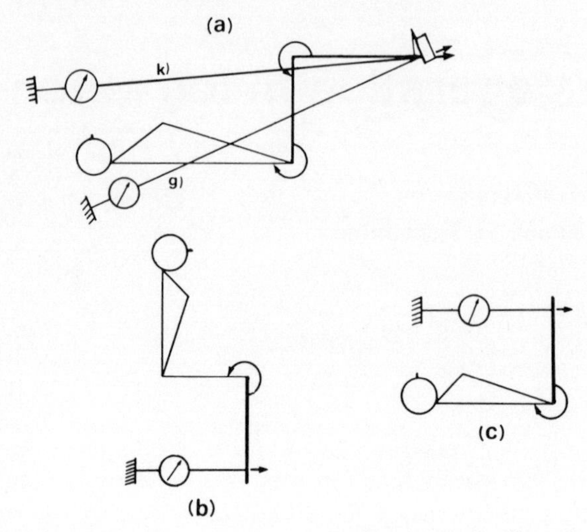

Figure 1. Experimental posture. (*a*) Leg extension: (*g*), the line of action of the force is from the ankle to the center of gravity of the body, (*k*), the line of action of the force is from the ankle to a point near the knee. (*b*) Knee extension alone. (*c*) Hip extension alone.

The isometric contractions were performed at several force grades from minimal to maximal in all tests. The value of integrated EMG for a second was measured in each force grade. Utilizing surface electrodes 10 mm in diameter, EMGs were recorded using a 13-channel multipurpose electroencephalograph (Sanei Sokki Co., EG-130). Muscles studied were the vastus lateralis, rectus femoris, gluteus maximus, and biceps femoris. EMG signal was integrated through a miller circuit.

RESULTS

As an example, integrated EMGs measured in each force grade of a subject are shown in Figure 2. Open circles, filled circles, and filled triangles show the values for knee and hip extension, leg extension, line *g*, leg extension, line *k*. The abscissa shows the percentage of maximal force exerted at each level and the ordinate, the value of the integrated EMG from the muscle tested at each level as a percentage of the EMG value for maximal force in each test. The vastus lateralis and rectus femoris were tested for knee extension alone and the gluteus maximus and biceps femoris for hip extension alone. In the case of the open circles (knee or hip extension alone) integrated EMG of every muscle increased in a nonlinear manner as the intensity of the force increased. A similar

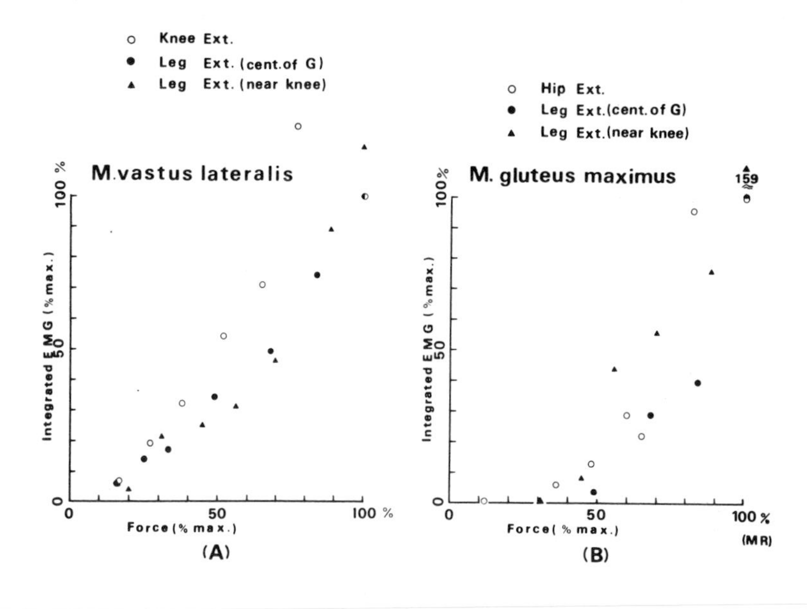

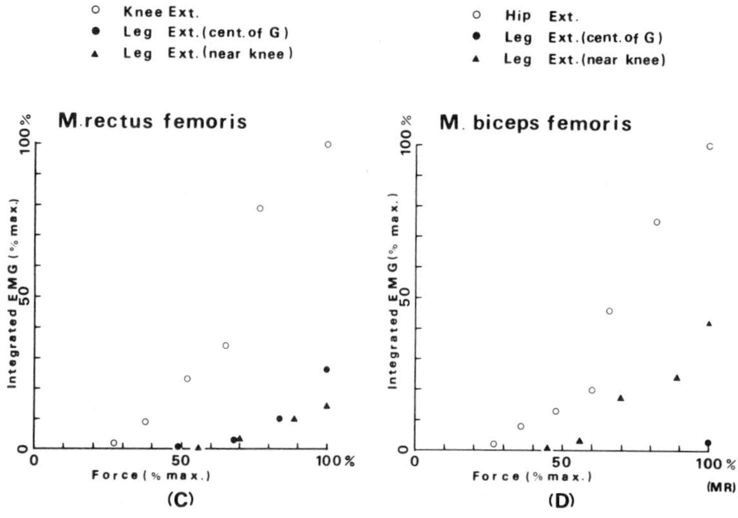

Figure 2. Relation between the integrated EMG and the resultant force. (*A*), vastus lateralis. (*B*), gluteus maximus. (*C*) rectus femoris. (*D*), biceps femoris. ○, Knee extension alone in *A* and *C*, and hip extension alone in *B* and *D*. ●, Leg extension exhibited along the line through the ankle and the center of gravity of the body. ▲, Leg extension exhibited along the line through the ankle and a point near the knee.

result was observed for leg extension in both directions (filled circles and filled triangles). However, this was not found in the cases of the rectus femoris and biceps femoris for both leg extensions where the slopes of the lines were considerably less than the ones for knee and hip extension alone.

DISCUSSION

Under the maximal effort of the leg extension, the joint that limited the resultant force shifted as the line of action changed according to Yamashita (1976). That is, when the force direction was through the center of gravity of the body and the ankle joint, the resultant force was limited by the force produced at the knee joint, but when the force was directed along a line from the knee to the ankle joint, the total force was limited by the hip joint. When a joint became the limiting factor, the single joint muscles that were participating in the joint movement exhibited near maximal activation. Cases were observed where the two-joint muscles were only submaximally activated as compared to the single joint muscles participating in the same motion, thus indicating that a graded activation of the muscles occurred even at maximal force production at a particular joint. From the present results for submaximal effort, the activation of both single joint muscles, vastus lateralis and gluteus maximus, was approximately proportional to changes in the total leg extension force provided the muscles were participating in the movement at the nonlimiting joint. While the values of integrated EMG of two-joint muscles, rectus femoris and biceps femoris, showed remarkably lower values in both cases of leg extension during maximal effort, which supports the findings of Yamashita (1976). In submaximal effort, the values for the two-joint muscles were also less. Thus, the slopes of the lines depicting the relationships between the integrated EMG and the force exerted were much lower than those for the knee or hip joint alone.

In summary, the relationship between the integrated EMG and the force exerted was observed during conditions in which the resultant force of isometric leg extension was systematically increased to maximum. The values for the integrated EMGs of the vastus lateralis, rectus femoris, gluteus maximus, and biceps femoris were approximately proportional to the forces exerted, even if these muscles were participating in the movements of the limiting or nonlimiting joint. Analysis of the slope of the line for the relationship between the measured integrated EMG and the force exerted revealed that single joint muscles produced the steeper slopes.

REFERENCES

Takagi, K., M. Kumamoto, and T. Okamoto. 1963. Kayakku no sutoroku ni tsuite. I. Kinryoku no dentatsukiko, II. Kinryoku to kyogiseiseki. (Studies on the strokes of Kayak. I. Mechanism of transmission of muscular force. II. Relation between muscular force and competition results. The Reports of Research Committee of Sports Science in Japan (in Japanese).

Yamashita, N. 1976. The mechanism of generation and transmission of force in leg extension. J. Human Ergology 4, 1: 41–50.

Effects of mass distribution and inertia on selected mechanical and biological movement components

C. H. T. Ward
Florida State University, Tallahassee

Under normal circumstances, the segments of the human body move in angular motion. Movements in work and play situations sometimes utilize implements and tools attached to the limbs. These elements are added to the neuromuscular-skeletal system and thus create a loaded situation. When body segments move in angular motion, the load that resists acceleration and deceleration is the moment of inertia. When objects are incorporated into or onto the moving segment they add to the original moment of inertia, so that the moment consists of both the segment plus the object. Since each individual is unique, it may even be said that within individuals, similar segments may be different. Constant loads will not provide the same resistive magnitude even within subjects, because of individual differences in limb segments. Hence, loads must be exclusive to each individual and perhaps even to each segment of that subject. Many investigators have not considered the moment of inertia as the resistive element in angular motion, nor have they created unique loads for each individual; either omission may seriously affect the validity of the study (Dern, Lavene, and Blair, 1947; Bouisset, 1964; Bouisset and Goubel, 1971). Since many investigations omitted the calculation of loads relative to individual characteristics, this study sought to generate loads that taxed subjects in proportion to personal mechanical properties, as did the study of Stothart (1973).

The moment of inertia consists of the summation of all the mass elements of a segment times the square of their distances from the axis

This research was carried out at the Biomechanics Laboratory, The Pennsylvania State University.

of rotation. Either the magnitude of the mass or the position of the mass relative to the axis of rotation can be manipulated to create mechanically identical moments of inertia. Thus, a load that consists of a large mass near the axis of rotation may be identical to a small mass located some distance from the axis. These were two of the elements under investigation in the present study: the effects of magnitude of inertia and equal moments of inertia created by different loads but concentrated near or far from the axis of rotation.

Anthropometric measurements were taken as well as volumetric measurements to estimate mass and calculate moment of inertia for the subject's right forearm and hand segments. Elbow extension in the horizontal plane was chosen as the movement used to investigate the variables of near and far loads. Each load was some multiple proportion of the individual subject's personal forearm and hand segment's moments of inertia.

Maximum angular acceleration, period of acceleration, and maximum velocity were the dependent variables used to describe the mechanical characteristics of movement under the various loads and conditions of load. In order to estimate the biological response to the treatment conditions, two measures of the electromyography (EMG) of the agonist muscle were used: integrated EMG/time to integrate to first movement and integrated EMG/time to integrate from first movement to maximum acceleration.

METHODOLOGY

The subjects were 16 male volunteers from the graduate and undergraduate programs as well as faculty and staff at The Pennsylvania State University. An apparatus was designed and constructed to allow the loading of the subject's forearm in multiples of the original moment of inertia without restriction of elbow flexion or extension. Weights were added to the apparatus at varying positions to manipulate the total moment of inertia and mass distribution.

A linear accelerometer was attached to the arm support apparatus for measuring the limb and apparatus acceleration. From the relationship between linear and angular motion angular acceleration was calculated from linear.

A potentiometer was used to measure the angular displacement of the forearm as a function of the voltage output from the variable resistance (elgon). For convenience the elgon was attached to the axis of rotation of the arm support apparatus.

Since forearm extension was the motor act investigated, the medial and lateral heads of the triceps brachii were monitored as the agonist muscle. The biceps brachii was considered the antagonistic muscle and its activity was also monitored. Surface electrodes (Narco-Bio System) were placed over the widest portions of the muscles. The muscle activity was amplified approximately 5,000 times by separate Tektronic amplifiers and displayed on a storage oscilloscope Model 564. The EMG signal was also amplified and output over a loudspeaker for subject auditory feedback.

The volumetric technique as described by Drillis and Contini (1966) was used to determine the volume of the hand and forearm for each subject. These were then summed to give the total mass of the forearm and hand (Dempster, 1955). This calculated mass value was used in the following equation for computation of the moment of inertia (I_0) of the subject's forearm and hand about the proximal end of the forearm:

$$I_0 = mk^2 + md^2$$

where m equals the total mass of the limb, d equals the distance from the axis of rotation to the center of gravity of the limb, and k equals the radius of gyration about the segment center of gravity.

Once the subject's forearm and hand moment of inertia values were known, the load conditions were calculated. The moment of inertia for the forearm, hand and support apparatus ($I1$) was then doubled ($I2$) and tripled ($I3$) to provide the appropriate load conditions.

The data for $I1$, $I2$, $I3$ were analyzed as a completely crossed design with repeated measures on the trials. Each subject was given ten trials to secure an average performance for that particular inertia treatment. After ANOVA, a series of linear contrasts were performed on the treatment means. The t ratios were then compared with tabular values to determine significance.

For the $I2$ and $I3$ conditions the data were compared between near and far load positions. A two $\times$ two factorial statistical design was utilized to compare inertial elements ($I2$, $I3$) and mass positions (near, far) using each subject as his own control.

RESULTS AND DISCUSSIONS

The results of the data analysis on the mechanical parameters demonstrated significant effects due to the inertial elements $I1$, $I2$, $I3$. Maximum acceleration between progressively greater resistance $I1$ and $I2$, $I1$ and

*I*3, *I*2 and *I*3 showed statistically significant decrements at the .05 level. Figure 1 graphically depicts the relationship between the mechanical dependent variable and the inertia and mass position treatment conditions.

The significant differences between the inertial loads were expected. As the load was increased or decreased the investigator expected to observe a corresponding variation in the mechanical and biological response. Stothart (1973), Bouisset (1964), Bouisset and Goubel (1971), and Dern, Levene, and Blair (1947) reported corresponding changes in the mechanical response to increased loads. In the present study, results indicated that from the *I*1 to *I*2 condition, a decrease in maximum acceleration of almost 40 percent occurred while from *I*2 to *I*3 a decrease of approximately 13 percent was observed.

The position-dependent variable was significantly different when using angular acceleration as the independent variable. The near position (large mass near axis) produced a significantly larger acceleration as compared with the far condition. Here the two positions for concentration of mass produced significantly different results even though the load was the same. For some reason, the position of mass concentration produced significantly higher maximum angular accelerations for the condition in which the load was nearest the axis of rotation.

The results for period of acceleration indicated that the *I*1 condition produced a mean time that was not significantly different from *I*2

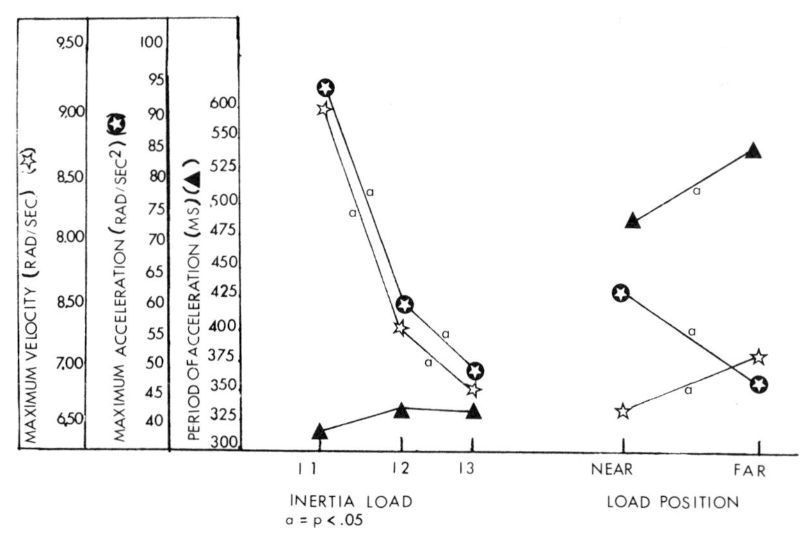

Figure 1. Results of mechanical analysis.

and *I*3. The inertia variable, however, did produce a significant mean difference for the *I*2 and *I*3 conditions, respectively. Again, this observation was expected since a larger load should take a longer time to reach constant velocity during maximal performance. Since the smaller load is accelerated more quickly than the larger load, it should also reach constant velocity sooner.

The position variable also produced significant results. The near loading condition produced a shorter period of acceleration, higher maximum acceleration but lower maximum velocity.

Figure 2 graphically depicts the relationship between biologically dependent variables and the inertia and mass position treatment conditions. The nonsignificant results for IEMG/T movement to maximum acceleration for the medial triceps and lateral triceps suggest that the biological response to the loads was nearly identical prior to movement. However, the significant difference in IEMG/time movement to maximum acceleration for medial triceps was interpreted as different between inertias, whereas the mechanical parameters were for the most part statistically different for load and position variables.

The biological system demonstrated general uniformity to the position and inertia variables. The data do not support the hypothesis of the investigator that a variation in the biological response would be observable with a change in load position.

The implication of the results of this research indicates that future research dealing with physical loading of limbs or segments must con-

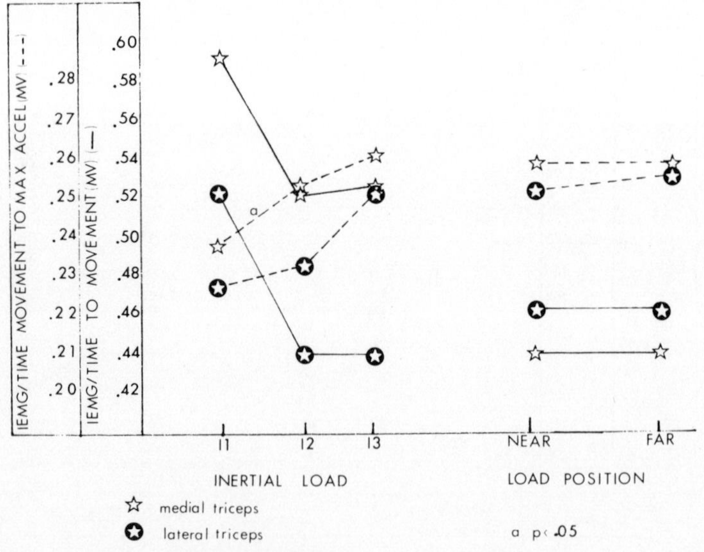

Figure 2. Results of biological analysis.

sider the distribution of mass variable. Simply calculating equal moments of inertia has been demonstrated in this study to be inadequate for creating identical mechanical response and serious questions have been raised regarding the biological responses to identical loads with mass concentrated at different points.

REFERENCES

Bouisset, S. 1964. Effet sur l'electromyogramme de la variation des facteurs definissant le moment de la charge deplacee. (Effect of definitive factors of energy variation on the electromyogram.) J. Physiol. 56: 303–304.

Bouisset, S., and F. Goubel. 1971. Interdependence of relation between integrated EMG and diverse biomechanical quantities in normal voluntary movements. *Activitas Nervosa Superior* (Praha) 13: 23–31.

Dempster, W. 1955. Space requirements of the seated operator. USAF, WADC, Technical Report. 55–159, Wright-Patterson Air Base, Ohio.

Dern, R. J., J. M. Levene, and H. A. Blair. 1947. Forces exerted at different velocities in human arm movements. Am. J. Physiol. 151:415–437.

Drillis, R., and R. Contini. 1966. Body segment parameters. Technical Report. 116603. New York University, New York.

Stothart, J. P. 1973. Relationship between selected biomechanical parameters of static and dynamic muscle performance. *In:* S. Cerquiglini, A. Vernerando, and J. Wartenweiler (eds.), Biomechanics III, pp. 210–217. S. Karger, Basel.

Relation between breaking and accelerating a movement with different loads

T. Bober and W. Erdmann
College of Physical Education, Gdansk

Movement in the opposite direction which appears as the preparation phase preceding the performance of the main task by concentric muscle work is a phenomenon that accompanies many human movements. It is of great importance in motions when the aim is to produce an extremely high final velocity. The relation between the impulse of the force needed to break the preparation phase and the one that accelerates the main movement determines, to some extent, the success of the movement. According to Hochmuth (1968), observing the right proportions between those forces is one of the basic principles of human movement. That phenomenon was studied by Hochmuth (1968) and Marhold (1963), who registered the force of reaction in the vertical jump, and by Kunz (1974), who concentrated on a similar problem. The aim of this study was to find the optimal proportions between the impulse of the force used to break and the one that was applied to accelerate the movement performed by a part of the body with a small mass and with different loads.

MATERIALS AND METHODS

The investigations were conducted on nine adult male subjects. Each of them performed a horizontal push of a weight with the upper limb. The weight was suspended on a rope put through a pulley. The examined subject acted on the weight by the rope, which he held in an extended hand (see Figure 1). Retraction of the hand caused the weight

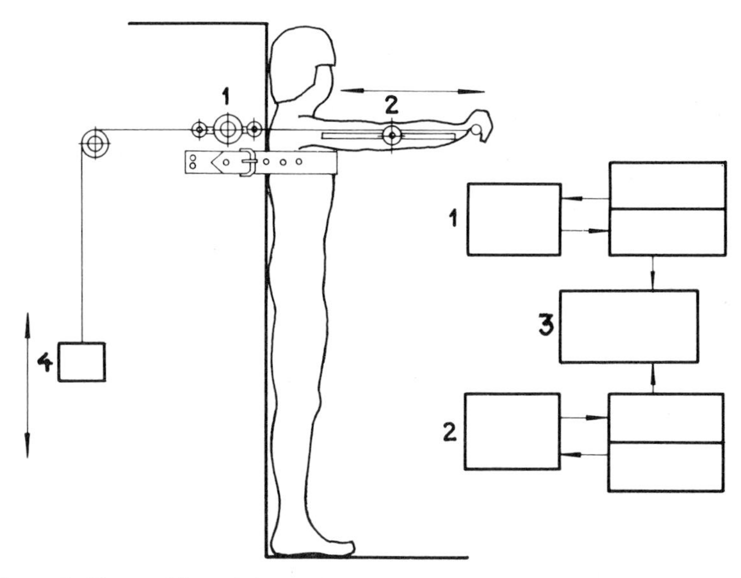

Figure 1. The position of the subject and the apparatus. Strain gauge (1), Elgon (2), recorder (3), and load (4).

to fall (negative work), which was immediately followed by a rapid extension of the arm (positive work). The body of the tested subject was stabilized in a standing position. The force was measured by a strain gauge (Figure 1, no. 1) set up in the rope and the angular changes by an electrogoniometer (Figure 1, no. 2). The results were recorded on an oscillograph while the distance of the hand retraction was measured mechanically. Each of the examined subjects performed five tests with loads of 2.5, 5, 10, and 12.5 kg. The surface contained under the curves of the force and time was the result of the push, in other words the impulse of the force (see Figure 2). The area P indicates the fall of the weight, area B equals area P and represents the breaking of the fall, and area A indicates the positive work that was done to overcome the resistance during the push. The area A represents the impulse of the force that determines the result of the push, and the relation of B to A indicates the quality of movement coordination.

RESULTS

Following the studies of other authors, the mutual dependence of the impulse of a force (that is, the result of the test) and of the relation of the impulse of the breaking force B to the impulse of the accelerating

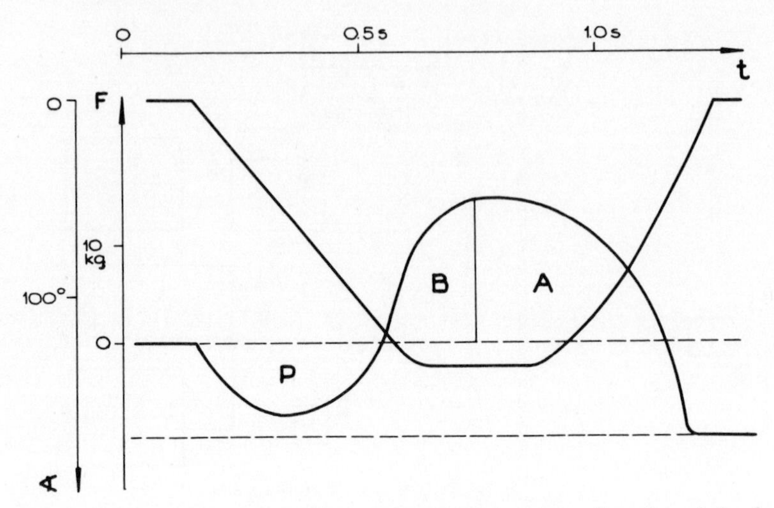

Figure 2. The example of force-time and angle-time curves. Impulse of the force of preparing phase (P), of breaking phase (B), and accelerating phase (A).

force A was examined (B/A). Figure 3 shows the distributions of mean results obtained for successive weights in five trials. The dot stands for the mean result obtained by all examined subjects. According to our expectations, the value of the impulse of a force increased as heavier weights were introduced. However, the larger the load, the higher were the proportions of breaking to accelerating forces. The correlation coefficient between the quantity of the pushed weight and the mean proportions in successive tests was $r = 0.96$, and the equation of weight regression to the B/A proportions was:

$$y = 0.034x + 0.87$$

The same phenomenon was studied in individuals where the relation between the results of successive trials and the B/A proportions was examined in the same subject. The tendencies observed were not different from those discussed above: in trials with heavier loads a greater impulse was achieved as well as higher B/A proportions. In addition it was discovered that at a given load the same subject obtained a better result at lower B/A proportions. That dependence increased along with the amount of the weight moved, which means that at larger loads, the correct movement performance had greater significance. This phenomenon can be utilized in the evaluation of movement coordination.

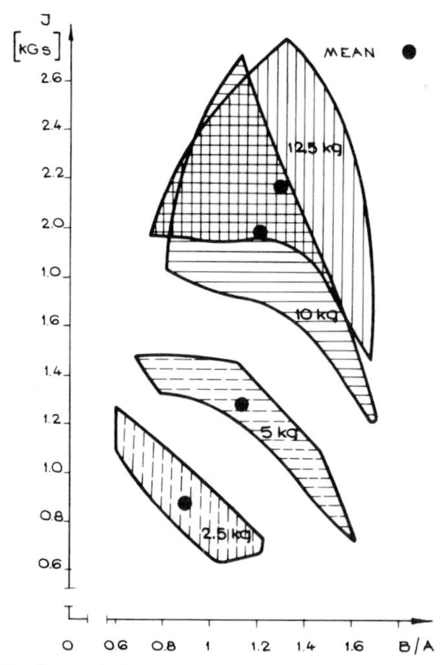

Figure 3. The distribution of the mean result of five trials performed by nine subjects with each of different loads.

Movement coordination, which is understood to be the use of potential possibilities, can be expressed as a relative value that was achieved by subdividing the impulse by the maximum isometric extension force of the upper limb (F_0) and thus obtaining an efficiency index (EI),

$$EI = \frac{I}{F_0} \times 1{,}000$$

Highly significant correlations between the efficiency index and the B/A proportions were found (see Table 1). Those correlations show that the lower the B/A proportions the better the result that can be achieved at the given load quantity. The tendency of the B/A proportions to increase in the tasks with heavier loads was still observed although it was not so distinct (see Table 1). The analysis of the factors accompanying the optimal breaking to accelerating proportions showed some regularities (see Tables 2 and 3). The performance time of the preparation phase seems to be of no significance at the loads utilized; however, a decrease of the angular velocity during that phase can be observed in the elbow joint. According to Hochmuth (1968) the period

Table 1. The optimal proportions B/A for the best relative results $[(I/F_o) \times 1{,}000]$ and correlation coefficients between these results and proportions B/A

Load (kg)	Optimal proportions	r
2.5	0.5–0.6	– 0.80
5	0.4–0.7	– 0.90
10	0.8–0.9	– 0.92
12.5	0.7–0.9	– 0.87

Table 2. Means of the time and distance of the trials with the best proportions B/A ($n = 9$)

Load (kg)	Time of phases (sec)			t_{break}/t_{acc}	distance of prep. phase (cm)
	t_{prep}	t_{break}	t_{acc}		
2.5	0.34	0.11	0.23	0.48	57
5	0.35	0.13	0.27	0.48	61
10	0.34	0.15	0.37	0.42	56
12.5	0.36	0.18	0.37	0.49	51

Table 3. Means of the angle and angular velocity in the elbow joint ($n = 9$)

Load (kg)	Angle (degrees) at:		Angular velocity of prep. phase (deg/sec)
	end of prep. phase	max. bend	
2.5	130	141	383
5	127	139	363
10	123	138	361
12.5	117	135	326

of time needed to break the preparation movement should be as short as possible. The conducted observations confirmed the above statement. For example, the mean times of the breaking phase in unsuccessful trials corresponded to the values of 0.13, 0.15, 0.17, and 0.19 sec for successive weights, which means that the subjects exceeded the mean times in the best trials (see Table 2). It was observed, however, that the rhythm of breaking and accelerating was not subject to any definite change in spite of the load increase. The times of these phases increased almost evenly. The amount of flexion at the elbow joint at the moment of completion of the preparatory movement and the maximum flexion in the unsuccessful trials were similar to the data obtained in the best test and are presented in Table 3. It can be seen, however, that the above quantities depend on the load. This fact indicates, although only an approximation as in the examined movement, that the whole kinematic chain was active. Thus, at higher loads the stretching of the extensor muscles of the elbow joint was less. This, however, is accompanied by an increased angle at which breaking took place (the difference between the maximum flexion and the angle at the end of the preparation phase).

CONCLUSIONS

Relating the observed proportions of the impulse of the breaking force and the impulse of the accelerating force to the absolute results, it can be stated that for upper limb movements the optimal proportions range from 0.6 to 1.3 and that they increase when a heavier load is utilized in the task. The results of the test expressed in relative values, that is, with respect to the isometric force (F_0), correlate significantly with the proportions of breaking to accelerating impulses, which means that this is the way to perform the movement. Carrying out trials with loads of different values showed certain relations between the optimal proportions of the analyzed movement phases and the parameters of kinematic movement structure such as the distance of the preparatory movement, the time of the breaking and accelerating phases, and the range of movement at the joint.

REFERENCES

Hochmuth, G. 1968. Biomechanische Prinzipien (Biomechanics Principles). *In:* J. Wartenweiler, E. Jokl, and M. Hebbelinck (eds.), Biomechanics I, pp. 155–160. S. Karger, Basel.

Kunz, H. 1974. Effects of ball mass and movement pattern on release velocity in throwing. *In:* Richard C. Nelson and Chauncey A. Morehouse (eds.), Biomechanics IV, pp. 162–168. University Park Press, Baltimore.

Marhold, G. 1963. Biomechanische Untersuchungen sportlicher Hochsprunge (Biomechanics investigation on sport high jump). Doctoral thesis, Leipzig. Cited by Hochmuth, 1968. *In:* J. Wartenweiler, E. Jokl, and M. Hebbelinck (eds.), Biomechanics I, pp. 155–160. S. Karger, Basel.

Biomechanical characterization of the behavior of an athlete's support-motor system under impact

A. A. Vain
University of Tartu, Tartu

In several events the activities of an athlete on a support surface are of an impact nature (in vaulting, broad jump, etc.). In the case of impact loads the acting forces acquire large values during a comparatively short period of time, causing the deformation of the athlete's support motor system. These deformations may be both of an elastic and a plastic character.

The energy released in the case of elastic deformations changes the velocity of the movement of parts of the body, while the restoration of plastic deformations may last for several days (Belyakova, Kuznetsova, and Kuznetsov, 1974). Plastic deformations are accompanied by the dispersion of energy and may cause injuries.

Several authors have studied the influence of mechanical loads on man in aviation (Diringshofen, 1932, 1942; Likhachov, 1935; Armstrong and Heim, 1938; Sergeyev, 1967; Gierke, 1964; Wittmann and Phillips, 1969; Savin, 1970; Stalnaker, Fogle, and McElhaney, 1971), but impact type exercises have not been investigated from a biomechanical standpoint. The biomechanical aspects of the athlete's skeletal muscular system make it possible to improve his technique to avoid micro-injuries and specify the methods of improvement of physical abilities in separate sport events.

METHODS

In laboratory and outdoor experiments five athletes performed take-offs with the aim of achieving maximum vertical velocity of the body. By means of two inductive three-coordinate accelerometers, the acceleration vector of the athlete's leg and body was registered on the paper of an oscillograph. The movements were studied by means of a 35mm camera that was synchronized with the paper oscillograph (see Figure 1).

RESULTS

The analysis of the experiments revealed that in spite of the same task of performance the absolute values of the accelerations in observed points and the character of their change were different in the case of one and the same athlete. Even in the case of one experiment, the maximum values of the two accelerometers did not coincide. The maximum values of the acceleration occurred in the braking phase on the leg. The load on the body was up to four times smaller. The oscillation frequency of the change in acceleration reached values up to 50 Hz. The gradient of the acceleration was the highest on the leg. The duration of the load in the braking phase and in that of take-off was approximately equal: 0.09 sec.

INTERPRETATIONS

On the basis of the recorded acceleration we can assert that in the case of impact loads, different forces fall on different parts of the body. The differences occur both in the absolute values of accelerations and in changes in time. On the basis of the study of the oscillation frequencies of the acceleration it is possible that when the oscillation frequency of the load and that of tensile muscle (Pakhomova, 1973) are close to the resonance frequency, microinjuries may occur.

Taking into account the inertial forces of the upper parts of body and gravitational forces, the values of the tension of the tendon calcaneus may be near the limit of the tensile strength of the mentioned sinew (Obysov, 1972).

CONCLUSION

In conclusion the body of an athlete during impact type movements must tolerate loads close to injury threshold levels. This causes the necessity for the biomechanical aspects to be considered in development of competition and training sites, equipment, and certain variants of sport technique. The aim is to elucidate possible origins of sport injuries, e.g., new sports equipment and variants of technique in order to exclude the danger of the injuries to the athletes.

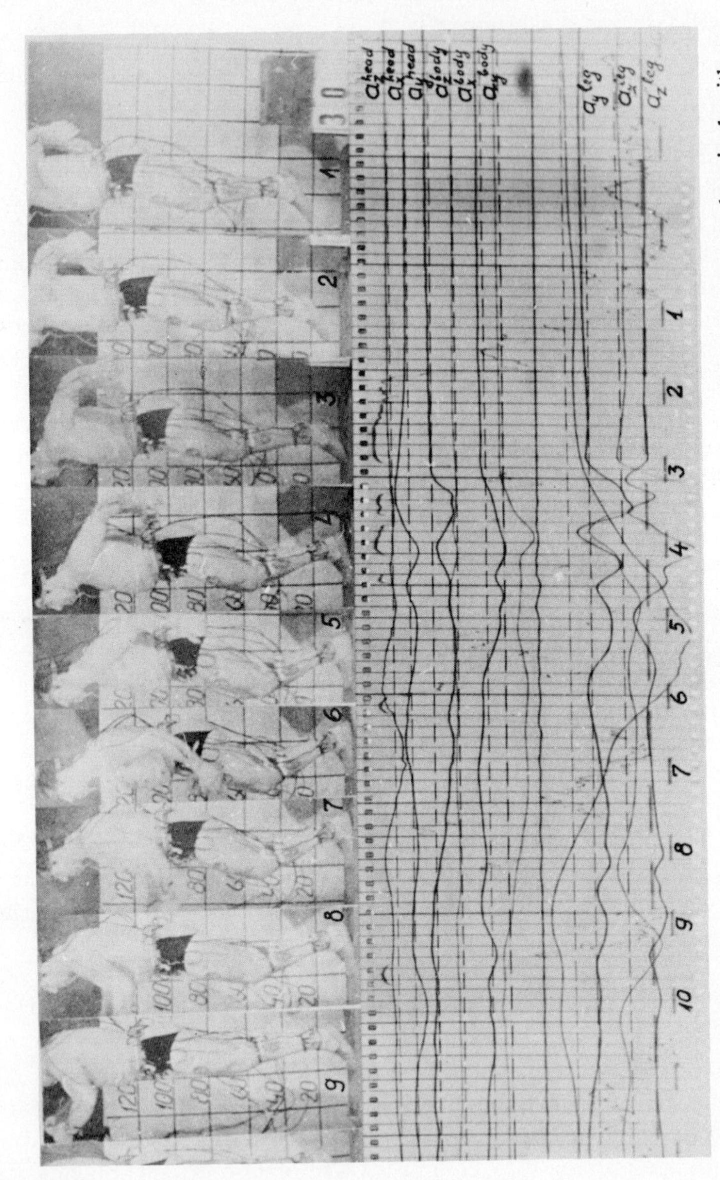

Figure 1. Acceleration vector of athlete's leg and body as shown by a 35mm camera synchronized with a paper oscillograph.

REFERENCES

Armstrong, H. G., and J. W. Heim. 1938. The effect of acceleration on the living organism. J. Aviation Med. 9: 199–215.

Belyakova, N. T., Z. I. Kuznetsova, and N. M. Kuznetsov. 1974. Influence of intensive loads on the apparatus locomotorius of young gymnasts. Gimnastika, coll. articles Vol. II, Moscow, "Fizkultura i sport" (in Russian).

Diringshofen, H. 1932. Die Bedeutung von hydrostatischen Druckunterschieden für den Blutkreislauf des Menschen bei Einwirkung hoher Beschleunigungen (On the significance of hydrostatic differences of pressure on blood circulation in the case of high acceleration). Z. Flugtechn. 23: 164.

Diringshofen, H. 1942. Die Wirkungen von Fliehkräften auf den Blutkreislauf des im Flugzeug sitzenden Menschen (On the influences of centrifugal forces on the blood circulation of men in a plane). Luftfahrtmedizin 6: 152.

Gierke, von H. E. 1964. Biodynamic response of the human body. Appl. Mech. Rev. 17, No. 12: 951–958.

Likhachov, A. A. 1935. On the influence of great accelerations on the organism. Proc. of All-Union conf. on stratosphere study, Moscow (in Russian).

Obysov, A. 1972. Die mechanische Festigkeit der biologischen Gewebe (Mechanical strength of biological tissues). Z. Mod. Med. S. 209–214.

Pakhomova, T. G. 1973. On the interdependence between elasticy, viscosity, force and bioelectrical activity of human muscles. Dissertation, Leningrad (in Russian).

Savin, B. M. 1970. The Effect of Acceleration and Functions of Central Nervous System. Nauka, Leningrad (in Russian).

Sergeyev, A. A. 1967. Physiological mechanisms of acceleration effects. Nauka, Leningrad.

Stalnaker, R. L., J. L. Fogle, and J. H. McElhaney. 1971. Driving point impedance characteristics of the head. J. Biomech. 4: 127–139.

Wittmann, T. J., and N. S. Phillips. 1969. Human body nonlinearity and mechanical impedance analyses. J. Biomech. 2: 281–288.

Jumping and running

Kinematic and kinetic correlates of vertical jumping in women

D. I. Miller and D. J. East
University of Washington, Seattle

By comparing the relative magnitudes of the body weight (W) and the vertical component of the ground reaction force (Rz), one can divide the Rz-time curve into three distinct phases that have functional significance for the standing vertical jump take-off (Figure 1). These phases are designated: preliminary unweighting, weighting, and final unweighting. Preliminary unweighting represents a major portion of the curve near the beginning of the take-off during which $Rz < W$ resulting in a negative (i.e., downward) acceleration of the jumper's center of gravity. For most individuals, this phase coincides with the initial downward motion of the body. The next phase, weighting, begins during the downward movement of the center of gravity and continues almost to the end of the upward motion preceding lift-off. Throughout this period, $Rz > W$ and the jumper accelerates positively with the maximum Rz often occurring when the body is at or near its lowest position. Final unweighting occupies a very brief time interval immediately before the jumper leaves the ground. Although the individual continues upward, the velocity is decreasing and, as a consequence, the acceleration is negative, thus making $Rz < W$. The role played by individual segments in producing this coordinated movement pattern has not been explicitly determined.

The present study was designed to investigate the vertical component of ground reaction force produced by female subjects during the performance of standing vertical jumps and to evaluate the contributions made by their body segments to the vertical impulse generated during the weighting phase of the take-off.

Research supported in part by Biomedical Sciences Support Grant 61–2300 NIH.

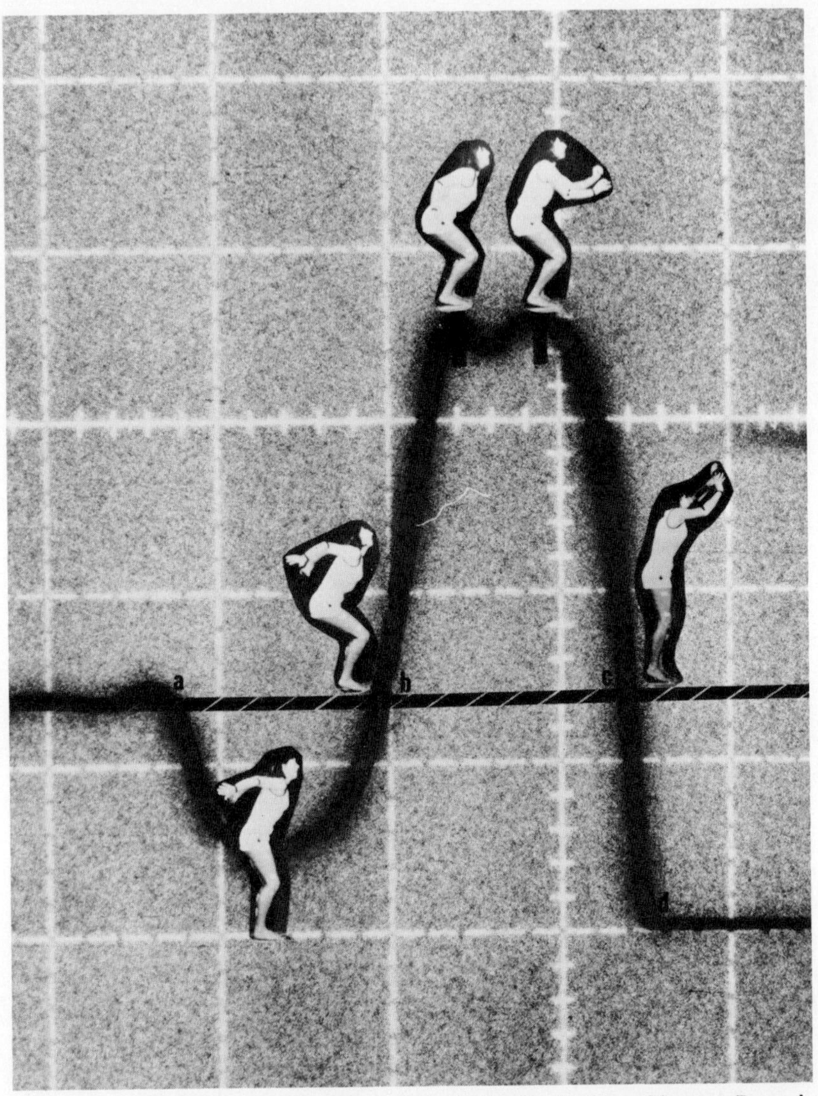

Figure 1. An Rz-time curve of subject three indicating body positions at Rz=min, W, max 1, max 2, and W, respectively, and the three major phases, preliminary unweighting (a to b), weighting (b to c) and final unweighting (c to d). The vertical scale is 500 N/div and the horizontal 0.2 sec/div.

PROCEDURES

After suitable orientation and practice periods, four physically active college women (ages 20–23 yr) performed eight standing vertical jumps for maximum height. The lateral view of these trials was filmed with a LOCAM camera operating at 98.4–98.9 frames/sec. Simultaneously, the ground reaction force was measured by a Kistler force platform (Type 9261A) and displayed on two Tektronix 5103N storage oscilloscopes. Permanent records were obtained by photographing the stored traces with a 35-mm Nikkormat camera. Time synchronization was accomplished by manually closing a switch to an oscillator, which triggered the horizontal sweeps of the oscilloscope beams, initiated a 25-Hz pulse train on the face of each scope, and activated one of the timing lights inside the camera.

The 16-mm films from four trials for each subject were analyzed on a Benson-Lehner digitizing system. Frames at 0.02–0.03-sec intervals throughout the take-off as well as those coinciding with critical points on the Rz-time curve were magnified 55 times and rear projected onto the analyzing surface. Coordinates of segmental endpoints were recorded on magnetic tape and subsequently processed by digital computer. Data analyses included determination of segmental and total body centers of gravity as well as body segment angles. Position-time relationships that corresponded to the weighting portion of the take-off were fitted through polynomial regression (least squares fit). In most instances, an eighth- or ninth-degree function described this phase of the vertical jump satisfactorily. The resulting smoothed curves were differentiated twice to determine linear velocities and accelerations. Integrals of segmental and total body inertial forces in the vertical direction were evaluated over the weighting phase.

A Hewlett-Packard 9820A Calculator and 9864A Digitizer were employed to integrate the force curves, determine maximum and minimum force values, and calculate the time required for the designated phases of the vertical jump.

RESULTS AND DISCUSSION

The analysis of the data was based upon three underlying assumptions. First, it was assumed that, during the weighting phase, each subject was moving her arms and legs symmetrically about her midsagittal plane; that her take-off was being performed in a single plane of motion; and that minor deviations of the segments from that plane did not necessitate employing three-dimensional cinematographic procedures. Second, it was assumed that the segmental mass distribution and mass center locations reported by Dempster (1955) and Dempster and Gaughran (1967) provided adequate estimates of these parameters for the subjects studied and that error introduced from this source would be small and thus not detract markedly from the results. Finally, the curve fitting technique utilized was considered to furnish reasonable estimates of position-, velocity-, and acceleration-time relationships. Visual inspection of the fitted curves and comparison of predicted force curves with those obtained from the force platform indicated that this was a logical assumption. Maximum forces calculated through double differentiation of the film data agreed within an average of 3.5 percent and force-time integrals within 4.3 percent of the force platform criteria. However, while characteristic intersubject differences in force patterns could be identified, subtle intrasubject variations were more difficult to discern.

Descriptive data on the average time durations of the three major portions of the force curve, maximum and minimum Rz values attained, and resultant vertical impulses [defined as $\int (Rz - W)dt$] elicited by each of the subjects are presented in Table 1. Times in the air are also indicated so that the relative efficiencies of the jumpers may be compared. Although not quantified in the analysis, a brief weighting period was displayed by subjects one, two, and four before their initial unweighting.

Analysis of the Rz-time curves indicated that subjects produced rather consistent yet individual vertical force patterns during the weighting phase. Subject one appeared to be somewhat restrained in her jumping pattern, employing a straight arm-swing throughout and exhibiting less hip flexion (91–99 degree hip angle) than the other subjects. Some of her force curves had a single apex, while the remainder had a 30–60 N dip at the top with the maxima on both sides being approximately equal in magnitude. Subject two displayed the greatest variability among the women tested. While a few of her patterns resembled a three-step ascending staircase, the most repeated had a minor peak followed by a dip and second peak about 150–220 N higher than the first. Subject three's weighting curve could be categorized as having predominantly one peak although there were minor perturbations at the apex (Figure 1). Her elbows flexed between 100 and 160 degrees and her minimum hip and knee angles were 89 and 82 degrees, respectively. Subject four, the most skilled jumper based upon time in the air, displayed the greatest hip flexion of the group (33 degree hip angle) and her weighting pattern was distinguished by two definite peaks of almost equal magnitude with the first being slightly larger. A drop of 300–500 N separated the two.

Table 1. Summary of resultant vertical force-time relationships during three phases of the vertical jump take-off*

Parameters	Subject			
	One	Two	Three	Four
Body weight (N)	548	615	655	576
Initial unweighting				
$\int Fz\, dt$ (N · sec)	−69 ± 5	−58 ± 6	−76 ± 9	−73 ± 7
Time (sec)	0.24 ± 0.01	0.41 ± 0.09	0.24 ± 0.01	0.39 ± 0.03
Min Rz (N)	50 ± 29	31 ± 5	16 ± 6	21 ± 4
Weighting				
$\int Fz\, dt$ (N · sec)	207 ± 6	220 ± 7	251 ± 7	251 ± 7
Time (sec)	0.30 ± 0.02	0.48 ± 0.07	0.30 ± 0.02	0.45 ± 0.03
Max Rz (N)	1451 ± 61	1358 ± 22	1790 ± 79	1399 ± 54
Final unweighting				
$\int Fz\, dt$ (N · sec)	−8 ± 1	−9 ± 1	−9 ± 1	−7 ± 2
Time (sec)	0.02 ± 0.00	0.02 ± 0.00	0.02 ± 0.00	0.02 ± 0.00
Free fall				
Time in air (sec)	0.49 ± 0.01	0.50 ± 0.01	0.51 ± 0.01	0.56 ± 0.01

*Values are means and standard deviations of eight trials per subject. Resultant vertical force $Fz=Rz-W$.

The percentage contributions of the segmental inertial forces to the weighting phase of the take-off are presented in Table 2. Since these forces were not constant, however, further insight into their roles was obtained by examining their magnitudes as functions of time (Figure 2). The trunk, because of its large mass, was responsible for the majority of the total impulse. Its inertial force-time curve was generally bimodal in all subjects. Although the present study investigated absolute rather than relative vertical forces, the contribution of the knee extensors to the upward acceleration of the trunk was inferred by examining the inertial force curves of the lower legs and thighs. In the trials of subjects one, two, and four and also in a few of those of subject three, these forces fluctuated near the zero line during the first half of the weighting phase and then exhibited a definite increase in a positive direction for the remainder of the period although they began to drop off rapidly at its termination. In these instances, it appeared that the initial maximum in the trunk force could be attributed principally to eccentric contraction of hip extensors since the jumper was still moving downwards. The second peak seemed to rely upon a large contribution from the concen-

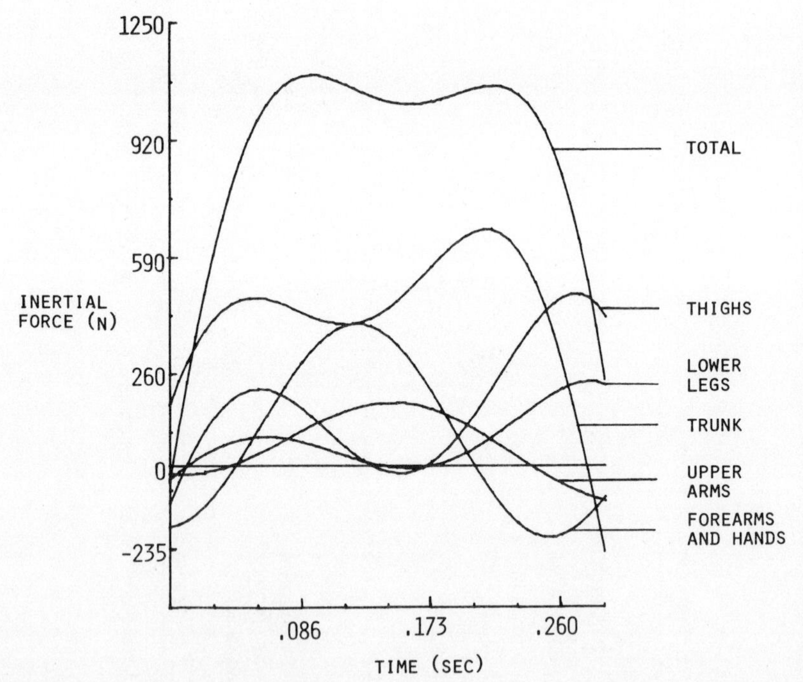

Figure 2. Segment and total body inertial forces during the weighting phase (subject three).

Table 2. Contributions of segment inertial forces to the vertical ground reaction force impulse during the weighting phase (percent)*

Segments	Subject			
	One	Two	Three	Four
Feet	0.8 ± 0.1	1.0 ± 0.1	1.3 ± 0.4	0.9 ± 0.3
Lower legs	5.0 ± 0.5	4.7 ± 0.3	7.2 ± 0.3	4.4 ± 0.3
Thighs	9.7 ± 1.3	9.1 ± 0.7	17.3 ± 0.6	7.1 ± 1.1
Trunk	54.5 ± 1.7	59.3 ± 2.8	46.6 ± 2.2	60.1 ± 0.6
Upper arms	9.6 ± 0.4	8.7 ± 0.4	7.8 ± 0.2	9.0 ± 0.7
Forearms–hands	11.5 ± 0.5	5.4 ± 2.8	10.8 ± 0.8	7.4 ± 0.6
Head–neck	9.6 ± 0.4	12.4 ± 1.6	8.0 ± 0.2	11.9 ± 0.3

*Values are means and standard deviations of four trials per subject.

tric action of the knee extensors. The trial of subject three, illustrated in Figures 1 and 2, however, implicated knee and hip extensors in both peaks since the inertial forces of the lower limbs in this case showed two comparable increases. In the women studied, the arms served to reduce the dip in the *Rz*-time function. Commonly, they exerted a negative force at the beginning of weighting, a positive force near the middle coinciding with the decrease in that of the trunk, and a negative force again at the termination of the weighting phase. Beyond these general observations that were noted in all subjects, individual patterns of segment contributions to the total impulse were also evident.

Future research will focus upon a relative force analysis and the determination of resultant muscular torques at the ankle, knee, and hip joints in an effort to gain additional information concerning this fundamental yet rather complex movement pattern.

REFERENCES

Dempster, W. T. 1955. Space requirements of the seated operator. WADC Technical Report. Aerospace Medical Research Laboratory, Wright-Patterson Air Development Center, Wright-Patterson Air Force Base, Ohio.

Dempster, W. T., and G. R. L. Gaughran. 1967. Properties of body segments based on size and weight. Am. J. Anat. 120: 33–54.

Polyparametric study of the vertical jump

M. Desiprés
University of the Orange Free State, Bloemfontein

Many researchers have used the vertical jump as described by D. A. Sargent (1921) as a test of "power." This was mainly due to L. W. Sargent (1924), who claimed it was a test of power.

However, after many years of application researchers applying the definition according to physics began testing the validity of this motor activity as a measure of power incorporating the essential components of displacement of the body weight as a function of time (Gray, Start, and Glencross, 1962; Davies, 1971; Cavagna et al., 1971; Adamson and Whitney, 1971; Payne, Slater, and Telford, 1968).

However, discrepancies regarding the findings among researchers have caused some concern. This can mainly be ascribed to a lack of standardization of performance procedures and methods of evaluation. The aim of this project was to make an intensive comparative study of the work done by previous researchers and to apply a multivariate study to determine the interaction among the following variables: displacement, velocity, and acceleration as a function of time; impulse, angle of knee flexion, angular velocity, and eventually power liberated during the vertical jump. In conjunction with these parameters the amount of positive and negative work done during the performance was determined electromyographically.

Research supported by the Human Sciences Research Council of Pretoria, Republic of South Africa.

METHODS AND PROCEDURES

Subjects ($n = 6$) were selected from three age groups, namely 13, 17, and 21 yr. Each age group was subdivided into sportsmen participating at provincial level and non-sportsmen.

The following parameters were researched.

Displacement of the greater trochanter making use of the Graph-Check sequence camera which was modified in order to give a pulse to the polygraph so that time elapsed between photos could be measured accurately. From the displacement-time graphs, velocity as a function of time was determined by drawing tangents at given time intervals. Acceleration was recorded directly using a Philips accelerometer placed over the greater trochanter. Knee flexion was recorded using an elgon. Vertical reaction force was recorded dynamographically from a simple force platform fitted with strain gauges. Muscle activity was recorded with Elema Schönander universal amplifiers using surface electrodes as sensors. The following muscles were examined: tibialis anterior, gastrocnemius, and soleus during the first jump and rectus femoris, biceps femoris, and vasti during the second jump. Electrode placement was standardized according to the method of Battey and Joseph (1966). Before each recording all sensors and recording equipment were recalibrated.

The subjects were not acquainted with the movement and were only allowed two to three practice jumps for orientation purposes before the test. Therefore, the influence of the learning factor was limited if not excluded. The impulse graphs recorded were used to compare movement patterns between subjects and within subjects. Power was determined according to the basic Newtonian laws of movement.

RESULTS AND FINDINGS

Evaluation of Polygrams (see Figure 1)

Vertical lines (A) were drawn from the camera pulses and used as reference points to be able to compare the various parameters during the performance. The paper speed was 100 mm/sec. Displacement time graphs (B) were plotted according to scale as well as velocity time graphs (C). Registration of reaction force (D) was recorded directly. Difference between baseline (E) and (F) represents body mass of the subject. The time difference between points (5) and (6) represents the

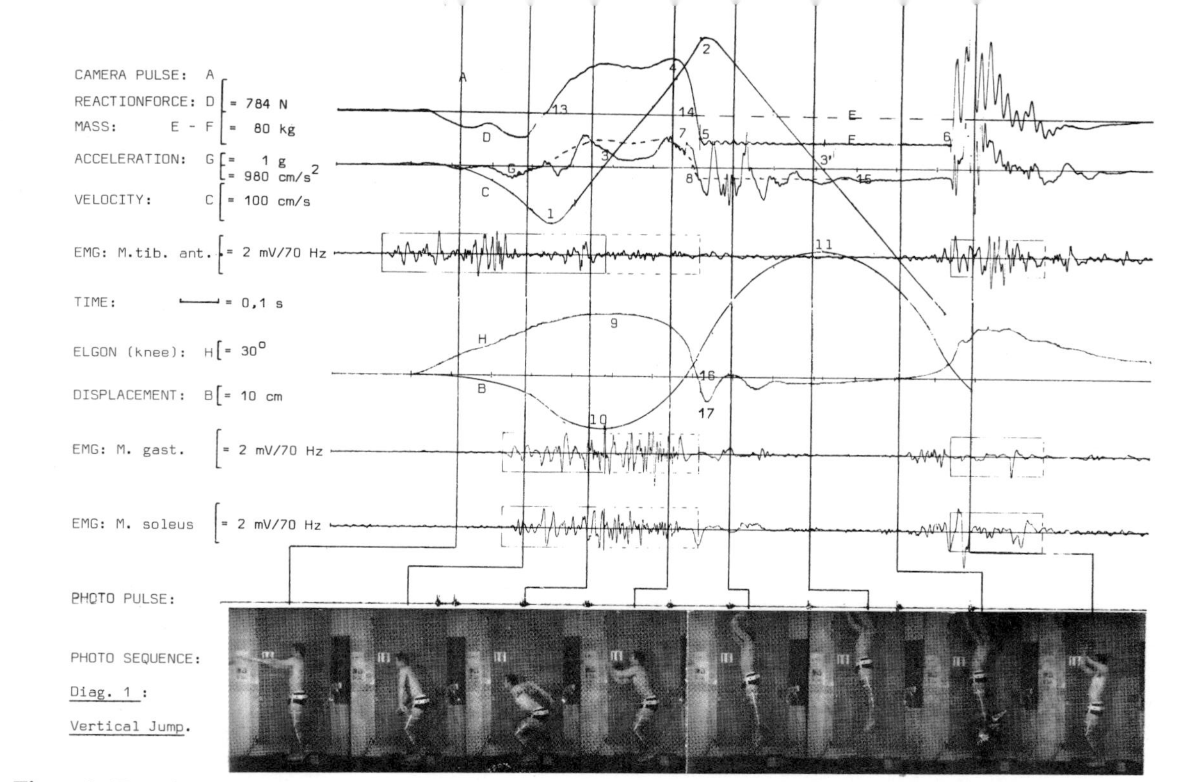

Figure 1. Experimental data of one sequence.

time of flight, i.e. after takeoff (5) and touch-down (6). Acceleration (*G*) was recorded directly. Angular acceleration in the knee joint was determined from the electrogoniograms (H) and time. During the vertical jump the applied muscle force is reflected by a ground reaction force (*D*).

During preparation for the jump, i.e., bending of the knees and swinging of the arms downwards (photo 2) the reaction force decreases and reaches a minimum. During this time eccentric contraction occurs in the leg muscles and after the turning point (10) the muscles perform positive work to bring about a positive reaction force (13), which coincides with maximum velocity in a negative direction (1). At point (3) velocity is zero, after which it increases rapidly until it reaches maximum at point (2) an instant before take-off (5). It is obvious that there is a decrease in velocity just before take-off. The outcome of this brings about a decrease in acceleration (*G* 7–8). At take-off (5) the reaction force is zero. Velocity of the body is in a negative direction reaching zero at point (3′), which is exactly in the middle of the flight, i.e., the turning point (11) on the displacement-time graph (*B*). The impulse graph (*D*) is depicted through points 13–4–14.

Figure 2 illustrates the impulses of reaction force of the different subjects during the two attempts, e.g., A_1 and A_2. The areas of impulse were determined planimetrically. It is clear that there are differences in areas with different age groups which possibly denote motor development. It should be noted that there is not a consistent increase in impulse for older subjects. Patterns of movement after repeat performances were very similar for each individual but differed significantly between subjects.

Regarding velocity and power it is interesting to compare results with other researchers (Table 1). It must be taken into consideration that different researchers did not use a standardized technique for performance. The values of Gerrish (1934) were high and it must be taken into account that this research project was done many years ago when instrumentation was perhaps not as sophisticated as today.

Table 2 is a comparison of ground reaction force obtained by different researchers.

Regarding angular velocity of the knee joint (Table 3) the author found that with an increase in age there was a decrease in the angular velocity during knee flexion and knee extension. The differences in angular velocities between age groups was, however, not statistically significant.

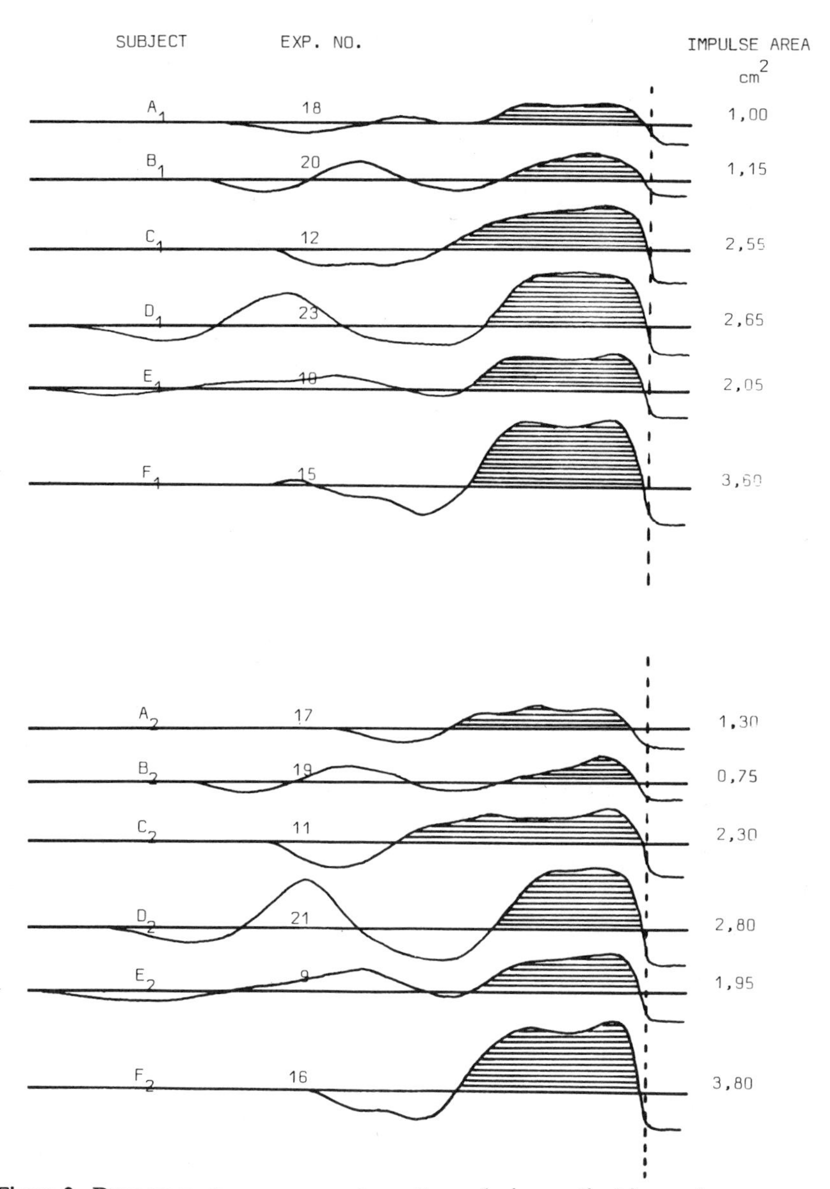

Figure 2. Dynagrams to compare motor patterns during vertical jump. Age group: *A* and *B*, 13 yr; *C* and *D*, 17 yr; *E* and *F*, 21 yr. Nonsportsmen: A, C, and E.

Electromyographic evaluation was done by enlarging the EMG's on a Kail reflecting projector. Amplitudes were measured and summated thereafter multiplied by the duration of the time of firing to give millivolt-second units (Desiprés, 1974). The rank order of biopoten-

Table 1. Comparison of findings of some researchers regarding velocity and power during the vertical jump

Researchers	Velocity (m/sec)	Power (kW)	Comments
Desiprés (1975)	2.90	2.05	Mean
Davies (1971)	3.40	3.86	Mean
Cavagna et al. (1971)	2.67	2.04	Mean
Gerrish (1934)	3.20	4.40	—
Gray et al. (1962)	?	1.22	No arms
Adamson and Whitney (1971)	2.20	?	—

Table 2. Comparison of findings of some researchers regarding reaction force

	N
Desiprés (1975)	1,244
Davies (1971)	1,647
Miller (1973)	1,249
Gerrish (1934)	1,468
Adamson and Whitney (1971)	1,147
Payne (1968)	1,302

Table 3. Angular velocity of knee joint

Age group	Flexion (rad/sec)	Extension (rad/sec)
21	2.49	5.36
17	2.73	6.26
13	2.76	6.71

tial liberated during positive and negative work can be seen in Table 4. The following correlations were determined:

Reaction Force/EMG: $r = 0.22$; $t = 0.702$ $(p \leq 0.20)$
Impulse/EMG: $r = 0.36$; $t = 1.235$ $(p \leq 0.20)$

Furthermore, it was found that the total biopotential liberated was 40 percent less in the group composed of active sportsmen as compared to the non-sportsmen, which might be ascribed to greater efficiency of movement. The amount of biopotential liberated during eccentric contraction was 37 percent less than during concentric contraction.

Table 4. Rank order of muscle potential liberated

Rank concentric contraction (positive work) (mV/sec)		Eccentric contraction (negative work) (mV/sec)	
1. Vasti:	5,088	Rectus femoris:	3,613
2. Gastrocnemius:	4,094	Tib. anterior:	3,557
3. Rectus femoris:	3,749	Vasti:	3,361
4. Biceps femoris:	2,915	Biceps femoris:	2,096
5. Soleus:	2,882	Soleus:	1,226
6. Tib. anterior:	2,586	Gastrocnemius:	1,124

This is in contradiction with Komi's (1973) findings. The conditions were different, however, and therefore this conclusion should be investigated further.

CONCLUSION

The conclusion reached in this study was that the application of the Newtonian laws of motion applicable to rigid bodies as applied to dynamic bodies in motion should be considered with caution. Mathematical models designed to simulate gross body movement to predict the outcome thereof must be designed with great care, always taking into consideration the multivariate factors that contribute to human motion.

REFERENCES

Adamson, G. T., and R. J. Whitney. 1971. Critical appraisal of jumping as a measure of human power. *In:* J. Vredenbregt and J. Wartenweiler (eds.), Biomechanics II, pp. 208–211. S. Karger, Basel.

Battey, C. K., and J. Joseph. 1966. An investigation by telemetering of the activity of some muscles in walking. Med. Biol. Eng. 4: 125–135.

Cavagna, G. A., L. Komarek, G. Citterio, and R. Margaria. 1971. Power output of the previously stretched muscle. *In:* J. Vredenbregt and J. Wartenweiler (eds.), Biomechanics II, pp. 159–167. S. Karger, Basel.

Davies, C. T. M. 1971. Human power output in exercise of short duration in relation to body size and composition. Ergonomics 14, No. 2: 245–256.

Desiprés, M. 1974. An electromyographic study of competitive road cycling conditions simulated on a treadmill. *In:* R. C. Nelson and C. A. Morehouse (eds.), Biomechanics IV, pp. 349–355. University Park Press, Baltimore.

Gerrish, P. H. 1934. Dynamic analysis of the standing vertical jump. Unpublished thesis, Columbia University, New York.

Gray, R. K., K. B. Start, and D. J. A. Glencross. 1962. A useful modification of the vertical power jump. Res. Quart. 33, No. 2: 230–235.

Komi, P. V. 1973. Measurement of the force velocity relation in human muscle under concentric and eccentric contraction. *In:* S. Cerquiglini, A. Venerando, and J. Wartenweiler, (eds.), Biomechanics III, pp. 428–433. S. Karger, Basel.

Miller, D. I., and R. C. Nelson, 1973. Biomechanics of Sport. Lea and Febiger, Philadelphia.

Payne, A. H., W. J. Slater, and T. Telford. 1968. The use of a force platform in the study of athletic activities. Ergonomics II, No. 2: 123–143.

Sargent, D. A. 1921. The physical test of a man. Am. Phys. Ed. Rev. 26 (No. 4): 188–194.

Sargent, L. W. 1924. Some observations in the Sargent test of neuro-muscular efficiency. Am. Phys. Ed. Rev. 29 (No. 2): 47–56.

Variation in horizontal impulses in vertical jumps

P. Tveit
The Norwegian College of Physical Education and Sport, Oslo

In this study vertical and horizontal forces in the vertical jump have been measured and corresponding impulses calculated in take-offs with and without a preparatory counter-movement, and before and after warming up.

The variation in vertical forces and in positive work in take-offs with and without a preparatory counter-movement has been extensively studied for many years. Among others Cavagna et al. (1971) and Asmussen and Bonde-Petersen (1974) have found that vertical forces and the positive work are greater in take-offs with than without a preparatory countermovement. In this study, however, the horizontal forces and impulses in different types of take-offs have been studied. The horizontal forces and impulses in take-offs from a semi-squatting position have been compared to corresponding forces and impulses in take-offs started with a preparatory counter-movement. Both kinds of take-offs have been studied before and after a standardized warm-up. The measurements in this study were made by means of a force-platform (Nigg, 1974), and the calculations were made by a computer to minimize errors and to make it possible to handle a larger amount of data (Tveit, 1975).

METHODS

About 320 children (14 yr of age) from different Norwegian schools were tested. The tests were arranged at their own schools.

The subjects made four vertical jumps. First they made one take-off (1A) from a semi-squatting position with 90° flexion at the knee-joint and one from a free position (1B) starting with a preparatory counter-movement. After these two jumps they all had a 5-min standardized warm-up on a bicycle ergometer, and then two jumps (2A) and (2B) of the same type as the former two. The children were asked to make all jumps as high as they could possibly manage and to land on the platform.

For each take-off the force-time graphs were printed by a UV-oscillograph and simultaneously recorded on an analog tape recorder (Tveit, 1975). The analog tape was later digitized by an A–D converter and recorded on a digital tape recorder. The digital tapes were then analyzed by a computer, and printed in suitable tables.

RESULTS

The force-time graphs were found to have two characteristic shapes as shown in Figure 1. The curves to the right are representative of take-offs started with a preparatory counter-movement.

The computer recorded the data from every take-off and printed the values indicated in Table 1. I SUM 2 is the sum of the numerical values of the horizontal forward and backward impulses in the part of the take-off with an upwards movement ($t > 0$). F. MAX is the largest

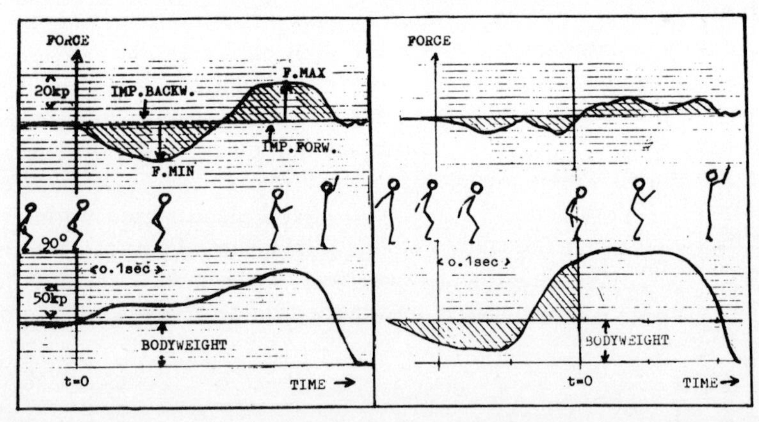

Figure 1. Examples of records from the force platform. *Ordinate*, force in kp; *abscissa*, time in sec. The records on top show the horizontal force and those at the bottom show the vertical force. The records to the left are representative for take-offs from a semi-squatting position, and those to the right are representative for take-offs started with a preparatory counter-movement. $t=0$ indicates where the upward movement is starting. F. MIN is the maximum force backward. F. MAX is the maximum force forward. The simple drawings in the middle indicate the different movements.

Table 1. Differences in horizontal forces and impulses in different types of take-offs. 1A, Take-offs from a semi-squatting position. Before warm-up. 1B, Take-offs from a standing position and with a preparatory counter-movement. Before warm-up. 2A; Like 1A but after warm-up. 2B, Like 1B but after warm-up. I SUM 2, Sum of the numerical values of the impulses backwards and forwards after $t=0$. F. MAX, Maximum force forwards. F. MIN, Maximum force backwards. BODY W, The average value of all bodyweights. MEAN, Arithmetic mean of all subjects. ST. D., Standard deviation for all subjects. N DIFF>0, Number of positive differences. N DIFF<0, Number of negative differences. % N DIFF>0, Positive differences given as percentage of the total N.

		$\frac{(1A—1B) \cdot \overline{BODYW.}}{BODY W.}$	$\frac{(2A—2B) \cdot \overline{BODYW.}}{BODY W.}$	$\frac{(1A—2A) \cdot \overline{BODYW.}}{BODY W.}$	$\frac{(1B—2B) \cdot \overline{BODYW.}}{BODY W.}$
I SUM 2 (Kp · sec)	MEAN	1.57	1.75	0.058	0.058
	S.D.	2.50	2.21	1.92	2.15
	N DIFF > 0	252	256	156	155
	N DIFF < 0	70	67	155	156
	% N DIFF > 0	78.0	79.3	50.2	49.8
	N	322	323	311	311
F. MAX (Kp)	MEAN	4.48	7.16	2.15	0.17
	S.D.	9.72	10.30	8.44	8.21
	N DIFF > 0	235	252	129	168
	N DIFF < 0	88	72	187	145
	% N DIFF > 0	72.8	77.8	40.8	53.7
	N	323	324	316	313
F. MIN. (Kp)	MEAN	4.42	4.19	0.87	0.81
	S.D.	9.95	8.85	7.57	8.73
	N DIFF > 0	228	219	172	170
	N DIFF < 0	93	106	146	144
	% N DIFF > 0	70.4	67.4	54.1	54.1
	N	321	325	318	314

force in the forward direction, and F. MIN is the largest force in the backward direction. For the take-offs from the semi-squatting position, $t = 0$ is the point where the upward movement is started. For take-offs with counter-movement, $t = 0$ is placed in a way that makes the two hatched areas at the bottom (to the right of Figure 1) equal.

The mean value of the maximal forces in the forward direction $(\overline{\text{F. MAX}})$ was found to be 26.1 kp and in the backward direction $(\overline{\text{F. MIN}})$ 23.4 kp. Corresponding values for the other kinds of take-offs were as follows: 20.9 kp and 18.8 kp for 1B, 24.4 kp and 26.1 kp for 2A, and 20.6 kp and 18.8 kp for 2B. The means of the numerical values of the forward and backward impulses in the part of the take-offs with upwards movement (I SUM 2) were as follows: 5.2 kp sec for 1A, 3.6 kp sec for 1B, 5.4 kp sec for 2A and 2.7 kp sec for 2B. To make the variations in the different kinds of take-offs more obvious, the differences in I SUM 2 were printed as shown in Table 1. The same was done for F. MAX and F. MIN in the same table. Table 1 is based on the actual *difference for every subject* and the mean difference is calculated from this.

The assumption has been made that there is a positive correlation between the body weight and the mass of muscles for each subject, and that there is also a positive correlation between the body weight and the horizontal impulses. To prevent errors due to these correlations, the impulses and forces for each subject have been divided by individual body weight and multiplied by the mean body weight of all subjects.

The corrected numbers using the following formula have been printed in Table 1.

$$\overline{\text{I SUM 2}} = \frac{1}{N} \sum \frac{[(\text{I SUM 2 in 1A}) \div (\text{I SUM 2 in 1B})] \cdot \overline{\text{BW}}}{\text{Body weight}}$$

Corresponding formulas have been used for $\overline{\text{F. MAX}}$ and $\overline{\text{F. MIN}}$, and for the different kinds of take-offs.

In Table 1 the subjects have been grouped in N DIFF > 0 and N DIFF < 0 as the impulses and forces are larger or smaller, respectively, without rather than with preparatory counter-movement in the take-offs.

DISCUSSION

In previous investigations Hochmuth (1967), Cavagna et al. (1971), and Asmussen and Bonde-Petersen (1974) have found that the vertical force is larger in take-offs started with a counter-movement than without.

In this study the same relations have been found for the vertical force, but the forces and the forward and backward impulses show the opposite relation: *the horizontal forces and impulses are smaller in take-offs with a preparatory counter-movement than without.*

The subjects were asked to jump as high as possible, and therefore there would be no value in horizontal impulses. The desired situation is, if possible, to direct the force vertically all the time without producing any horizontal components. When using the extensor muscles in the hip joint the total force would point backwards, while separate use of the extensors in the knee joint would make the total force point forward.

Among others, Hochmuth (1967) has maintained that the muscular coordination is improved when the movement is started with a preparatory counter-movement. The present study has verified these theories using an alternative method and given measurable expressions for the coordination. The mean values for both F. MAX and F. MIN have throughout the data indicated that the force is generated more directly and more steadily in the vertical direction in take-offs started with a preparatory counter-movement. The force components in the forward and backward directions are producing impulses in the corresponding directions, but to make the jump vertical these impulses must neutralize each other. The horizontal forces and impulses are wasted because they do not influence directly the height of the jump.

From Table 1 one will find that 78.0 percent of the subjects have the largest impulses in take-offs made without counter-movement when the take-offs are made before the warm-up. In take-offs after warm-up the corresponding ratio is 79.3 percent. From this one finds that both with and without warm-up *the horizontal impulses are smaller or the coordination is better* in take-offs started with a preparatory counter-movement.

The data show a similar tendency for F. MAX and F. MIN even though one will find that the difference in F. MAX with and without counter-movement increased considerably after warm-up. This may be due to the fact that this special warm-up on a bicycle ergometer had a larger effect on the extensors in the hip joint.

In take-offs before and after the warm-up there were only small variations in F. MAX, F. MIN and I SUM 2. The largest variations have been found in the take-offs from the semi-squatting position, where the warm-up seems to have a negative influence upon the coordination. Only 40.8 percent of the subjects have a smaller I SUM 2 after warm-up. This lack of influence or even negative influence was not expected because theories are teaching that coordination is improved as a result of warming-up.

CONCLUSIONS

1. The less the horizontal impulses the better the coordination in executing the vertical jump.
2. Coordination is better in the vertical jump take-offs started with a preparatory counter-movement than in take-offs without such a preliminary movement.
3. Coordination is not improved in vertical take-offs after a standardized warm-up on a bicycle ergometer.

REFERENCES

Asmussen, E., and F. Bonde-Petersen. 1974. Storage of elastic energy in skeletal muscles in man. Acta Physiol. Scand. 91: 385–392.

Cavagna, G. A., L. Komarek, G. Citterio, and R. Margaria. 1971. Power output of the previously stretched muscle. *In:* J. Vredenbregt and J. Wartenweiler (eds.), Biomechanics II, pp. 159–167. S. Karger, Basel.

Hochmuth, G. 1967. Biomechanik sportlicher Bewegungen (Biomechanics of Sport movement). Sportverlag, Berlin, pp. 187–209.

Nigg, B. M. 1974. Sprung-Springen-Sprünge. (Jump, Jumping, Jumped). Juris verlag, Zuerich, pp. 45–49.

Tveit, P. 1975. Methods and instrumentation when measuring forces in biomechanics. NORA 30, NIH, Oslo.

Reexamination of the mechanical efficiency of horizontal treadmill running

R. W. Norman, M. T. Sharratt, J. C. Pezzack, and E. G. Noble
University of Waterloo, Waterloo

The term mechanical efficiency is frequently used to describe the amount of work done as a proportion of the energy expended to do it. To perform the calculation one needs a measure of mechanical work, the oxygen cost of performing the work, and the caloric equivalent of the metabolic substrate assumed to be the primary energy source. In spite of an apparently simple calculation there is a disconcerting divergence of efficiency values for similar tasks that can be accounted for, not by individual differences of the subjects, but by the variety of methods used in the estimation of both the mechanical "work" done and the physiological cost. Depending on the techniques used to measure the mechanical efficiency, values range from 25 to 50 percent (Cotes and Meade, 1960; Cavagna, Saibene, and Margaria, 1964). This discrepancy relates to the different techniques used by these researchers to calculate mechanical work in the context of human motion. Their calculations seem to imply that the external work can be estimated by modeling the human body as if it were a point mass rather than a linked segment system. Furthermore, the rotation of the limbs, including the legs, are dealt with in a manner that gives one the feeling that their movements are extraneous to the motion of the center of gravity rather than the cause of it.

PURPOSE

The purpose of the study was to reexamine the concept of mechanical efficiency in horizontal treadmill running.

MATERIALS AND METHODS

The mechanical pseudowork output of each segment of a 12-member model was partitioned into three components, the kinetic energy of translation, the potential energy, and the kinetic energy of rotation.

The model developed is a rigid planar linkage model, similar to that of Fenn (1930) but including the head and neck combined, upper arms, forearms and hands combined, trunk, thighs, shanks, and feet. The instantaneous work done on each segment was taken as the change in energy level from frame to frame in the motion. The total pseudowork done was estimated as the sum of the absolute changes of the instantaneous energy, thus accounting for a continual physiological cost of both positive and negative work:

$$\text{TPW} = \sum_{j=1}^{n}\sum_{i=1}^{12} \left| m_i g h_{ij+1} - m_i g h_{ij} \right| + \left| \tfrac{1}{2} m_i V_{i\,j+1}^{2} - \tfrac{1}{2} m_i V_{i\,j}^{2} \right| + \left| \tfrac{1}{2} I_{cgi}\omega_{i\,j+1}^{2} - \tfrac{1}{2} I_{cgi}\omega_{i\,j}^{2} \right|$$

where

TPW = total pseudowork per stride
n = number of frames per stride
12 = number of segments in the model
i, j = segment number, frame number
m_i = mass of segment i
g = gravitational constant, 9.8 m/sec^2
I_{cgi} = segment moment of inertia about mass center
h_{ij} = vertical position of segment
ω_{ij} = segment absolute angular velocity
V_{ij} = segment resultant translational velocity including treadmill velocity

Subjects

Three trained subjects were used, one of whom (AT) is the World Master's Marathon Champion. Body weight variations among the subjects amounted to only 3.2 kg.

Procedures

Following an initial determination of $M\dot{V}O_2$ from a level treadmill test, the runners performed runs at 50, 66, and 100 ± 5 percent $M\dot{V}O_2$. Gas was collected 2 min before a 5-min run, during the last minute, for 15 min following, and for a final minute after recovery. The physiological cost of each run was calculated in three ways: (1) the gross caloric work rate with no adjustment for pre-exercise O_2 levels; (2) the net aerobic work rate (gross — pre-exercise O_2 level); (3) the net combined aerobic + anaerobic work rate. The average recovery O_2 consumption per minute was added to the per minute aerobic cost.

A 16-mm Locam camera, operated at 100 frames per second, was used to film the last 5 sec of the third minute of each run. Seventeen pairs of coordinates were located in each frame analyzed on a C.E.C. "digitable." Segment lengths were measured and their masses, mass moments of inertia, and centers of mass calculated using data from Dempster (1955). The displacement data were smoothed using a two-pass, digital filter (Winter, Sidwall, and Hobson, 1974) at a cutoff frequency of 7 Hz, and absolute angular and translational velocities of each segment were determined via finite differences from the filtered displacement data.

RESULTS AND DISCUSSION

Mechanical Efficiency Variation

Table 1 is a matrix of mechanical efficiency values calculated in several ways. The number derived varied appreciably depending upon what combination of numerator and denominator was used in the ratio of pseudowork output to work input. The table reveals differences in mechanical efficiency values of up to 8 percent by altering the value of the physiological work rate factor alone while much larger percent deviations were seen in altering the mechanical pseudowork term. Without knowledge of the numerator and denominator, even consistent values for mechanical efficiency would be difficult to interpret.

Interpretation of Mechanical Efficiency

Row A of Table 1 shows the total mechanical pseudowork rate as a percentage of the net aerobic physiological work rate. Considering only the mechanical efficiency number, runner AT was the most efficient of the three at any treadmill speed regardless of how this number was calculated. Runner AS was the least efficient. However, if one considers only the mechanical work term (numerators of row B), runners AS and DP are more efficient than runner AT. For example, they can maintain a treadmill speed of about 19 km/hr at a pseudomechanical work rate of about 75 cal/kg/min, a speed that requires a work output of about 90 cal/kg/min on the part of AT. On the other hand, AT must be considered most efficient physiologically because his net aerobic energy cost for this run was only 264 cal/kg/min, while AS and DP required an input of 291 and 271 cal/kg/min, respectively (denominators of row B).

High efficiency is normally reflected by a large work output at a low physiological cost. Accordingly, AT had an efficiency value over 30

Table 1. Mechanical efficiency of horizontal treadmill running

Subj.	AS			AT			DP		
% $M\dot{V}O_2$	50	66	100	50	66	100	50	66	100
Speed (km/hr)	9.8	13.0	18.7	11.0	14.4	19.6	9.6	12.5	19.2
					(percent)				
A. TPW/ NAerW	18.8	20.4	26.1	33.7	31.7	33.8	22.4	29.4	27.4
B.*	26.4 / 140.1	35.3 / 173.2	75.9 / 290.9	41.4 / 122.8	54.7 / 172.3	89.1 / 263.8	26.9 / 120.0	45.4 / 154.2	74.2 / 271.2
C. TKW/ NAerW	13.3	15.6	22.6	25.2	25.4	28.9	14.7	22.6	23.0
D. TPW/ NAer + AnW	—	19.0	24.1	—	29.0	31.9	22.2	26.9	26.1
E. TKW/ NAer + AnW	—	14.5	20.9	—	23.1	27.3	14.7	20.7	21.9
F. TPW/ GAerW	17.5	18.3	24.1	27.2	28.7	31.5	18.7	25.8	25.7
G. TKW/ GAerW	12.3	14.0	20.8	20.4	22.9	27.0	12.4	19.8	21.6

*Ratios of total mechanical pseudowork/net aerobic work (expressed as a percent in row A).
TPW, total mechanical pseudowork rate.
TKW, translatory kinetic work rate.
NAerW, net aerobic work rate.
NAer + AnW, net aerobic + anaerobic work rate.
GAerW, gross aerobic work rate.
All work rates calculated in cal/kg/min.

percent. However, if the speed is fixed, a low mechanical work output combined with a low physiological cost is desired. For example, if AT could combine his physiological delivery system with the mechanical output of AS or DP or vice versa, one could speculate that they would be "more efficient" runners. The new ratio derived for AT and AS calculated on a per kilometer basis is 15.6/55.7 Kcal/km = 28 percent. Paradoxically, this "efficiency" value is not impressively high, although a "mechanically" and "physiologically" more efficient hybrid runner has been produced than runner AT who had originally achieved a mechanical efficiency rating of 34 percent (row A, Table 1, 19.6 km/hr).

Since subjects AS and AT are most divergent, Table 2 has been produced to facilitate comparison of the runners segment by segment for the purpose of identifying sources of their differences in pseudowork output.

Mechnical Pseudowork Partitioning

The partitioned work is presented in two ways in Table 2, on an absolute scale, (Part *A*) (joules/stride), and as a percentage of the total mechanical work done (Part *B*). The subjects are very similar in proportions of the type of work performed although the absolute values are different. Translatory kinetic energy changes of the body segments account for about 85 percent of the total pseudowork, while rotational kinetic energy changes and potential energy changes account for about 5 and 10 percent, respectively. From 30 to 35 percent of the total work is accounted for by changes in motion of the trunk and head, about 10 percent by the arms and 55–60 percent by the legs. On a percentage basis, segment by segment, the two subjects are very similar. On an absolute basis the main difference between the two runners is the amount of translational kinetic work done by the thighs, shanks, and upper arms. The elevated mechanical work of AT is possibly associated with the energy changes necessary to accommodate his slightly longer stride (3.1 versus 2.9 m) in roughly the same time as that available to AS. A joint torque analysis would be necessary to confirm tthis.

The Model

The linked segment model used in determining the mechanical pseudowork has the advantage over point-mass types of models in that the work of the limbs, particularly that of the legs, becomes an integral part of the work output. Admittedly there are many assumptions made for a linked system just as there are for point-mass models. The largest problem is the assumed rigidity and thus constancy of the location of the

Table 2. Partitioning mechanical work per stride for two subjects running at 100% of $M\dot{V}O_2$

(joules/stride)

A	TWK		RKW		PW		Segment Total	
Subject	AT	AS	AT	AS	AT	AS	AT	AS
Upper arms	13.1	9.1	1.0	0.24	1.5	1.3	15.6	10.6
Lower arms	9.2	9.0	0.2	0.12	1.2	1.0	10.6	10.1
Head and neck	5.9	6.2	0.1	0.03	1.4	1.0	7.4	7.2
Trunk	55.7	53.2	1.6	1.16	8.9	7.6	66.2	62.0
Thighs	45.3	34.3	4.1	3.02	5.1	4.2	54.5	41.5
Shanks	45.8	36.6	3.1	2.32	2.8	1.6	51.7	40.5
Feet	23.9	20.7	0.9	0.76	2.0	1.2	26.8	22.7
Work type								
Total	198.9	169.1	11.0	7.65	22.9	17.9	232.8	194.6

(percent)

B	TKW		RKW		PW		Segment Total	
Subject	AT	AS	AT	AS	AT	AS	AT	AS
Upper arms	5.7	4.7	0.42	0.12	0.7	0.7	6.8	5.5
Lower arms	4.0	4.6	0.09	0.06	0.5	0.5	4.6	5.2
Head and neck	2.5	3.2	0.03	0.02	0.6	0.5	3.1	3.7
Trunk	24.0	27.3	0.70	0.60	3.8	3.9	28.5	31.8
Thighs	19.5	17.7	1.75	1.55	2.2	2.2	23.5	21.5
Shanks	19.7	18.8	1.34	1.19	1.2	0.8	22.2	20.8
Feet	10.3	10.7	0.37	0.39	0.9	0.6	11.6	11.7
Work type								
Total	85.7	87.0	4.70	3.93	9.9	9.2	100.3	99.7

TKW, translational kinetic work.
RKW, rotational kinetic work.
PW, potential work.

mass center and the length of the radius of gyration of each link. However, this system is closer to reality in describing human work output than the point-mass systems. Including the "negative work" as a physiological energy consumer rather than ignoring it, as others have done, is arbitrary. On the one hand, a slight overestimation of the potential energy during the falling portion of the airborne phase is made by including this value. This is not serious since the total potential energy change is small anyway. On the other hand, ignoring it results in an underestimation of the mechanical pseudowork done during a step, particularly when only the motion of the center of gravity is accounted for.

The total pseudo-mechanical work is the term that should be used as the work output. The net aerobic + anaerobic physiological work is the term that should be used as the work input. However, the method of partitioning the anaerobic component still needs refinement (Di Prampaero, 1970). One is compelled to agree with Whipp and Wasserman (1969) that "mechanical efficiency" is an inaccurate and misleading term. In the absence of knowledge of both the numerator and denominator of the ratio, one is at a loss to know whether the value cited is because the subject is mechanically inefficient, physiologically inefficient, or both, and a distinction between these terms is necessary.

REFERENCES

Cavagna, G. A., F. P. Saibene, and R. Margaria. 1964. Mechanical work in running. J. Appl. Physiol. 19: 249–256.

Cotes, J. E., and F. Meade. 1960. The energy expenditure and mechanical energy demand in walking. Ergonomics. 3: 97–119.

Dempster, W. T. 1955. Space requirements of the seated operator. WADC Technical Report 55–159, Wright-Patterson Air Force Base, Ohio.

DiPrampaero, P. E., C. T. M. Davies, P. Cerretelli, and R. Margaria. 1970. An analysis of O_2 debt contracted in submaximal exercise. J. Appl. Physiol. 29: 547–551.

Fenn, W. O. 1930. Work against gravity and work due to velocity changes in running. Am. J. Physiol. 93: 433–482.

Whipp, B. J., and K. Wasserman. 1969. Efficiency of muscular work. J. Appl. Physiol. 26: 644–648.

Winter, D. A., H. G. Sidwall, and D. A. Hobson. 1974. Measurement and reduction of noise in kinematics of locomotion. J. Biomech. 7: 157–159.

Inhibitory effect of long distance running training on the vertical jump and other performances among aged males

M. Ono
Tokyo Gakugei University, Tokyo

M. Miyashita
Tokyo University, Tokyo

T. Asami
Tokyo University, Tokyo

The fact that exercise does not always increase all aspects of motor ability and/or strength may well be expected from the limitations in the effects of isometric training, as Gardner (1962) pointed out. Nevertheless, many researchers seem to support the facilitative effects of exercise, based on the neural overflow theory of Davis (1900) and the cross-education effect by Hellebrandt (1962). In contrast, Henry and Smith (1961) proved the undeniable inhibitory effect of neural overflow, which was also suggested by the results reported by the current investigator in 1963. In the current study, experimental results show that long distance running training caused a deterioration of vertical jump performances.

EXPERIMENTAL RESULTS AND DISCUSSION

The Japanese Ministry of Education presented the results of a physical fitness test that was administered to a group of healthy males of ages between 50 and 54 yr. The subjects were a random sample from the national population. The test battery was composed of grip strength, vertical jump, repeated side-step, zig-zag dribble, and 1500 meter run-walk. The statistical analysis included the mean values of the total test score for five different groups, classified according to the amount of work done in their daily lives. According to the results, as shown in Figure 1, the highest test scores were obtained by the subjects in the most sedentary group, and the poorest by the most active group. The Ministry published the statistical results but did not make any comments at all about them.

The result of the current study seems to complement some of the reasons behind the above-mentioned statistics.

Running 4 km in 20 min on the average once every day was the assigned exercise for 18 wk to 36 healthy males ranging in age from 30 to 71 years. A physical fitness test was administered to the subjects, before and after 9 wk, and upon completion of 18 wk of exercise. Results of three tests were analyzed in order to obtain mean values for each test item for various age groups.

The distance covered in the 12-min run and the time spent to reach all-out effort by progressive loading, as shown in Figure 2, steadily increased with exercise in each age group. Among the four items demonstrated in Figure 3, grip and back strengths showed certain increasing trends, while vertical jump and back hyperextension recordings showed

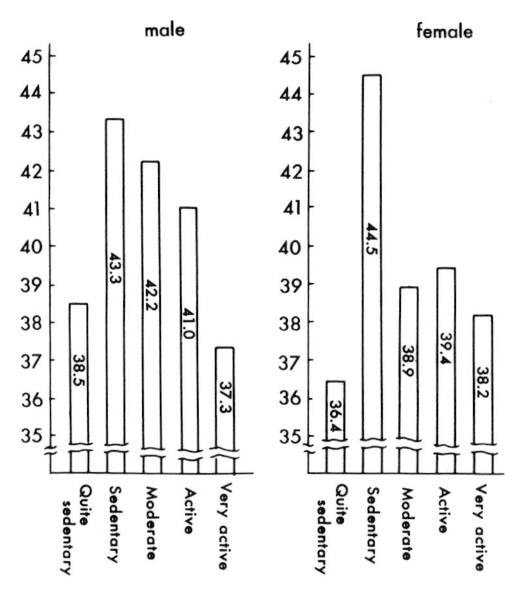

Figure 1. Total physical fitness test scores among middle-aged Japanese classified by work intensity (50–54 yr).

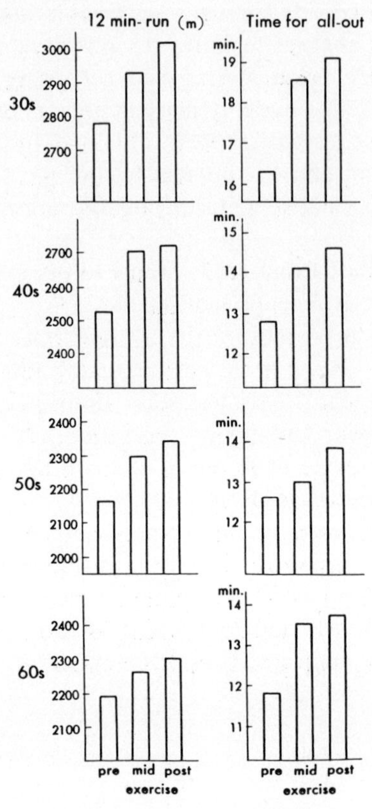

Figure 2. Changes in 12-min run and time to reach all-out on the time course along with training.

decreasing trends with age. A tendency for poorer performances was observed to occur with aging, particularly as the age advanced up to 50 yr.

When the vertical jump scores were analyzed according to the length of time of daily running, i.e., less than 10 min, between 10 and 25 min and over 25 min, the shortest running time resulted in largest decrease in jumping records. When other test items in each group were examined, as shown in Figure 4, the group with the greatest decrease in vertical jump scores had its largest increase in VO_2 max/wt and in running time to reach all-out fatigue. Hence, running speed during training exercise in this group was assumed to be quite fast. Since the vertical jump performances decreased in the 30-yr-old age group less than other age groups, age was judged not to be the primary cause.

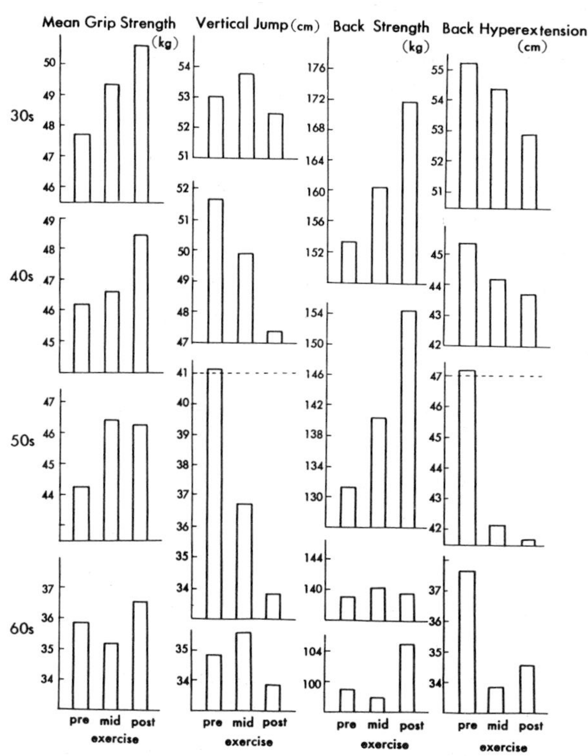

Figure 3. Changes in mean grip strength, vertical jump, back strength, and back hyperextension according to ages.

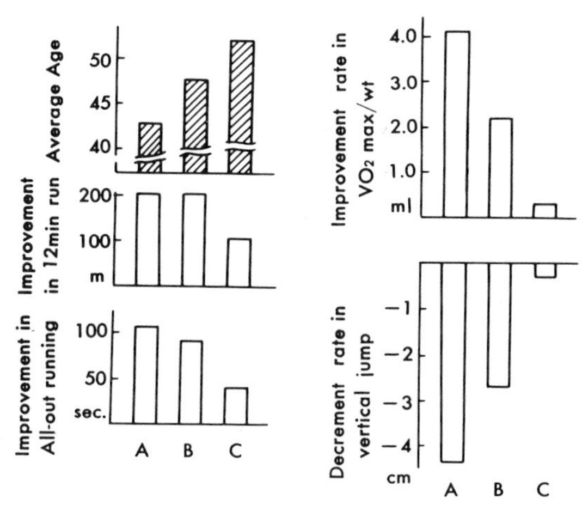

Figure 4. Test results classified according to amount of running time. *A*, less than 10 min; *B*, 10–25 min; *C*, over 25 min.

Running exercise of approximately 10 min with considerable speed and effort every day, therefore, might cause restraining of explosive muscular contractions such as needed in vertical jumping, for such activity is contrary to endurance running.

In Figure 5, the relationship between training time length and rate of decrease in back hyperextension is shown; here, the decreased rate was greatest in the group running less than 10 min. As has been pointed out repeatedly by this writer, the decrease in back hyperextension is suspected to be due to maturation or degeneration in the muscles. Should this premise be accepted, the decrease in vertical jump may be due more to changes in muscles themselves than to neuromuscular coordination or the intraneural system.

This reason is probably not the overwhelming cause but rather one of several causes, as one may see from Figure 6. For, no relationships were found between decreased scores in the vertical jump and in back hyperextension.

In Figure 7, the results of repeated side-step test, or test of agility, are shown. The running training apparently produced significant improvement in this test; the group running less than 10 min recorded an especially remarkable increment. Thus, running exercise is judged not to produce a restraining effect on neuromuscular coordination.

The positive effect on back strength is seen in Figure 8. The maximum isometric strength was exerted in attempting to pull the grip bar with the trunk flexed at 30 degrees in this test. The back strength increased the most among those whose running speed was the greatest.

The motor units, included in a bundle of muscles, do not always fire simultaneously; rather, certain motor units provided for excitation at

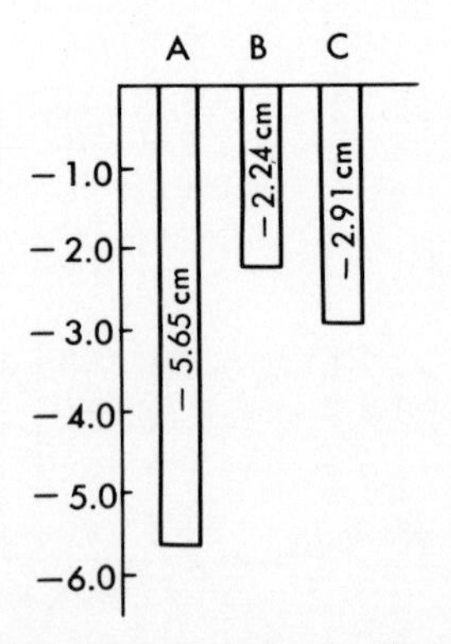

Figure 5. Decrement of back hyperextension classified according to daily running training time. *A*, up to 10 min; *B*, 10–25 min; *C*, over 25 min.

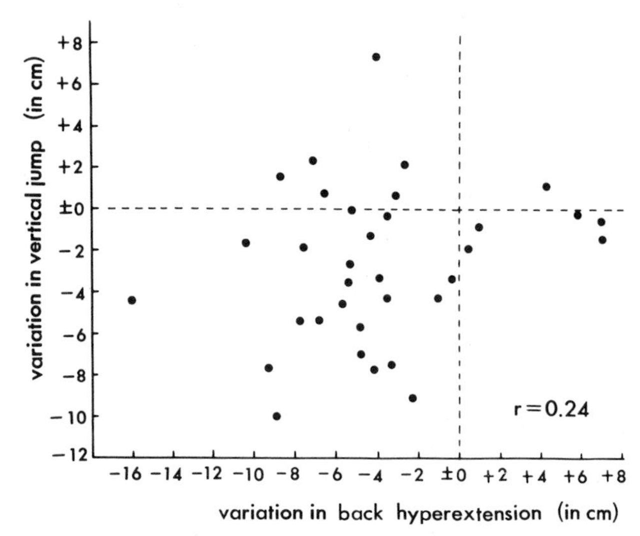

Figure 6. Correlation between the variation in vertical jump and back hyperextension.

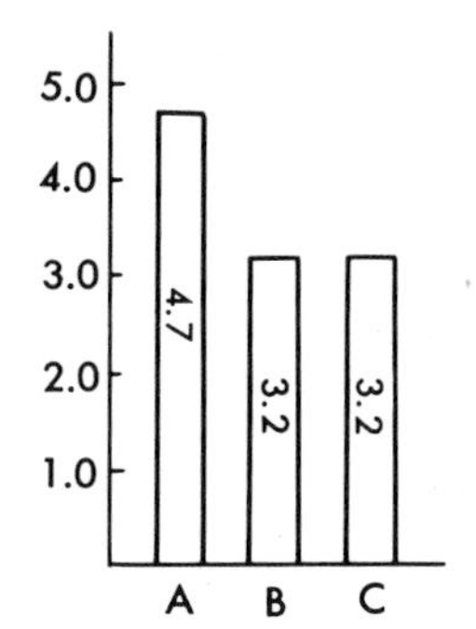

Figure 7. Increment of side-step score classified according to daily running training time. *A*, up to 10 min; *B*, 10–25 min; *C*, over 25 min.

certain joint angles are fired to participate with strength exertion at the respective joint angle. Therefore, an assumption may be made that the motor units to be mobilized for pulling the bar of the back strength dynamometer with the trunk tilted 30 degrees forward are apt to be strengthened by running exercise.

Such an assumption may well be applied to the vertical jump; i.e., running exercise may restrain only the group of motor units that are related to major muscles involved in vertical jump.

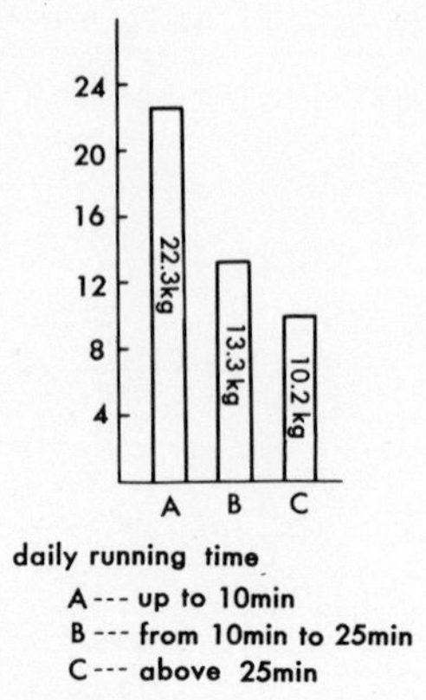

Figure 8. Variation in back strength according to daily running training. *A*, up to 10 min; *B*, 10–25 min; *C*, over 25 min.

CONCLUSION

In the case of aged males, prolonged running exercise might cause restraining of explosive muscular contraction such as vertical jump and a decrease in back hyperextension. No relationship was found between lowered scores in the vertical jump and in back hyperextension.

REFERENCES

Davis, W. W. 1900. Research in cross education. Studies Yale Psychol. Lab. 8: 64–108.

Gardner, G. W. 1962. Specificity of strength changes of exercised and non-exercised limb following isometric training. Res. Quart. 34 (1): 98–101.

Hellebrandt, F. A. 1962. The scientific basis of weight training. *In:* F. D. Sills, L. E. Morehouse, and T. L. De Lorme (eds.), Weight Training in Sports and Physical Education, Chapter 3, pp. 25–34. AAHPER, Washington, D.C.

Henry, F. and L. E. Smith. 1961. Simultaneous vs. separate bilateral muscular contractions in relation to neural overflow theory and neuromotor specificity. Res. Quart. 32: 42–46.

Ono, M. 1963. Physical exercise and its effect in middle and elder age. Report 3. Muscular strength and trunk flexibility in relation to exercise habits. Bulletin of the Phys. Fit. Research Inst., The Meiji Life Foundation of Health and Welfare. 1: 35–43.

Other movements

Distribution of body masses of a sportsman

V. N. Tichonov
Moscow Institute of Cooperation, Moscow

The initial attempts to determine moments of inertia of different parts of a body in a man were undertaken by Braune and Fischer in 1889. These two scientists determined moments of inertia of limbs in a frozen dead body by method of swinging.

Methods of determining the moments of inertia of limbs in living men were described by Drillis, Contini, and Bluestein (1964), and Gurfinkel and Saphronov (1969). Pohl (1930), Page (1969), Hochmuth (1960), Nubar (1962), Du Bois and Santschi (1962), Korsakov et al. (1970), and Bouisset and Pertuzon (1967) have determined moment of inertia of the whole body in the alive man.

PURPOSE

The purpose of this study was to determine axial moments of inertia and static moments (mr), and on the basis of these data to calculate the frequency of oscillations of limbs of an athlete in relation to the axis of proximal joints.

METHODS

Moments of inertia were measured with a device constructed by the author under the guidance of Professor M. M. Gernet. It was a one-thread suspended device, the action of which was based on the utilization of torsion oscillations (Figure 1). The system was suspended on a steel pivot and fixed to the support. The whole system was balanced with the

Figure 1. A one-thread suspended device for measuring moments of inertia.

help of the weight so that the two points coincided. The first point was fixed on the device and the other one on the floor.

An athlete stood on this device and the system was balanced. The moment of inertia of the whole system in relation to the central axis was determined according to its period of torsion oscillations using the formula

$$I_1 = \frac{T^2}{4\pi 2} \cdot C - I_2$$

where I_1 = moment of inertia of the whole body in relation to the central axis of the device,
C = coefficient of torsion of the steel bar pivot,
T = period of the torsion oscillation of the whole system,
I_2 = moment of inertia of the device, which is known.

Taking into consideration the moments of inertia of the balancing weight and the weight of the athlete's body we determined the location of the center of gravity of the athlete's body in relation to the central axis of the device and recomputed the moment of inertia of his body in relation to his center of gravity.

To determine the moment of inertia of the arm or leg in the first experiment the limb was located along the central axis of the device. In this position its moment of inertia is approximately equal to zero (Figure 2). In the second experiment the limb was located in a horizontal position while the whole body remained in the same position (Figure 3). The moment of inertia of the limb was determined as the difference between these two measurements. The miscalculation of the measurement of a moment of inertia, for example, of an arm, is very small and equal to 1–2 percent.

Technique and procedures have been published (Gernet and Tichonov, 1967; Ivanov, Stepanov, and Tichonov, 1970).

RESULTS

Results of the investigation are given in Tables 1 and 2. The study of static moments of limbs in relation to the axis of proximal joints showed their high variability, particularly in man (Table 3). The weight of the

Figure 2. Moment of inertia when the limb was located along the central axis of the device.

Figure 3. Moment of inertia when the limb was located in a horizontal position while the whole body remained in the same position.

Table 1. Moments of inertia (I) of the limbs obtained on the living man (sportsman) (in kg-cm sec^2)

Sex	Limbs	Mean	Standard deviation	Range
Men	Arm			
(20)	Left	4.32	1.27	5.90– 2.04
	Right	4.04	1.00	5.66– 2.53
	Leg			
	Left	25.35	5.25	36.29–15.50
	Right	6.27	6.01	37.95–15.79
Women	Arm			
(13)	Left	2.70	0.68	3.54– 1.66
	Right	2.51	0.70	3.69– 1.07
	Leg			
	Left	18.83	4.28	23.88–13.16
	Right	19.65	4.31	27.78–13.17

limb multiplied by the distance between the center of gravity of the limb and the proximal joint was determined. Afterwards the frequency of oscillations of these limbs was calculated according to the formula of mechanics (Table 4).

Table 2. Moments of inertia (I) in relation to central axis* (in kg cm sec^2)

Sex	Axis	Mean	Standard deviation	Range
Men (20)	Longitudinal	17.20	3.88	28.26– 13.47
	Sagittal	128.48	20.78	185.83–105.25
	Frontal	161.81	27.41	231.73–134.30
Women (13)	Longitudinal	16.11	2.50	19.18– 11.95
	Sagittal	98.84	14.03	118.01– 82.67
	Frontal	119.86	17.22	147.20– 85.26

*During measurement in relation to sagittal and frontal axis with the athletes' arms up.

Table 3. Static moments (GR) of limbs in relation to the axis of proximal joints (men, 12)

Particulars	Weight (kg)	Stature (cm)	Static moments (kg cm)	
			Arm	Leg
Mean	66.6	176.4	90.7	561.2
Standard Diviation	6.3	6.4	22.5	100.6

Table 4. The frequency of oscillations (K) of the limbs (sec^{-1}) in connection with the level of experience

$$(K = \sqrt{\frac{G \cdot R}{I}})^*$$

Limb	Sportsman of		
	High qualification	Medium qualification	Beginner
Arm (left)	4.6	5.4	4.5
Arm (right)	3.3	4.9	4.2
Leg (left)	4.4	4.5	5.3
Leg (right)	4.9	4.6	5.2

G, weight; R, distance from the center of gravity of the limb to the proximal joints; $G \cdot R$, the static moment of limbs.

CONCLUSION

On the basis of our experiments we believe that indices of mass distribution can be used for selection of athletes and for quantitative estimation of motor activity.

REFERENCES

Bouisset, S. and E. Pertuzon. 1968. Experimental determination of the moment of inertia of limb segments. *In:* J. Wartenweiler (ed.), Biomechanics I, pp. 106–109. S. Karger, Basel.

Braune, W., and O. Fischer. 1889. Bestimmung der Tragheitsmomente des menschlichen Körpers und seiner Glieder Abh.d Kgl. S.G.d.W. Bd. XVIII.

Drillis, R., R. Contini, and M. Bluestein. 1964. Body Segment Parameters. A. Survey of Measurement Techniques Artificial Limbs, Vol. 8, No. 1, pp. 44–66.

Du Bois, J., and W. R. Santschi. 1962. The determination of the moment of inertia of the living human organism. Paper read at the International Congress of Human Factors in Electronics, Institute of Radio Engineers, Long Beach, Calif., May, 1962.

Gernet, M. M., and V. N. Tichonov. 1967. Experimental determination of moments of inertia of human body and its limbs. Theory and Practice of Physical Culture, 11: 27–30.

Guzfinkel, V. S., and V. A. Saphzonov. 1969. The apparatus for measure of moments of inertia of different segments of human body. The Author's Certificate, no. 255483, USSR.

Hochmuth, G. 1960. Das biomechanische Prinzip der Impuls erhaltungen. Theorie and Praxis der KörperKultur, No. 8, 683.

Ivanov, V. V., V. I. Stepanov, V. N. Tichonov. 1970. The apparatus for determination of axial moments of inertia of human body and its limbs. The Author's Certificate, No. 343685, USSR.

Korsakov, S. A., O. A. Romodanovsky, L. A. Tsherbin, and A. V. Maslov. 1972. The question of the position of general center of gravity and moments of inertia in dependence on the posture of a man. The model of wounds of head, chest and vertebral column. The materials of the conference 30 January, 1970, Prof. Gzomov (ed.), Moscow, p. 40.

Nubar, Y. 1962. Rotating platform methods of determining moment of inertia of body segments. Unpublished report, Research Division, College of Engineering, New York University.

Page, R. L. 1969. Moments of inertia of the human body about a vertical axis. J. Phys. Ed. 4, No. 4: 221–223.

Page, R. L. 1970a. Moments of inertia of the human body—1. Br. J. Phys. Ed. 1, No. 5.

Page, R. L. 1970b. Moments of inertia of the human body—2. Br. J. Phys. Ed. 1, No. 6.

Pohl, R. W. 1930. Einführung in—DIE Mechanik und Akustik. Verlag von Julius Springer. Berlin.

Contribution of body segments to ball velocity during throwing with nonpreferred hand

T. Hoshikawa and S. Toyoshima
Aichi Prefectural University, Nagoya

In a previous study (Toyoshima et al., 1974) the contribution of each segmental action leading to the final ball velocity was investigated during throwing with the preferred hand. In this study, a corresponding investigation of throwing with the nonpreferred hand was undertaken and was compared with the previous research on the preferred hand. Furthermore, a longitudinal study consisting of 15 wk of training was conducted on nonpreferred hand throwing.

PROCEDURES

Four male adults were employed as subjects in this experiment, including two left-handed persons. Some characteristics of the subjects are presented in Table 1. The subjects were asked to throw a hard rubber ball (weight, 250 g) with maximum effort using four different throwing patterns as follows: pattern 1, a normal overhand throw; pattern 2, an overhand throw without movement of the leading foot; pattern 3, an overhand throw with the upper body immobilized (without trunk and hip rotations); and pattern 4, an overhand throw with the upper arm placed on the chair and immobilized (throwing with forearm alone).

More than ten trials were administered for each hand in each of the four conditions. The ball velocity at the moment of its release was measured electrically through a phototransistor system (Toyoshima et al., 1974). For each condition only the throw with the highest velocity was used for analysis.

Table 1. Physical characteristics of subjects

Subject	Height (cm)	Weight (kg)	Strength of elbow extensors (kg)		Power of elbow extensors (kg.m/sec)		
			Pref.	Nonpref.	Before tr., pref.	Before tr., nonpref.	After tr., nonpref.
T.H.	162	57	13.5	12.0	7.49	7.20	7.20
S.T.	165	61	15.0	14.0	8.14	7.20	7.49
K.E.*	157	60	14.0	13.0	7.80	7.35	7.49
U.M.*	167	51	13.0	12.0	6.83	6.37	6.58

*Left-handed subject.

In order to make a comparison of the sequence and timing in the segmental action between the preferred and nonpreferred hand, throwing form was filmed at rate of 64 frames per second by means of a Bolex motion picture camera and analyzed through a Graf-Pen system. Furthermore, EMGs of bilateral muscles, not only of the upper extremities but also of the trunk and lower extremities, were recorded simultaneously, using bipolar surface electrodes.

The training program consisted of 15 wk of nonpreferred hand throwing three times a week for approximately 50 throws during each training session.

RESULTS

Ball Velocities in Throwing with Preferred and Nonpreferred Hands

Before Training The values obtained in this study are summarized in Table 2. As the subjects threw with pattern 1, the mean ball velocity for all subjects was 27.5 m/sec in the preferred hand and that of the nonpreferred hand was 19.4 m/sec. The percentage of the velocity using nonpreferred to the preferred hands was 70.5 percent. For pattern 4 these figures were 10.2 m/sec, 8.8 m/sec, and 86.2 percent, respectively. The difference between the preferred and the nonpreferred hand was greatest in pattern 1. Thus the more segments of the body that are brought into action, the greater the difference in the ball velocity between the preferred and nonpreferred hand.

After Training Following 15 wk of training with the nonpreferred hand, the ball velocity increased for all subjects and throwing patterns (P). The percentage of the nonpreferred to the preferred at each condition was 76.4 (P-1), 81.5 (P-2), 93.5 (P-3), and 98.5 (P-4) percent, respectively. After training, the differences between the preferred and nonpreferred hand were progressively reduced with increased number of immobilized segments.

Contribution of the Body Segments to Ball Velocity

The contribution of the body segments to the ball velocity may be seen from the decrease in the ball velocity resulting from successively immobilizing adjoining segments and is indicated by percentages of the respective maximum for pattern 1.

Before Training Ratios expressed as a percentage of each condition to pattern 1 were 82.4 (P-2), 50 (P-3), and 36.9 (P-4) percent, respectively, in the case of the preferred; corresponding values of the nonpreferred were 87.6 (P-2), 59.8 (P-3), and 45.1 (P-4) percent, respectively.

Table 2. Contribution of body segments to ball velocity indicated by percentage of the respective maximum for pattern 1 and the percentages of the nonpreferred hand to the preferrd hand.

Pattern	Subject	Preferred (m/sec)	Before training Nonpreferred (m/sec)	Before training n/p (%)	After training Nonpreferred (m/sec)	After training n/p (%)
1	Right-handed	27.9	19.5	69.9	21.6	77.4
	Left-handed	27.1	19.2	70.8	20.4	75.3
	Average	27.5	19.4	70.5	21.0	76.4
2	Right-handed	22.5	16.7	74.2	18.6	82.7
	Left-handed	22.8	17.3	75.9	18.3	80.3
	Average	22.7	17.0	75.1	18.5	81.5
	Pattern 2/Pattern 1	82.4%	87.6%	—	87.9%	—
3	Right-handed	13.8	11.3	81.9	13.2	95.7
	Left-handed	13.8	11.9	86.9	12.5	91.2
	Average	13.8	11.6	84.4	12.9	93.5
	Pattern 3/Pattern 1	50.0%	59.8%	—	61.2%	—
4	Right-handed	10.2	8.6	84.3	10.3	101.0
	Left-handed	10.1	8.9	88.1	9.7	96.0
	Average	10.2	8.8	86.2	10.0	98.5
	Pattern 4/Pattern 1	36.9%	45.1%	—	47.6%	—

It should be mentioned that the contribution of the peripheral segments, shoulder and elbow, to the velocity of the throw is relatively greater for the nonpreferred than for the preferred hand.

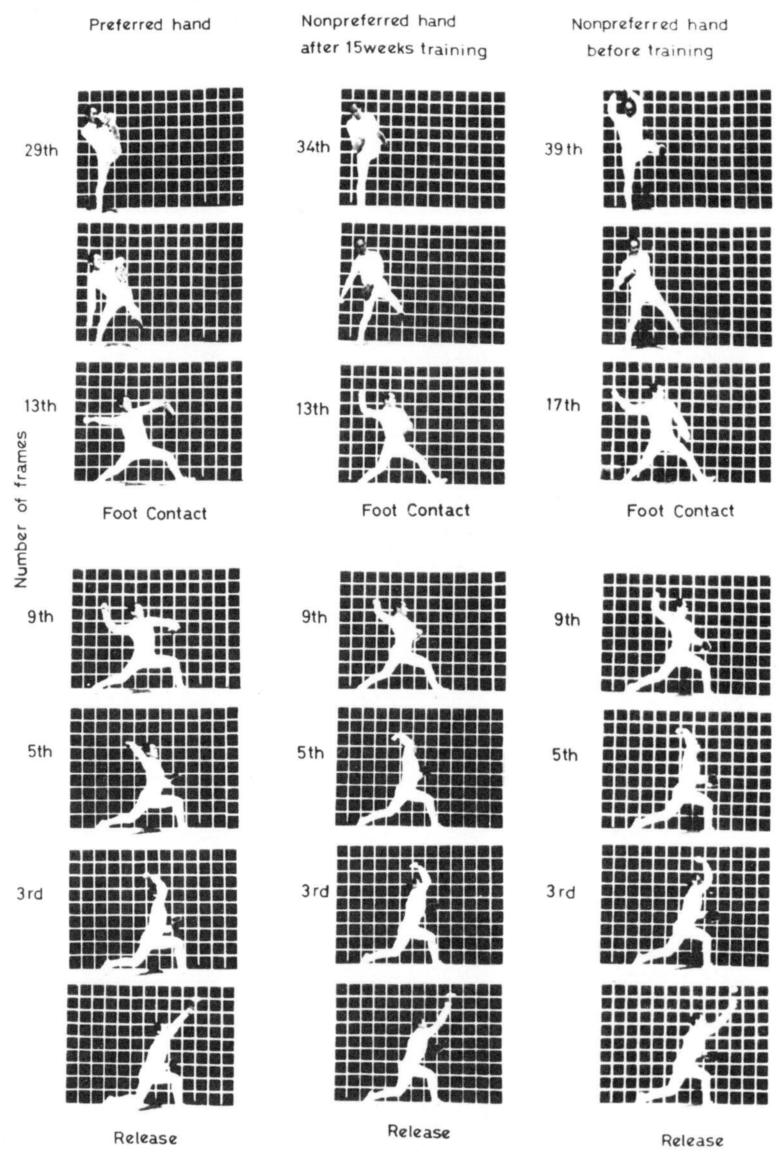

Figure 1. Serial pictures for the preferred, nonpreferred and untrained, and nonpreferred throwing after 15 wk training. In order to make comparison with the preferred hand the photographs for the nonpreferred hand were printed in reverse.

After Training Figures shown in percentages following training were 87.9 (P-2), 61.2 (P-3) and 47.6 (P-4) percent, respectively. These values were about the same as those before training. Even following training the more effective contribution to the acceleration of the ball in throwing is provided by the peripheral levers.

CINEMATOGRAPHIC FINDINGS

Figure 1 shows the cinematographic data obtained from the film at a speed of 64 frames per second. Many distinctive differences among the series of pictures of each throw are evident. It is of special interest that the difference in the three types of throwing seemed to be less in the movement pattern of the distal segments, shoulder, elbow, and wrist, which were considered as main contributors to ball velocity, rather than in the movement pattern of the proximal segments which would exert a large force during throwing. This tendency also was found in EMG recordings.

From the figures it also appears that there are discrepancies in displacements, temporal patterns, and velocities of each segment. Changes in the horizontal velocity for the hip, shoulder, and elbow joints and ball are presented in Figure 2.

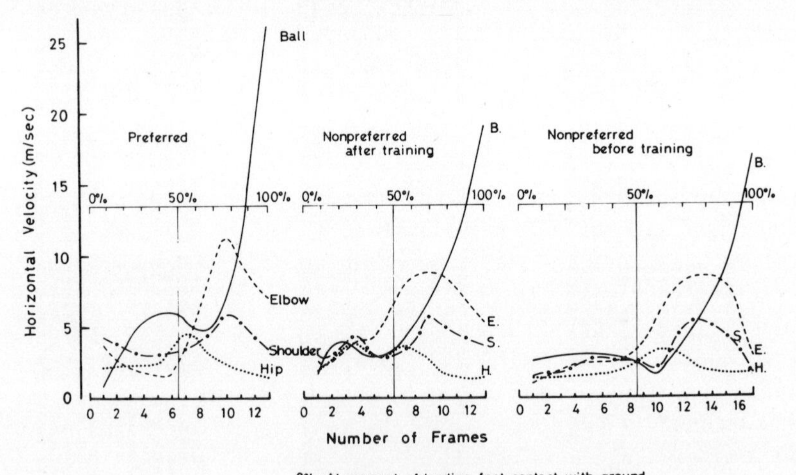

Figure 2. Patterns of sequence and timing of segmental movement viewed from changes in horizontal velocity for the preferred, nonpreferred and untrained, and nonpreferred throwing after 15 wk training. The zero reference position is the time when the leading foot makes contact with the ground, and the 100 reference is the time when the ball is released.

In throwing, the levers of the body should operate so that each contributes effectively to final ball speed. As shown in Figure 2 there are distinctive differences in the sequence and timing of segmental actions between throwing with the preferred and nonpreferred hands. For instance, when using the preferred hand, the slower but more powerful levers, the hip and trunk, begin their forward movement prior to action in the faster, but relatively weak, elbow and wrist segments.

ELECTROMYOGRAPHIC FINDINGS

A clear difference was also found in muscle activity, from an analysis of the EMG recordings, between the preferred and nonpreferred hand. In throwing with the preferred hand, muscles of the opposite hand and trunk recorded stronger potential as compared with throwing with the nonpreferred hand. However, no difference in the muscle action potentials for the throwing hand was found.

DISCUSSION

It is impossible to measure exactly the force exerted by each segment in whole body throwing. Therefore, the contribution of each body segment to ball velocity was estimated from the reduction due to the immobilized segment.

The values obtained in this study tend to agree with those of previous reports (Broer, 1969, and Toyoshima et al., 1974), using the same experimental procedure as this study. Broer reported that approximately 50 percent of the velocity of the overhand throw resulted from the step and the body rotation and the remainder came from the action of the distal segment. Toyoshima et al. obtained 53 percent for the step and the body rotation. In this experiment the result was 50 percent and agreed well with Broer's report. The contribution of the body segments in throwing with the nonpreferred hand was 12 percent for the step and 40 percent for the step and body rotation; the remainder came from the shoulder and elbow actions. It may be considered that the action of the distal segments is dominant in throwing with the nonpreferred hand. This tendency did not change following training.

Although there are a number of factors with respect to bilateral asymmetry, conceivable factors in low performance of throwing with the nonpreferred hand are the cross-sectional area of muscles to which strength exerted is proportional (Hettinger, 1961; Fukunaga, 1969), and neuromotor coordination. Although the cross-sectional area of the bilateral muscles would be nearly the same, there is a difference in muscle strength between both arms, generally stronger for the preferred (Bookwalter et al., 1950). Consequently, a discrepancy of ball velocity between both types of throw, preferred and nonpreferred, might be attributable to the neuromotor function. This was confirmed from the results in this study which show that the ball velocity for the nonpreferred hand increased following training, perhaps without hypertrophy in the muscles.

Since throwing is an ontogenic movement and special preference for one hand is relatively high in motor learning (Lotter, 1960; Singer, 1966), throwing skill should be learned posteriorly. In all conditions, the percentage of ball velocity for the nonpreferred hand to the preferred increased with progressive reduction of the number of the segments taking part in throwing. The contribution of the distal segments to ball velocity was relatively greater in throwing with the nonpreferred hand than with the preferred. These results indicate that there is a lack of integrative neuromotor coordination in throwing with the nonpreferred hand. This was supported also from kinematic and electromyographic analysis.

CONCLUSION

The contribution of body segments to ball velocity in throwing with the nonpreferred hand was estimated from the reduction in velocity due to successively immobilizing segments, and these contributions were compared with those previously determined for throwing with the preferred hand. For the nonpreferred throw approximately 40 percent of the ball velocity of the overhand throw resulted from the step and trunk actions and remainder came from the shoulder and elbow rotations. Compared with the preferred throwing, the action of the distal segments was relatively dominant in throwing with the nonpreferred hand. This tendency did not change following 15 wk of training.

REFERENCES

Bookwalter, K. W., L. Hoffman, P. Newquist, D. Rains, and M. Rousey. 1950. Grip strength norms for males. Res. Quart. 21: 249–273.

Broer, M. R. 1969. Efficiency of Human Movement. W. B. Saunders, Philadelphia.

Fukunaga, T. 1969. Calculation of muscle strength per unit cross-sectional area of human muscle by means of ultrasonic measurement. Res. J. Phys. Ed. (Japan) 14: 28–32.

Hettinger, Th. 1961. Physiology of Strength. Charles C Thomas, Springfield, Illinois.
Lotter, W. S. 1960. Relationships among reaction times and speeds of movement in different limbs. Res. Quart. 31: 156–162.
Singer, R. N. 1966. Interlimb skill ability in motor skill performance. Res. Quart. 37: 406–410.
Toyoshima, S., T. Hoshikawa, M. Miyashita, and T. Oguri. 1974. Contribution of the body parts to throwing performance. *In:* R. C. Nelson and C. A. Morehouse (eds.), Biomechanics IV, pp, 169–174. University Park Press, Baltimore.

Biomechanical structure of movements under conditions of resilient speed-force interaction with athletic apparatus

J. Povetkin
USSR Council of Ministers, Moscow

Knowledge of conditions under which the interaction between the athlete and the apparatus proceeds appears to be of great significance for the preparation of high-class athletes in technically complicated types of sport. Weight lifting happens to be a sport where force interactions between the athlete and the athletic apparatus attracted attention of researchers only in recent years (because of the appearance of weights with an elastic bar). This lag in research led in turn to contradictions between the techniques of particular movements and the competition rules of refereeing. This again resulted in instances of highly subjective estimations of the press movement, while the international federation of weight lifting came out to combat subjectivism by mere cancellation of this movement. Our investigations are dedicated to the structural study of the press and push movements with both arms under the conditions of resilient speed–force interaction with the athletic apparatus.

MATERIALS AND METHODS

To study the techniques and the structure of movement a six-channel telemetric device was used. The radius vector, its two vertical deviations angles α and γ, the acceleration of the major disk mass of the barbell in three reciprocally perpendicular planes were measured in the process of investigation. Acceleration sensors were located on the athletic apparatus in the area of concentration of the major bulk of barbell disks in a specially designed box protected from occasional pushes and shocks by a shock absorbing device. The relative coordinates sensor was located on the floor and connected with the major bulk of disks by a 5-mm-diameter steel cable. While the barbell was lifted the steel cable unwound from a reel connected with a potentiometer. The constant load added by the steel cable was 1.6 kg and there were also the loads of the acceleration sensor, its hull, and that of the shock absorbing device, all of which were balanced by a barbell disk weighing 10 kg. Registration of shifts and accelerations of the barbell was performed on the paper band of a light-beam oscillograph. In addition, investigations were performed to find out the rigidity of athletic apparatus in a static regime. Also determined were the potential possibilities of athletic apparatus in oscillatory regime, their time and amplitude characteristics. Oscillograms of more than 1,000 movements were registered and the dynamic properties of 10 barbells investigated.

RESULTS

Results of the analysis of the press and push movements indicated that the force interaction between the athlete and the barbell causes a characteristic disk oscillation relative to the center of force application. The frequency and amplitude of disk oscillations depend on the rigidity of the fulcrum-muscular apparatus of the athlete, the grip width of the bar and its elasticity, the proper weight of the barbell, the weight of disks in assembly, and their assembly order on the barbell's ends.

In the case of an ideally rigid fulcrum (the barbell is fixed in its center on a metallic stand), any vertical impact produced on the disks causes prolonged proper oscillations of an amplitude directly proportional to the disks' mass and inversely proportional to the bar's rigidity. In the case of an elastic fulcrum, a singular impact does not produce oscillations at all because of the low durability of such an oscillatory system. The rigidity of the athlete's fulcrum-muscular apparatus happens to be intermediate between these boundary conditions. It depends on morphological features of the organism, its level of training and technical preparedness. It was discovered that the majority of high-class athletes employ the elastic properties of the bar when lifting the weight up to the breast level and further pushing it upwards. Thus the phases of upward jerk and pushing are stressed in the process of weight lifting according to the time parameters of the oscillatory process. Here two or three oscillatory periods of disks are employed.

Athletes are particularly effective in using disk oscillations when lifting the barbell from the chest by press movement. When executing a push there is no necessity in this because the athlete straightens his arms by lowering his body under the barbell. When executing a press movement, there is no such possibility and athletes are forced to employ all potential reserves of the dynamic system "athlete-apparatus." It turned out that the press movement is significantly more dynamic than the push movement and possesses various specific features of technical application of the oscillatory parameters pertaining to the barbell. The analysis of results scored by ten strongest athletes of the world in different weight categories in the push and press movements showed a characteristic dependence between the percentage ratio of the results' difference to their average value, which along with the growth of achievements and weight categories decreases accordingly from 12 percent down to 0.8 percent. This in turn proves that the bar's elasticity progressively influences the result along with increasing weight of the barbell, while the percentage ratio curve shape indicates that there may come a moment when press results will be higher than push results. This is quite natural since the biostructure of the press movement is more adequate to the lifting of great weights and the fact that there are no weak links in the fulcrum-muscular apparatus (while in the push movement there is the bent knee and coxal joints) permits the athlete to create lifting efforts much greater than those produced by the pushing movement. The grade of interaction between the athlete and an elastic athletic apparatus may reach an optimal ratio on the condition that there occurs a resonance effect in the system "athlete-apparatus." This in turn ensures conditions for a maximum employment of potential possibilities of the system.

CONCLUSIONS

Employment of the dynamic qualities of the athletic apparatus according to its amplitude-frequency characteristics within the structure of the speed–force interaction in the system "athlete-apparatus" allows the athlete to obtain maximum results in weight lifting. Resonance–force interactions are most promising when one strives to achieve high results in weight lifting, but they require the athlete to possess techniques ensuring adequate conformity of his speed–force capacities with oscillatory parameters of modern athletic apparatus.

Stability of standing posture against an external force

K. Kobayashi and H. Matsui
University of Nagoya, Nagoya

Electromyographic studies on postural muscles have been carried out by Joseph and Nightingale (1952), Portnoy and Morin (1956), Basmajian (1962), and by a number of other investigators. The main topic in these papers was the maintenance of the human upright position against the force of gravity. On the other hand, Houtz and Fischer (1961) and Kobayashi et al. (1974) discussed muscular response to external forces other than gravity. In the present study, activities of the postural muscles required to maintain balance in a standing posture were investigated for two different types of external force applied to the trunk.

METHODS

The subjects were 20 healthy adults (16 men and 4 women) ranging in age from 19 to 32 yr. The mean body height of the male subjects was 171.4 cm and the mean body weight was 70.0 kg. These values for the female subjects were 155.3 cm and 49.0 kg, respectively, One end of a rope was secured to the chest (or to the body's center of gravity) of the standing subject, while the other end was attached to known weights. These weights were attached to the rope as illustrated in Figure 1. The subjects were asked to hold a static upright standing posture with heels together and toes separated by about 20 cm, against this additional force. Approximately 10 sec later, the attached weight was suddenly removed from the rope by means of an electromagnetic system. The subject tried to maintain his initial position by overcoming the effects of the sudden change of force. Electromyograms of the lower extremities were recorded throughout the experiments. Integrated electrical activities

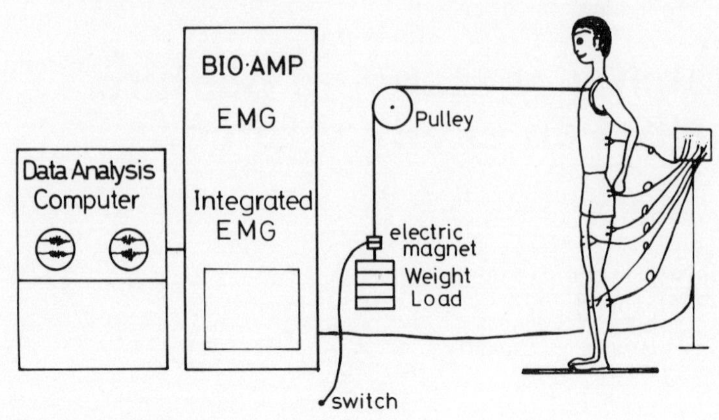

Figure 1. Instrumentation of the experiment.

(IEMG) in the static resisting condition were investigated for various magnitudes of the external force and the muscular reaction time to a sudden directional change of the force was investigated. The time interval between the removal of the weight and the beginning of the response of the postural muscles was measured using the data analysis computer ATAC 501–20.

RESULTS

A given subject could maintain an upright static position for 10 sec, provided the force applied to the chest did not exceed a certain maximum amount. The mean values of these maximum forces were 13.6 kg (range, 11–18 kg) for the male subjects and 8.5 kg (range, 7–10 kg) for the female subjects. On the other hand, when the force was suddenly released, the mean maximum values were 6.8 kg (range, 4–14 kg) for the male and 4.3 kg (range, 4–5 kg) for female subjects. The coefficients of correlation between body weight and the maximum force that could be overcome were 0.49 ($p < .05$) under the static resisting condition, and 0.77 ($p < .001$) under the impulse condition in male subjects.

Considerable individual differences were seen in the patterns of muscular activity required to hold an upright posture for ten seconds statically, when the force was applied to the body. There was a tendency in all subjects for the integrated muscular activity (IEMG) to be larger the greater the force. The tibialis anterior or the gastrocnemius or both of all subjects showed marked activity in approximate proportion to the magnitude of the external force (Figures 2, 3, and 4). For all subjects, the integrated value of the electromyogram of the tibialis anterior or of

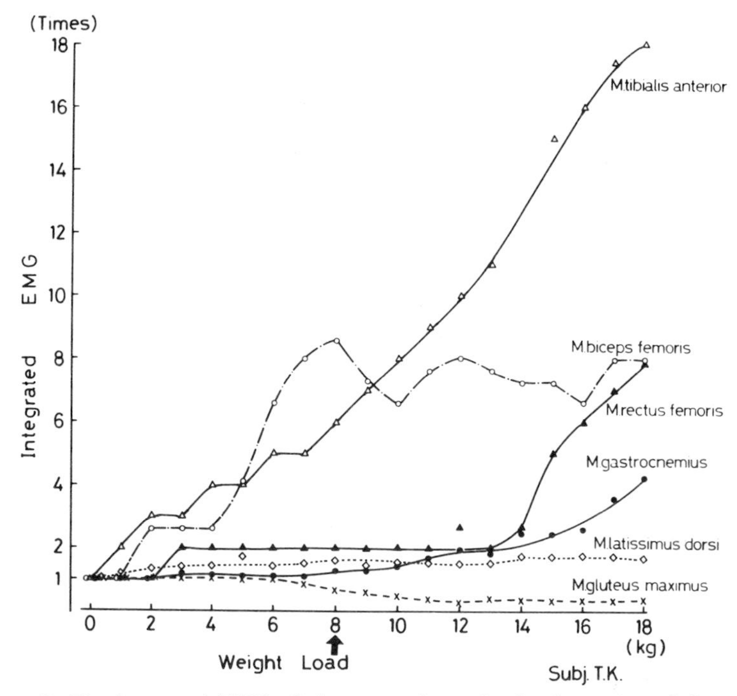

Figure 2. The integrated EMG of the postural muscles in the static resisting condition, when the external force was applied to the chest of the standing subject. *Ordinate*, multiples of the integrated EMG for 10 sec for the resting standing posture. *Abscissa*, the weight loads from zero to its maximum. *Arrow*, maximum level against which the subject could hold the standing posture when the external force was suddenly changed. (Subject is a Judo man, 173 cm, 86 kg.)

the gastrocnemius increased 6–18 times when the static force was increased from zero to its maximum values.

The IEMG of the rectus femoris increased remarkably when the force was increased to its maximum level for each subject. If the IEMG of the tibialis anterior or the gastrocnemius did not exceed 5–6 times the IEMG for the resting standing posture, then the subjects were able to maintain the standing posture when they resisted a sudden directional change of the force. When the force was suddenly released, of all the muscles in the lower extremities the tibialis anterior reacted most sensitively. The average reaction time of the tibialis anterior for all subjects is shown in Table 1 for different magnitudes of the force. The reaction time became shorter as the force increased. The difference in the reaction times of the tibialis anterior for weight loads of 1 and 5 kg was statistically significant ($p < .01$). The same was also true for the difference observed for loads of 5 and 10 kg. The reaction time of the tibialis ante-

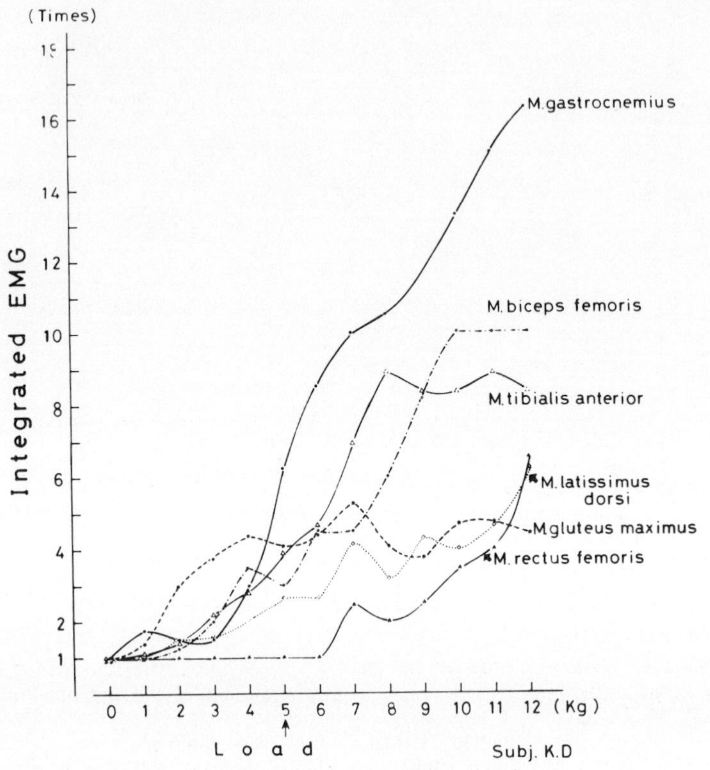

Figure 3. Subject is an ordinary male student, 168 cm, 59 kg.

rior was slower when the external force was applied to the approximate location of the body's center of gravity than when the same force was applied to the chest.

DISCUSSION

The magnitude of the external force that can be overcome seems to indicate a degree of stability of an upright posture to the external force. The standing posture in the impulse resisting condition is more unstable by about 50 percent than in the static resisting condition. Joseph and Nightingale (1952) concluded that when a human being is standing at ease, the soleus and the gastrocnemius show marked activity but the tibialis anterior is quiescent. As previously indicated, the gastrocnemius and the tibialis anterior showed marked activity, especially when the external force was relatively large. This was because the heels of the

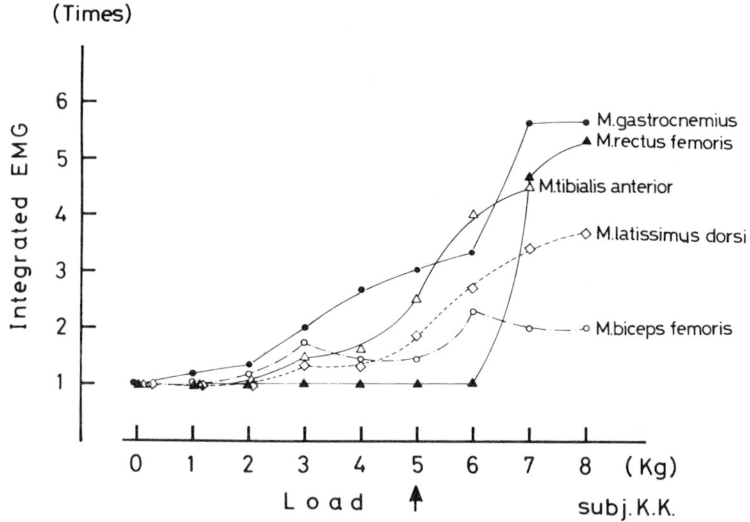

Figure 4. Subject is a female student, 158 cm, 45 kg.

subjects were raised slightly when resisting the external force. When a human being lifts his body weight in this way the tibialis anterior and the gastrocnemius become active even if an external force is not applied (Basmajian, 1962).

In this study, two different types of resisting posture have been observed. In one type, the hip was drawn backward and the trunk was moved slightly forward. In the other the hip joint was extended and the trunk was inclined backward. In the former type, the IEMG of the tibialis anterior was very noticeable, and in the latter type the gastrocnemius was more active. It has been found by many researchers that the

Table 1. Reaction time of the tibialis anterior to a sudden directional change of the force applied to the body ($N = 20$)

Magnitude of the force (weight load) (kg)	Body part to which the force was applied: Chest		Body's center of gravity	
	Mean (msec)	SD	Mean (msec)	SD
1	117.2	34.1	113.6	13.9
3	90.1	12.6	101.8	9.0
5	78.9	8.5	95.3	6.9
7	71.1	4.0	90.3	7.7
10	68.0	4.5	87.5	7.3

activity in the muscles of the thigh are surprisingly slight during relaxed standing (Basmajian, 1962).

In this study, the activity of the biceps femoris was extensive and the rectus femoris reacted intensively as the external force became greater. It is interesting that the human body will become more unstable to a sudden change of the force when the IEMG of the tibialis anterior or the gastrocnemius exceeds 5–6 times the IEMG for the resting standing posture. This fact shows that the initial activity in these muscles must be relatively low if the standing posture is to be maintained when the direction of the external force is suddenly changed.

As for the reaction time of the muscles, the movement of the body's center of gravity must be considered. When the force is suddenly released, the body's center of gravity will be accelerated. The relationship between the reaction time of the muscles and the acceleration of the body's center of gravity will be investigated more precisely in future experiments.

CONCLUSIONS

The tibialis anterior and the gastrocnemius are remarkably active in maintaining a standing posture when the subject is resisting an external force.

The tibialis anterior is most sensitive to a sudden directional change of the external force.

The reaction time of the muscles to a sudden change of the force becomes shorter as the size of the change in load increases.

REFERENCES

Basmajian, J. V. 1962. Muscle Alive. Their Functions Revealed by Electromyography. Williams and Wilkins, Baltimore.

Houtz, S. J., and F. J. Fischer. 1961. Function of leg muscles acting on foot as modified by body movements. J. Appl. Physiol. 16:597–605.

Joseph, J., and A. Nightingale. 1952. Electromyography of muscles of posture: leg muscles in males. J. Physiol. 117: 484–491.

Kobayashi, K., M. Miura, Y. Yoneda, and K. Edo. 1974. A study of stability of standing posture. *In:* R. C. Nelson and C. A. Morehouse (eds.), Biomechanics IV, pp. 53–59. University Park Press, Baltimore.

Portnoy, H., and F. Morin. 1956. Electromyographic study of postural muscles in various positions and movements. Am. J. Physiol. 186: 122–126.

Mechanics and muscular dynamics of rising from a seated position

D. L. Kelley, A. Dainis, and G. K. Wood
University of Maryland, College Park

Examples of early twentieth century interest in muscular mechanics can be found in the works of Lombard (1903), Fischer (1906), and Lombard and Abbott (1907). Lombard and Abbott were concerned with the contributions of lower extremity muscles to hip and knee joint movements in the frog. Lombard (1903) argued that those muscles which cross both hip and knee joints have better leverage as extensors than as flexors, and in sit-to-stand movements, the concurrent extensions of hip and knee joints result from the hamstrings prevailing across the back of the hip joint and the rectus femoris across the front of the knee joint. Thus the phenomenon of complex co-contractions has been recognized for some time as being a normal function under certain conditions. Although there has been scant interest since then in studying sit-to-stand movements, interest in co-contraction and the mechanics of two-joint muscles has prevailed up to the present time: Landsmeer (1961), Molbech (1965), Carlsöö and Molbech (1966), and Carlsöö, Fohlin, and Skoglund (1973). This study was undertaken in order to investigate the nature of variations in the execution of the sit-to-stand task as performed by male and female subjects of differing body sizes, and the function of selected one- and two-joint muscles crossing the hip and knee joints.

PROCEDURES

A prescribed rising movement, demonstrated by three men and three women (each group consisted of a tall, a medium, and a short subject), was investigated through synchronized cinematography and electromyography. Figure 1 illustrates the prescribed starting position. An adjusta-

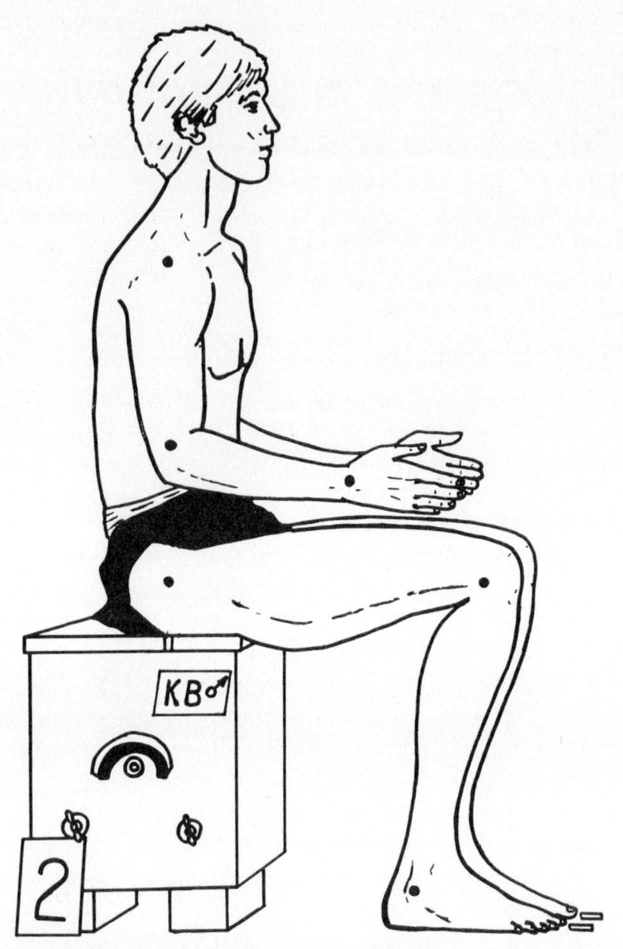

Figure 1. The initial position before the start of movement.

ble seat made it possible for all subjects to start from a seated position in which the thigh was oriented horizontally, with the knee–thigh angle at 75 degrees. The specified rising time was 2 sec, and was controlled by metronome pacing. Each subject practiced the movement before data gathering began until highly consistent performances were obtained. Records were taken of five trials for each of the six subjects.

Four superficial muscles were monitored electromyographically. They were: rectus femoris, biceps femoris (long head), vastus medialis, and gluteus maximus.

Cinematographic and electromyographic records were synchronized via simultaneous event markers, a light mounted on the adjustable seat, and an event pen on the Beckman recorder. These were actuated when the subject lost contact with the seat.

A five-segment simple-link model, as described by Dainis (1975), was used for the numerical analysis. The ankle joint was established as the fixed point about which the body moved. The positions of the joint centers were read from the film and segment angles with respect to the horizontal were calculated. Smoothing of the data was achieved by a least squares fit of quadratic functions of time to every group of five points.

RESULTS AND DISCUSSION

As might be expected of a movement innate to almost everyone, many similarities among subjects were noted in the execution of the rising task. Using the results from one of the male subjects, Figures 2 and 3 summarize the similarities that are generally representative of the be-

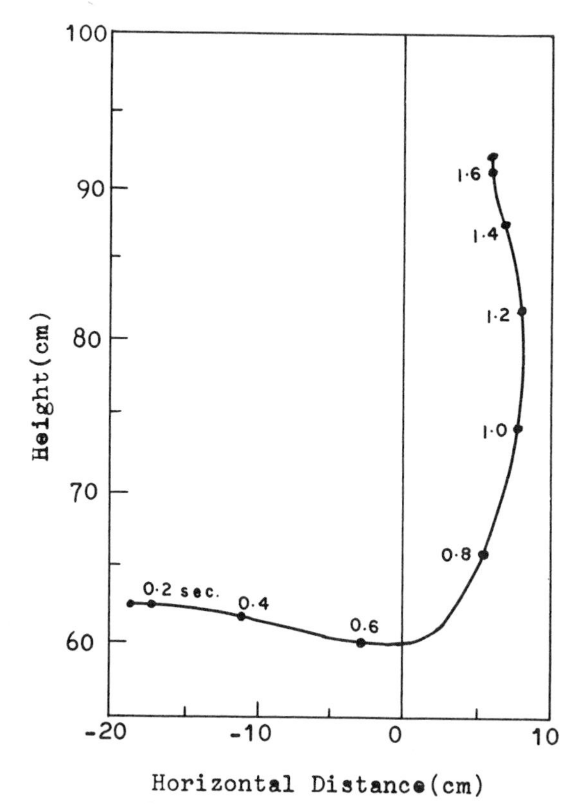

Figure 2. The height and horizontal position of the whole body center of mass in relation to the ankle joint.

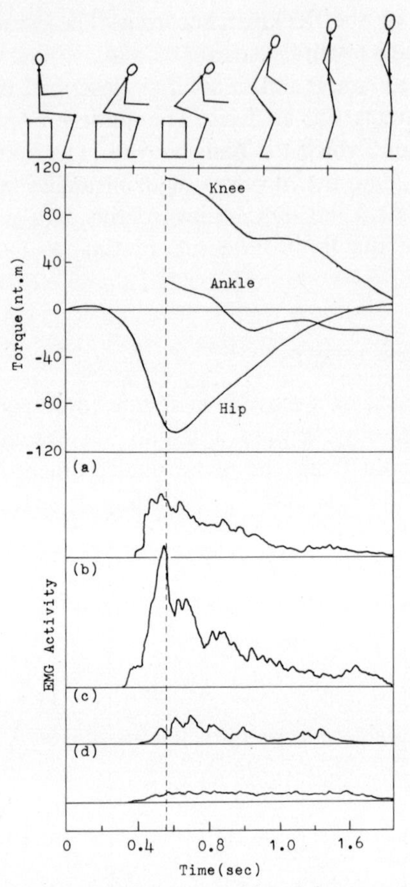

Figure 3. The relationship between body position, joint torques, and EMG activity for a typical subject. Positive torques act to decrease the anterior joint angles, and (*a*), (*b*), (*c*), and (*d*) show the EMG activity of the vastus medialis, rectus femoris, biceps femoris, and gluteus maximus, respectively.

havior of four of the six subjects. Stick figures indicate sequential body positions (Figure 3) as they relate to internal joint torques and action potential records over a common time base. Figure 2 shows a plot of a typical whole body center of mass (CM) path, beginning at the first identifiable movement (hip flexion) and ending in the erect standing posture.

The movement was initiated by a forward pivoting of the upper body and a forward thrust of the upper extremities. This caused the CM to move forward and slightly downward until the vicinity of seat separation (light-on) was reached. With small variations in timing, the vertical component of the CM velocity reversed its direction at that point but the CM continued to move forward to a maximum of approximately 7.5 cm in front of the lateral malleolus. The final part of the CM upward movement was accompanied by a backward motion of about 2.5 cm.

Generally smooth torque curves for the three joints of the lower limbs were noted, Figure 3. A very slight, positive hip torque at the outset depicts the small muscular turning action required to begin the forward pivoting of the upper body. The very short duration of this torque indicates that once begun, hip flexion proceeded as the result of gravitational action, and also that it was completed before the rectus femoris muscle was activated. Immediately thereafter, hip torque became negative and increased in magnitude rapidly to a maximum shortly after the body was free of its support. Hip extensor EMG appears to be timed for a maximum contribution to hip extension with some early eccentric action assisting in the control of the upper body as it approaches its lowest position. The absence of hip extensor activity along the torque curve for about 0.2 sec indicates that muscles other than hamstrings and gluteus maximus were the sources of the early negative torque. Although they were not monitored, an adductor group involvement is suspected.

Figure 2 demonstrates that the CM was several centimeters behind the ankle joint axis at light-on. Following light-on, ankle joint torque was positive, indicating that initially, dorsiflexion accompanied the forward movement of the CM. Knee torque was very large in response to the transfer of weight from the seat to the lower extremities and the knee extensor muscles showed peak activity at this point. It appears that the early biceps femoris activity could not contribute to knee extension through the co-contraction mechanism described by Molbech (1965) because the knee joint angle is too small. The horizontal component of the ground reaction force through the ankle joint became strongly negative about this time, which correlates well with the large positive knee torque and the steeply rising EMG envelopes of the knee extensor muscles. Since hip extensor–knee flexor torque would act to produce a positive ground component, it appears that the earlier and more impulsive contractions of the quadriceps group override the action of the hamstrings, particularly at the knee joint. The horizontal component remained negative until very near the completion of the movement, where it became positive for a short duration. For the latter part of the motion, the knee and hip torques diminished smoothly as the joints approached complete extension, and action potentials of the four muscles at the end of the movement were essentially at baseline level, indicating an easy standing position.

Certain major differences were noted in the way some subjects executed the movement. For example, the amounts of forward arm thrust and forward pivoting of the upper body varied substantially. Maximum shoulder flexion varied from 11 to 53 degrees, and maximum hip flexion varied from 19 to 36 degrees. However, all subjects completed these flexions in the vicinity of light-on.

Hip angle versus knee angle plots demonstrated generally linear characteristics (Figure 4). In five of the six subjects, the graph had a continually increasing slope, as shown by curve (*b*), indicating that knee extension lagged behind hip extension. The mean angle of the knee joint was 115.8 degrees (S.D. = 2.6 degrees) when the hip angle was 120 degrees. One subject, the tall male, demonstrated a continually decreasing slope, curve (*a*). In this case, knee extension proceeded at a higher rate than hip extension, the angle exceeded 140 degrees when the hip angle was 120 degrees. His data also showed that the biceps femoris did not become active until about 0.3 sec after light-on although he did display gluteus maximus activity similar to the other subjects. This result indicates that in the majority of subjects, early hamstrings (biceps femoris) activity tended to augment hip extension while at the same time inhibiting knee extension. In the case of the remaining subject,

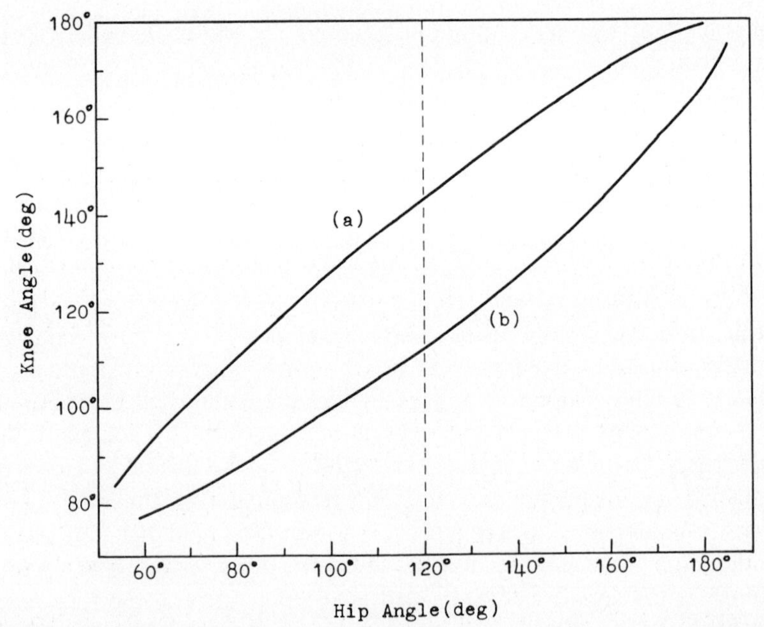

Figure 4. Knee angle versus hip angle graph for an atypical case (*a*), and a typical case (*b*).

hamstrings activity was considerably delayed in comparison to the others, and no early resistance to knee extension was offered. This allowed the knee to extend faster than the hip. One other notable exception was found. One subject consistently performed the movement without once activating the gluteus maximus. However, his other data did not indicate an atypical movement pattern.

SUMMARY AND CONCLUSIONS

Many more similarities than differences were noted in the EMG and motion patterns. Once the body left the seat, co-contractions of the quadriceps and hamstrings were observed throughout the movement in all but one subject, and with another exception, the gluteus maximus was also active throughout the entire period of hip extension. The major observable movement pattern differences occurred in the first part of the movement, these being the variations among subjects in hip and shoulder flexion before loss of seat support. In the six subjects investigated, no size- or sex-related variations were detected.

This preliminary work indicates that a more detailed study is warranted to determine the causes of individual differences in accomplishing the rising task. Additional study of the Molbech model (1965) and horizontal ground reaction components may serve to explain the functions of the various muscle groups.

REFERENCES

Carlsöö, S., L. Fohlin, and G. Skoglund. 1973. Studies of co-contraction of knee muscles. *In:* J. E. Desmedt (ed.), New Developments in Electromyography and Clinical Neurophysiology, Vol. 1, pp. 648–655. S. Karger, Basel.

Carlsöö S., and S. Molbech. 1966. The functions of certain two-joint muscles in a closed muscular chain. Acta Morphol. Neerlando-Scand 6: 377–386.

Dainis, A. 1975. Analysis and synthesis of body movements utilizing the simple n-link system. *In:* R. C. Nelson and C. A. Morehouse (eds.), Biomechanics IV, pp. 513–518. University Park Press, Baltimore.

Fischer, O. 1906. Theoretische Grundlagen Fur eine Mechanik der Lebenden Korper mit Speziellen Anwendungen auf den Menschen sowie auf einige Bewegungsvorgange an Maschinen. (Theoretical bases for the mechanics of the living body with special applications to man as well as several movement processes in machines). Teubner, Leipzig.

Landsmeer, J. 1961. Studies in the anatomy of articulation. II. Patterns of movement of bi-muscular and bi-articular systems. Acta morphol. Neerlando-Scand. 3: 304–321.

Lombard, W. 1903. The action of two-joint muscles. Am. Phys. Ed. Rev. 8: 141–145.

Lombard, W., and F. Abbott. 1907. The mechanical effects produced by the contraction of individual muscles of the thigh of the frog. Am. J. Physio. 20: 1–60.

Molbech, S. 1965. On the paradoxical effect of some two-joint muscles. Acta morphol. Neerlando-Scand. 6: 171–177.

Energy efficiency of ball kicking

**T. Asami, H. Togari, T. Kikuchi,
N. Adachi, K. Yamamoto, K. Kitagawa, and Y. Sano**
University of Tokyo, Tokyo

Robb (1972) defined a skilled movement as one in which a predetermined objective is accomplished with maximum efficiency and a minimum outlay of energy. Studies along this line have been made on the relation between energy expenditures and performance in such cyclic movements as running, swimming, and bicycling, but not in sport skills such as throwing, jumping, and hitting.

In this study, with its larger energy output and ease of evaluating its skill level, the mechanical efficiency of the instep kick used in soccer was measured and a comparison made between skilled and unskilled soccer players.

METHOD

Four soccer players (the skilled group) and four non-soccer players (the unskilled group) were used as subjects. Each subject was asked to kick a placed ball at several predetermined ball velocities aiming at the target, which was drawn on a canvas hung from the ceiling 5 m away from the ball.

The ball velocity was measured electrically using an apparatus with photoelectric tubes, designed by Toyoshima and Miyashita (1973).

After each kick, the subjects were informed of the resultant ball velocity and asked to adjust their kicks so as to obtain the designated velocity. Each subject performed 15 consecutive kicks at the same designated ball velocity at 10 sec intervals. The predetermined ball velocities were 10, 12, and 14 m/sec and continued on at 2 m/sec increments up to the subjects' maximal effort.

Longitudinal and transverse lines were drawn on the canvas at 30 cm intervals and the third central square from the bottom was painted as the bull's-eye of the target. All squares were numbered and the number of the square hit by the ball was recorded to determine the accuracy of the kick. The ball used was an official soccer ball weighing 400 g.

Expired air was collected into Douglas bags during the exertional period and the following recovery period in sitting position for 10 min. O_2 and CO_2 concentrations were analyzed by an Expired Gas Analyzer 1HO2 (Sanei-Sokki). The ventilatory volume was measured by Respiration Gasmeter (Max Planck Institute).

The net energy expenditure for each kick was calculated by dividing by 15 the difference between the gross energy expenditure and the energy expenditure for each subject without kicks for the same period of time.

Kinetic energy of a kicked ball was calculated from the weight of the ball and the average ball velocity of 15 kicks, using the formula $\frac{1}{2}mv^2$. Thus the mechanical efficiency was obtained by the following formula:

$$\text{Mechanical efficiency} = \frac{\text{Kinetic energy of kicked ball}}{\text{Net energy expenditure of ball kicking}}$$

In these calculations, 1 liter of oxygen was assumed to correspond to 5 Kcal and 1 Kcal to 427 kg.

RESULTS AND DISCUSSION

Figure 1 shows the relationship betwen the ball velocity and the energy expenditure, or mechanical efficiency, in a representative case from both the skilled and unskilled groups. Other cases showed similar tendencies. Energy expenditure increased exponentially with increasing ball velocity as in the relationship between energy requirement and speed obtained by Sargent (1926) and Yamaoka (1965) in running, and Karpovich and Pestrecov (1939), Yamaoka (1965), and Klissouras (1968) in swimming.

The relationship between mechanical efficiency and the ball velocity showed a bell-shaped quadratic curve. A similar relationship was reported in swimming by Klissouras (1968). Average efficiency at the ball

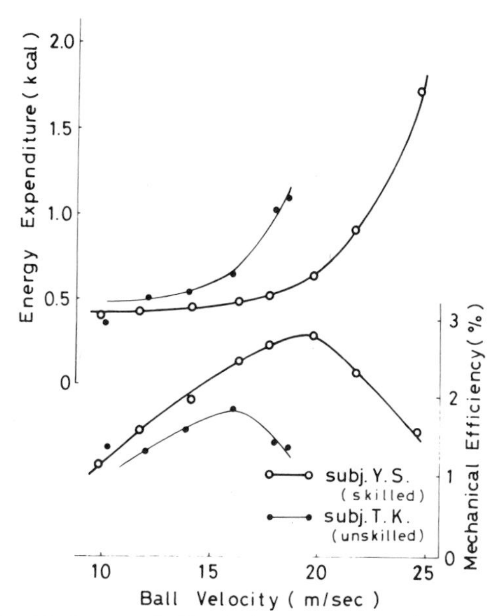

Figure 1. Relationship between ball velocity and energy expenditure or mechanical efficiency.

speed of 10 m/sec was 1.38 percent (1.20–1.46) in the skilled group and 1.21 percent (1.00–1.39) in the unskilled. The ball velocity achieved with maximal effort averaged 22.9 m/sec (21.8–24.6) in the skilled group and 18.0 m/sec (17.1–19.8) in the unskilled. The efficiency at these two levels was 1.86 percent (1.58–2.65) and 1.16 percent (0.77–1.38), respectively.

Figure 2 shows the relationship between the percent of the maximum ball velocity and the efficiency achieved by each subject. A definite difference in efficiency was apparent between the skilled and the unskilled groups. The skilled players were found to kick harder with less energy. A similar relationship was reported in running by Yamaoka (1964, 1965) and in swimming by Karpovich and Millman (1944). Yoshizawa (1973) observed the likelihood of improvement in efficiency through improving skill by training in his bicycle ergometer studies.

The maximum efficiency appeared to be achieved at 70–85 percent of the maximum ball velocity in each subject. In the skilled group this was somewhat lower than 80 percent, but higher in the unskilled. The average maximum efficiency was 2.78 percent (2.30–3.30) in the skilled group and 1.76 percent (1.55–1.95) in the unskilled.

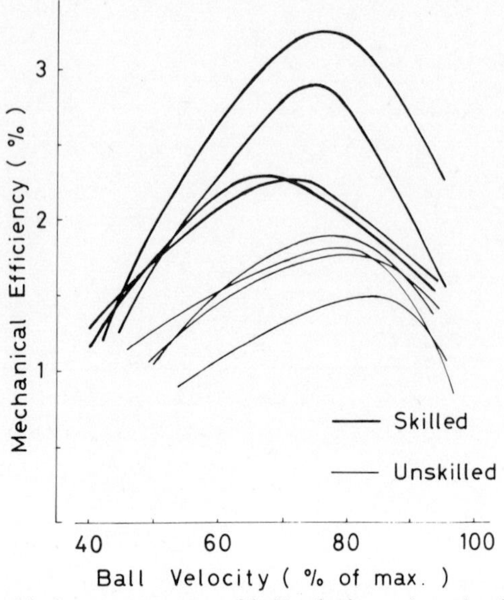

Figure 2. Relationship between percent of ball velocity and mechanical efficiency.

Comparison of these values with those so far reported on the mechanical efficiency of physical movement showed that these were much lower than 22.7 percent (Fenn, 1930) and 37.7 percent (Furasawa, Hill, and Parkinson, 1927) in sprint running, and 11.4–21.5 percent (Dickinson, 1929) in bicycling, and that similar to 0.5–2.2 percent (Karpovich and Pestrecov, 1937), 1.71–3.99 percent (Adrian, Singh, and Karpovich, 1966) and 1.0–3.2 percent (Klissouras, 1968) in swimming.

To further delineate these factors, the relationship between the degree of accuracy and the efficiency was analyzed. The accuracy of the kick was assessed by the extent of deviation of the ball from the bull's-eye of the target. Such deviation was the least in most subjects at a velocity of approximately 80 percent of the maximum velocity. Thus it was found that when the highest efficiency was attained to the energy output, the direction of the kick was also the most accurate.

Figure 3 shows the relationship between the maximum efficiency and the accuracy in each subject. The accuracy of kick was expressed as the average of distances from the bull's-eye for all kicks performed by each subject. There was a very high negative correlation, $r = -0.96$. Therefore, it can be said that the higher the efficiency is, the more accurate is the kick.

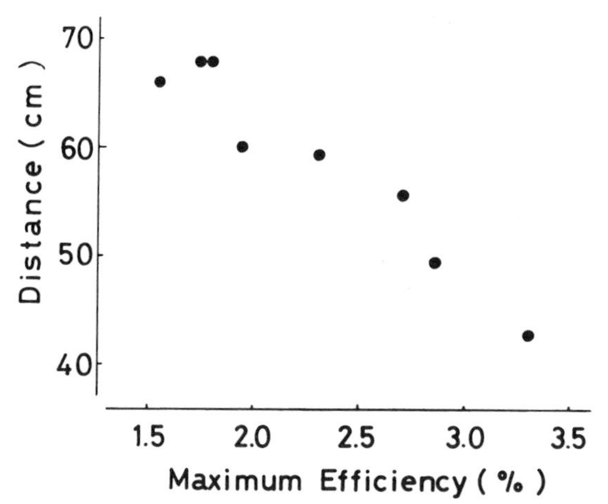

Figure 3. Relationship between maximum efficiency and accuracy of direction of kicked ball.

CONCLUSION

On the basis of the results described it is concluded that mechanical efficiency can be used as a reasonable scale for an objective assessment of the skill of physical movement.

REFERENCES

Adrian, M. T., M. Singh, and P. V. Karpovich. 1966. Energy cost of leg kick, arm stroke, and whole crawl stroke. J. Appl. Physiol. 21: 1763–1766.

Dickinson, S. 1929. The efficiency of bicycle-pedalling as affected by load. J. Physiol. 67: 242–255.

Fenn, W. O. 1930. Frictional and kinetic factors in the work of sprint running. Am. J. Physiol. 92: 583–611.

Furusawa, K., A. V. Hill, and J. L. Parkinson. 1927. The energy used in "Sprint" running. Proc. Roy. Soc. B 102: 43–50.

Karpovich, P. V., and K. Pestrecov. 1939. Mechanical work and efficiency in swimming crawl and back strokes. Arbeitphysiol. 10: 504–514.

Karpovich, P. V., and N. Millman. 1944. Energy expenditure in swimming. Am. J. Physiol. 142: 140–144.

Klissouras, V. 1968. Energy metabolism in swimming the dolphin stroke. Int. Z. Angew. Physiol. Einschl. Arbeitphysiol. 25: 142–150.

Robb, M. D. 1972. The Dynamics of Motor-skill Acquisition, p. 3. Prentice-Hall, Englewood Cliffs, N. J.

Sargent, R. M. 1926. Relation between oxygen requirement and speed in running. Proc. Roy. Soc. B 100: 10–22.

Toyoshima, S., and M. Miyashita. 1973. Force–velocity relation in throwing. Res. Quart. 44: 86–95.

Yamaoka, S. 1964. Speed and Oxygen. *In:* E. Hisamatsu and M. Ikai (eds.), Sports Medicine (in Japanese), p. 180–182. Taiikunokagakusha, Tokyo.

Yamaoka, S. 1965. Studies on energy metabolism in athletic sports. Jap. Res. J. Phys. Ed. 9(3): 28–40.

Yoshizawa, S. 1973. The studies on the effect of training on aerobic work capacities in adolescents. Rep. Res. Cent. Phys. Ed. 1: 14–23.

Biological rhythms in the development of human locomotions in ontogenesis

V. K. Balsevich
Institute for Physical Culture, Omsk

This research is proposed to study the age development of biomechanical systems of locomotion in connection with the evolution of morphological and functional systems of the human organism. With the help of such an approach to the problem we tried to determine the main regularities of development of the system-structural complex, realizing the human locomotor function.

METHODS

The information about the development of biomechanical elements of locomotion was obtained by means of the tensiometric recording of the main parameters of walking and running of 950 subjects ranging in age from 5 to 65 yr. The data on the morpho-functional development were obtained by anthropometry and testing. Our conclusions about the structural relations have been based on the results of correlative analysis.

RESULTS

The analysis of the research materials has shown that one of the main regularities of the ontogenesis of human locomotor function is the oscillating character of its development with age. This regularity, as revealed by the author, was reflected in all those levels of the locomotor system that were studied. In the ontogenesis of biomechanical elements it was displayed by the distinct periods of faster and slower development.

Fluctuations in the biomechanical systems as a whole are characterized by the rhythmic changes in the periods of intrasystem structural consolidation of elements, and by the periods of their dissociation or discord. The age dynamic of intersystem relations in motorics has also the definite rhythm of development as a distinctive feature, which is expressed in the consecutive changing of processes of increasing and decreasing degrees of coordination between the biomechanical, morphological, and functional systems we investigated (Figure 1).

The characteristic feature of cooperation of the oscillating processes in the development of various levels of the locomotor system is the synchronism of fluctuations in the evolution of separate elements. At the same time these same fluctuations are asynchronous in their structural complexes (systems).

Thus, the evolution of locomotor function is characterized by the strongly pronounced cyclic recurrence, which forms the biologic rhythm of the development of locomotions.

The essence of the biologic rhythm of age evolution of locomotor function lies in the consequent changing of periods of accelerated development of separate elements of the motor system and by the periods of their structural consolidation.

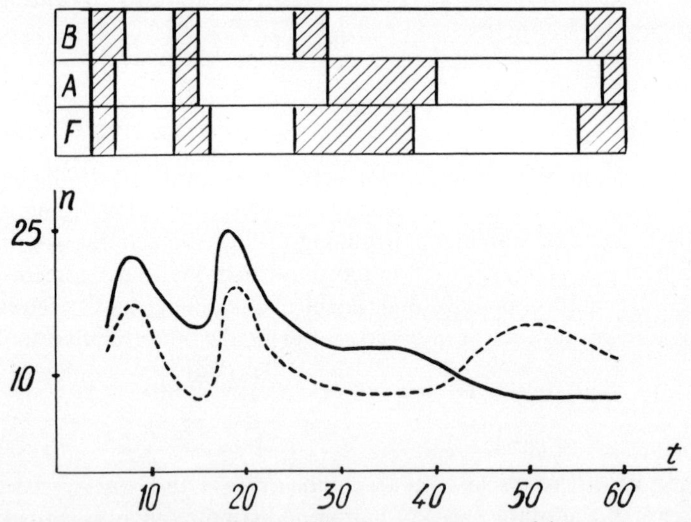

Figure 1. The age development of different components of locomotor system. n, the number of reliable correlations in the biomechanical system; t, age in years; *solid line*, females; *broken line*, males. The hatched fragments of the diagram placed above the curves show the periods of intensive development of the elements of biomechanics (B), anthropometry (A), and physical fitness (F) in the same scale of time.

The cycles of development including the phase of primary development of motor elements and the phase of their organizational, structural arrangement in the integral system are the components of the discovered biologic rhythms. The given cycles and phases of the development are not closed and discrete. The limit of the organizational regulations of the system in every cycle is conditioned by the achieved level of the development of separate elements. Therefore, as soon as the possibilities of structural organization approach this limit, the new cycle of the system development begins. The separate elements of motorics evolve with an acceleration again against a background of structural dissociation, and this discord is changed into the phase of consolidation, the increased level of the system organization.

Thus the process of the development of locomotor system goes in a spiral; both phases of biologic rhythm are present on each coil of this spiral. The main peculiarity of this process is the qualitative difference in each of the next periods of the development from the previous one.

DISCUSSION

The physiologic mechanism of the forming of cycles of the locomotor development can be understood in the light of Uchtomsky's study about the dominant. In this connection it is natural to explain the oscillating character of the age evolution of biomechanical elements and system by the asynchronous speed of development of different levels of motorics. Besides, the homogeneous dominants are repeated at the latter cycles of development, reflecting each time the new character of current adaptive behavior of the system.

In conforming with the author's ideas it seems to be possible to suggest the following hypothesis for discussion.

The program of the development of locomotor function is expressed in the hereditary information which is formed in the philogenesis and conditions of the algorhythms of the development of locomotor systems in ontogenesis. In the process of initial stages of individual evolution, this program is transformed into the complex of connections in the nervous system. These connections form the structure of the model or the organ of control of the development of locomotor function.

The organ of the control of the development can be imagined as a neurodynamic structure uniting the great number of elements of the nervous system, interacting in the great number of combinations.

The given organ is dynamic, i.e. it is able to self-correct under the influence of exo- and endogenous factors. The level of the determination of various aspects of the model's functioning is different, that is, determined by the degree of vital importance of various components to the process of development for the organism.

Thus the biologic rhythms and the sequence of the development of separate elements of motorics and their system-structural complexes are strongly determined. At the same time the level of the development of the system of locomotor acts on the whole and the separate elements of the system depends to a considerable extent on the influence of the environment and inner medium and is determined by them.

The different degree of determination of separate aspects of the functioning of the organ of control of the development is confirmed by the results of comparative research of the rhythm of the age development of biomechanics of running and jumping of children who participate or do not participate in sports, where the distinct synchrony of the speeding up and slowing of development of separate elements and systems of motorics of sportsmen and non-sportsmen was noted (this research was carried out at our laboratory). Besides, the absolute mean scores of sportsmen were higher than those of non-sportsmen. Even such an intensive influence on motorics as sports does not change the biorhythm of its development, although the development itself goes on at a higher level.

As far as the revealed biorhythms condition the periods of the speeding up and slowing down of development of different system-structural components of motorics, in the process of the governing of the development of the locomotor function it is necessary to realize the age differentiation of means and methods in the training of sport locomotions and their improvement.

The author thinks that the effectiveness of the governing of the process of locomotor improvement will be significantly higher, if the accents of the pedagogic influence will coincide its character with the natural accelerations in the development of separate elements and structures of motorics. In this case the training information appeared to be adequate to "tuning" of the instructed student and will be actively mastered by him. If the accents of pedagogic influence do not coincide with the primary trend of the development of locomotor system, the training information will appear in the role of a "putting out" factor, i.e. the factor breaking the natural algorhythm of the development, coordinated with the evolution of all the other systems of the whole organism. Naturally, the organism will resist the action of such information. But the possibilities of such information in this relation are not unlimited. If the current of "inopportune," "unwilling" information is intensive enough, the "break-down" in the process of development, the discord in the whole system, the break-up of the normal course of the evolution of its separate groups can happen.

In this connection it is necessary to realize the age differentiation of means and methods of the governing of the process of the locomotor improvement with regard for the specific peculiarities of the motor development in different periods of the individual evolution. Therefore, the author can formulate the principle of the correspondence of the character of pedagogic influence to biologic rhythm of the age motor development.

The realization of this principle in practice of governing the development of the locomotor function is possible on the basis of the regard for the revealed regularities of the age evolution of movements and the usage of the vast arsenal of means and methods of physical and sport training being at the disposal of sport teachers and coaches.

Biomechanics of sport

Introductory paper

Biomechanics of sports from the viewpoint of methodological advances

M. Miyashita
University of Tokyo, Tokyo

It was pointed out by Plagenhoef (1971) that by combining anatomical data, principles of mechanics, data-collecting equipment, and the computer, more specific and accurate data can be obtained on human movement than ever before. Furthermore, data on neuromuscular function should be added to the four factors stated above, for the study of learning and development of movement.

Among these five factors, anatomical data such as lengths, weights, and shapes of numerous body segments, and principles of mechanics such as velocity, acceleration, force, work, etc., are already well known. On the other hand, many physiologists are still contributing to the study of neuromuscular function, including the mechanisms controlling an individual muscle in the various levels of the central nervous system, with contractile properties from histochemical and biochemical aspects. Moreover, recent developments in data-collecting equipment and computers have contributed to obtaining more precise recording and analysis of human movement.

The present report deals with studies of various sports activities from the viewpoint of methodological advances. For this purpose, a brief biomechanical composition of sports activities is illustrated (see Figure 1). Motion is the prime element in most sports. Performance refers to motion, while motion is created only through force. There are two forces acting on the human body; a positive one generated by muscular contraction and a passive one resulting from weight, volume, shape, and velocity. The positive force is completely controlled, while

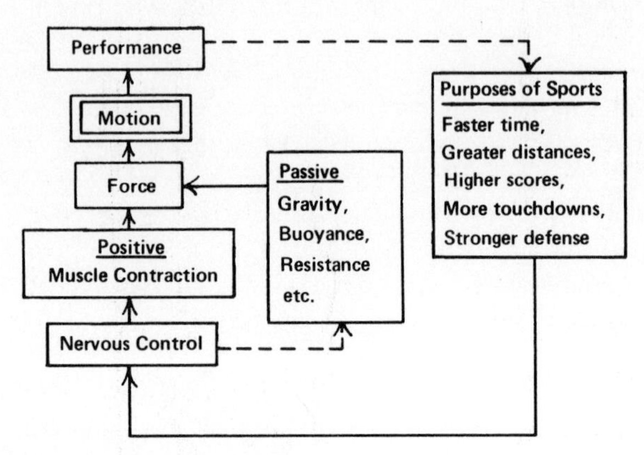

Figure 1. Biomechanical composition of sports activities.

the passive force is only partially controlled by the nervous system according to the individual purpose of the sport.

Early in the history of biomechanics, most researchers approached only one component. In the case of performance, movement speed was first measured in the locomotor type sports. Running speed was measured by Furusawa, Hill, and Parkinson (1928) using a magnetic coil system and by Ikai (1968) with photosensitive cadmium cells. Swimming speed was measured by Karpovich with a revolving cylinder Natograph (1930) and magnetic tape Natograph (1970), and by Miyashita using a photoelectric cell (1971). Similarly ball velocity was measured using a light-sensitive field (Nelson et al., 1966; Toyoshima and Miyashita, 1973).

As for motion, since Muybridge (1901) published his famous book *The Human Figure in Motion*, many investigators have tried to analyze human movement using the motion camera. Hellebrandt (1961) observed the jumping behavior of children of different ages. Broer (1966) indicated that the basic body movements are very similar in three different activities such as the basketball throw, tennis drive, and batting. Many motor patterns in various sports such as batting in baseball (Race, 1961; Watkins, 1963), pole vault (Fletcher, Lewis, and Wilkie, 1960; Steben, 1970) and ski jump (Komi, Nelson, and Pulli, 1974) have been investigated by means of cinematography.

The electrogoniometer is also useful to describe the pattern of motion. With the electrogoniometer, Gollnick and Karpovich (1964) recorded the angular motion of each joint during locomotion and some athletic movements.

Force can be indirectly estimated from the changing position of the body or its parts. That is, by using the photographic method, not only the minute detail of the various movements but also the force generated during movement have been investigated. Fenn (1930) and Cavagna, Saibene, and Margaria (1964) tried to measure speed variations of individual body segments during a running stride, and Miyashita (1974) measured the horizontal displacement of the swimmer's waist cloth while swimming breast stroke. These authors estimated work done and power exerted during running and swimming, based on speed variation. This kind of study will be more popular with the help of the new intermittent high-speed motion camera, highly sensitive film, and the modern film analysis systems.

Direct measurements of force have been conducted by many authors. In order to measure and evaluate the forces exerted between the feet and the ground during walking, running, and jumping, pneumatic, mechanical, and electric methods of recording have been used (Drillis, 1958). Recently Payne (1974) devised a platform that is capable of measuring forces in all six modes (vertical and two horizontal directions and moments of forces about these axes). The amount of propulsive force developed by the kick was measured by Cureton (1930) using spring scales. Propelling force during tethered swimming was measured in the four competitive swimming styles by Mosterd and Jongbloed (1964) with a spring and by Magel (1970) with an electrical force transducer. Wittmann (1974) placed strain-gauge force transducers between ski and normal release-binding elements in order to determine the resultant forces generated during alpine skiing. An accelerometer has also been used to measure the influences of force on the body in skiing (Neukomm and Nigg, 1974). The recent progress in these force transducers has made it possible to evaluate the forces more precisely and more conveniently.

Passive force was directly measured by means of various kinds of dynamometers. For instance, air resistance to a runner was measured by Hill (1928) using a balance, and that of a skier was measured by Watanabe and Ohtsuki (1975) with a semiconductor transducer, both in wind tunnels. Water resistance was also measured in relation to speed by Karpovich (1933), towing the swimmer by means of a spring connected to an electric motor. Holmer (1974) utilized strain-gauge methods for measurements in the swimming flume.

As a joint transforms rectilinear movement of the muscle into circular movement of the limb segment, positive force affecting human motion varies instantaneously. Because of this functional structure, it is very difficult to evaluate the force during movement. However, it has been reported that there is a linear or curvilinear relationship between electrical activities of muscle and force, both in static and dynamic contractions (Lippold, 1952; Kuroda, Klissouras, and Milsum, 1970; Bigland-Ritchie and Woods, 1974). That is, EMG can be used as an indication of the force exerted by the muscle momentarily (Rau and Vredenbregt, 1973). Ikai, Ishii, and Miyashita (1964) recorded EMG during swimming and reported that the extensors of the arm and trunk contracted more vigorously to move the body forward than the flexors. Kitzman (1964) investigated the relationship of the action potentials of the muscles and the various phases of the baseball batting swing with EMG.

However, for the purpose of applying these research results more significantly to sports activities, human movements should be investigated in some other aspects. For this investigation it is essential to use a combination of methods. For example, the motion–force relationships in jumping were identified by Murray, Seireg, and Scholz (1967), combining high-speed motion photography and a force transducer. The motion–positive force relationships in swimming the crawl stroke were studied by Clarys et al. (1974), combining a motion picture camera and EMG apparatus. Moreover, telemetry was developed so that electrical activities of muscles, angular movements at each joint, and the acceleration of body segments could be recorded without wires. Consequently, the trend to apply the combination of methods to the study of sports activities seems to be increasing rapidly.

Recent advances in data-collecting equipment and computers surely promise that in the near future, human movement will be analyzed simultaneously in several aspects. In other words, since methodology in biomechanics has become better established in recent years, individual sport activity can now be analyzed more quantitatively and systematically than ever before.

As the quantity of research increases, knowledge of the biomechanics of sport is expected to provide advantageous change in the teaching techniques of various sports. Furthermore, Hay (1973) states, "A knowledge of the biomechanical principles involved might well enhance the performance of an already skilled athlete." However, the problems associated with how one acquires skill are numerous and complex. Although the direction that the biomechanics of sport will take is still uncertain (Nelson, 1973), it should contribute to the solution of many of those numerous and complex problems. For this purpose, in the field of biomechanics, biological aspects must be taken into consideration literally, in addition to mechanical aspects. As previously described, instrumentation systems and research methodology are almost fully developed for the mechanical approach. On the other hand, the organismic variables that affect performance are still unrevealed since they depend mostly upon the neuromuscular function. Therefore, the research results concerning the mechanism of neuromuscular function must be considered and at the same time the proper methodology must be developed. As a consequence of those efforts, biomechanics of sport will be established as one of the interdisciplinary and applied sciences.

REFERENCES

Bigland-Ritchie, B., and J. J. Woods. 1974. Integrated EMG and oxygen uptake during dynamic contractions of human muscles. J. Appl. Physiol. 36(4): 475–479.

Broer, M. R. 1966. Efficiency of Human Movement (2nd Ed.). W. B. Saunders, Philadelphia.

Cavagna, G. A., F. P. Saibene, and R. Margaria. 1964. External work in running. J. Appl. Physiol. 19(2): 249–256.

Clarys, J. P. J., Jiskoot H. Rukin, and P. J. Brouwer. 1974. Total resistance, in water and 15 relation to body form. *In:* R. C. Nelson and C. A. Morehouse (eds.), Biomechanics IV, pp. 187–196, University Park Press, Baltimore.

Cureton, T. K. 1930. Mechanics and kinesiology of the crawl flutter kick. Res. Quart. 1: 87–121.

Drillis, R. 1958. Objective recording and biomechanics of pathological gait. Ann. N. Y. Acad. Sci. 74: 86–109.

Fenn, W. O. 1930. Frictional and kinetic factors in the work of sprint running. Am. J. Physiol. 2: 583–611.

Fletcher, J. G., H. E. Lewis, and D. R. Wilkie. 1960. Human power output: The mechanics of pole vaulting. Ergonomics 3: 30–34.

Furusawa, K., A. V. Hill, and J. L. Parkinson. 1928. The dynamics of sprint running. Roy. Soc. Proc. B 102: 29–42.

Gollnick, P. D., and P. V. Karpovich. 1964. Electrogoniometric study of locomotion and of some athletic movements. Res. Quart. 35: 357–369.

Hay, J. G. 1973. The Biomechanics of Sports Techniques. Prentice-Hall, Englewood Cliffs, N.J.

Hellebrandt, F. A. 1961. Physiological analysis of basic motor skills. Am. J. Phys. Med. 40: 14–25.

Hill, A. V. 1928. The air-resistance to a runner. Proc. Roy. Soc. B. 102: 300–385.

Holmer, I. 1974. Physiology of Swimming Man. Acta Physiol. Scand. Suppl. 407.

Ikai, M., K. Ishii, and M. Miyashita. 1964. An electromyographic study of swimming. Res. J. Phys. Ed. 7: 47–54.

Ikai, M. 1968. Biomechanics of spring running with respect to the speed curve. *In:* J. Wartenweiler and E. Jokl (eds.), Biomechanics II, pp. 282–290, S. Karger, Basel.

Karpovich, P. V. 1930. Swimming speed analyzed. Sci. Amer. 142: 224–225.

Karpovich, P. V. 1933. Water resistance in swimming. Res. Quart. 4: 21–28.

Karpovich, P. V., and G. P. Karpovich. 1970. Magnetic tape natography. Res. Quart. 41: 119–122.

Kitzman, E. W. 1964. Electromyographic study of batting swing. Res. Quart. 35: 166–178.

Komi, P. V., R. C. Nelson, and M. Pulli. 1974. Biomechanics of ski jumping. Studies in Sport, Physical Education and Health. University of Jyväskylä Report No. 5.

Kuroda, E., V. Klissouras, and J. H. Milsum. 1970. Electrical and metabolic activities and fatigue in human isometric contraction. J. Appl. Physiol. 29(3): 358–367.

Lippold, O. C. J. 1952. The relation between integrated action potentials in a human muscle and its isometric tension. J. Physiol. (London). 117: 492–499.

Magel, J. R. 1970. Propelling force measured during tethered swimming in the four competitive swimming styles. Res. Quart. 41: 68–74.

Miyashita, M. 1971. An analysis of fluctuations of swimming speed. L. Lewille and J. P. Clary's (eds.), First International Symposium on Biomechanics in Swimming, pp. 53–58. University Libre de Bruxelles, Brussels.

Miyashita, M. 1974. Method of calculating mechanical power in swimming the breast stroke. Res. Quart. 45: 128–137.

Mosterd, W. L., and J. Jongbloed. 1964. Analysis of highly trained swimmers. Int. Z. angew. Physiol. einschl. Arbeitsphysiol. 20: 288–293.

Murray, M. P., A. Seireg, and R. C. Scholz. 1967. Center of gravity, center of pressure and supportive forces during human activities. J. Appl. Physiol. 23(6): 831–838.

Muybridge, E. 1901. The Human Figure in Motion. Dover, New York, reprinted 1955.

Nelson, R. C., G. Larson, C. Crawford, and D. Brose. 1966. Development of a ball velocity measuring device. Res. Quart. 37: 150–155.

Nelson, R. C. 1973. Biomechanics of sport. *In:* S. Cerquiglini, A. Venerando, and J. Warteweiler (eds.), Biomechanics III, pp. 336–341. S. Karger, Basel.

Neukomm, P. A., and B. Nigg. 1974. A telemetry system for the measurement, transmission, and registration of biomechanical and physiological data applid to skiing. *In:* R. C. Nelson and C. A. Morehouse (eds.), Biomechanics IV, pp. 231–235. University Park Press, Baltimore.

Payne, A. H. 1974. A force platform system for biomechanics research in sport. *In:* R. C. Nelson and C. A. Morehouse (eds.), Biomechanics IV, pp. 502–509. University Park Press, Baltimore.

Plagenhoef, S. 1971. Patterns of Human Motion. Prentice-Hall, Englewood Cliffs, N.J.

Race, D. E. 1961. Cinematographic and mechanical analysis of the external movements involved in hitting a baseball effectively. Res. Quart. 34: 394–404.

Rau, G., and J. Vredenbregt. 1973. EMG-force relationship during voluntary static contractions (M. Biceps). *In:* S. Cerquiglini, A. Venerando, and J. Wartenweiler (eds.), Biomechanics III, pp. 270–274. S. Karger, Basel.

Steben, R. E. 1970. A cinematographic study of selective factors in the pole vault. Res. Quart. 41: 95–104.

Toyoshima, S., and M. Miyashita. 1973. Force-velocity relation in throwing. Res. Quart. 44: 86–95.

Watanabe, K., and T. Ohtsuki. 1975. Postural changes and aerodynamic forces in alpine skiing (in preparation).

Watkins, D. L. 1963. Motion pictures as an aid in correcting baseball batting faults. Res. Quart. 34: 228–233.

Wittmann, G. 1974. Biomechanical research on release bindings in alpine skiing. *In:* R. C. Nelson and C. A. Morehouse (eds.), Biomechanics IV, pp. 243–249. University Park Press, Baltimore.

Track and field

Computer techniques to investigate complex sports skills: application to conventional and flip long jump techniques

R. Mann, M. Adrian, and H. Sorensen
Washington State University, Pullman

Within recent years, numerous cinematographic procedures, both two- and three-dimensional, have been developed to aid in studying the mechanics of human motion. Because of the enormous number of calculations necessary to analyze such data, the computer has become a much relied upon tool.

Drawbacks within the cinematographic field have arisen from the trade-offs between cost and accuracy. While low-cost, single-camera planar analysis is adequate for a majority of human activities, it cannot provide reliable data in those activities where nonplanar body rotation is an important element.

However, the only reliable three-dimensional techniques proposed have problems of precise multiple camera positioning and synchronization, as well as an equipment cost that the average researcher just cannot afford (Miller and Petak, 1973).

In the use of the computer, programs that have been developed to aid in data analysis have, in many cases, been either too limited in their usefulness or written in a language that is not readily available at most computer installations.

The purpose of this research was to develop a program, termed FILMDATA, that would eliminate many of these problems. It was decided that a planar data processing procedure would be utilized, with an approach that would allow expansion to three-dimensional analysis at a future time. This decision was made to accommodate the researcher whose only cinematographic tool is a single camera.

To ensure that FILMDATA was as compatible as possible to most computer systems, it was written in FORTRAN IV. The problems that this language presented were outweighed by the widespread use and adaptability of the language.

To make FILMDATA as useful as possible, broad options concerning both data processing and presentation were implemented, giving the user control over what is considered important in each individual project.

FILMDATA DESCRIPTION

Data Collection

FILMDATA was developed to accept data cards of *X* and *Y* coordinates of up to 12 body points or external references. Once input, the data are smoothed by a least squares procedure to moderate human data collection errors (Wylie, 1966).

Data Manipulation

From the data points, FILMDATA performs the following calculations and prints the results: (a) smoothed displacement (*X* and *Y*), (b) velocity (*X* and *Y*), (c) acceleration (*X* and *Y*), (d) resultant displacement, (e) resultant velocity, and (f) resultant acceleration.

From these data bases, the following options may be utilized:

1. A subroutine that accepts the arrays of any three identified points and calculates the following joint results: (a) angular displacement, (b) angular velocity, and (c) angular acceleration.
2. A center of gravity (COG) routine that utilizes the available body points to calculate the following: (a) segment COG displacement, (b) segment COG velocity, and (c) segment COG acceleration. If sufficient body points are available, the total body COG may be calculated. In all calculations Dempster's (1955) body data were utilized.
3. A moment of inertia (MI) routine that utilizes the available COG values to generate the following: (a) MI of the entire (left, right) leg around the (left, right) hip, (b) MI of the entire (left, right) arm around the (left, right) shoulder, and (c) MI of the total body around the body COG. As in the COG routine, Dempster's (1955) body data were used to generate the results.

Units of FILMDATA

FILMDATA was written to work in either the SI or the English system of units.

Output of the Data

All generated results are available in printed, formatted form. FILMDATA also includes the option to plot any or all of the results in any manner the researcher deems valuable. A further option allows results to be carried across data sets so that plotted comparisons of different performances can be generated.

With the addition of plotters to many computer centers, the plot routine can be a great help in both analyzing and presenting data.

FILMDATA APPLICATION

(Biomechanical Comparison of the Conventional Long Jump and Flip Jump Styles)

The conventional long jump style with its many variations, has been analyzed and modified to the point where every major mechanical advantage has been invoked. Until recently, this basic style was deemed unsurpassed as the method of attaining the maximum results from the given effort.

With the recent advent of the flip style long jump (flip jump), the question of superiority has come under serious question. Conventional long jumpers cling to past performances for support, and wait apprehensively while the proponents of the flip jump predict new performance frontiers.

The two styles differ radically, which not only increases the controversy, but serves to make direct comparison very difficult. From the time the take-off leg touches down, to the time of landing in the pit, the conventional style is utilized to oppose the body rotation produced during the plant-and-drive phase of the jump. In contrast, the flip jump style strives to eliminate the braking force and utilize the body rotation to execute a complete flip while airborne.

In the most critical phase of the jump, the take-off, both styles exhibit strong points. The conventional stylists concede significant horizontal braking action to oppose rotation, but point to the critical lift potential generated by the limbs at take-off. The flip innovators claim similar lift generation, coupled with the fact that a braking effort is not necessary since the rotation is channeled into the flip.

Analyses of these types of movements are excellently suited for film analysis and interpretation by a program such as FILMDATA.

An Olympic decathlete skilled in both jumping styles was filmed at 200 frames per second with the take-off as the focal point. Data points were then generated from the film with a computerized graphic tablet system, and then coupled with Program FILMDATA. Jumps of similar distances were compared to determine any differences in the techniques of the two styles.

RESULTS

The extreme technique differences of the two styles can easily be seen in the moment of inertia values around the body center of gravity shown in Figure 1. In the conventional long jump style, where rotation is being opposed, the moment of inertia remains high throughout. In the flip jump, after take-off, the moment of inertia is quickly minimized to aid in body rotation.

Because of the forward body action initiated by the flip, the upper body projection in both the horizontal and vertical directions was superior in the flip jump style. However, the major deviation in the two

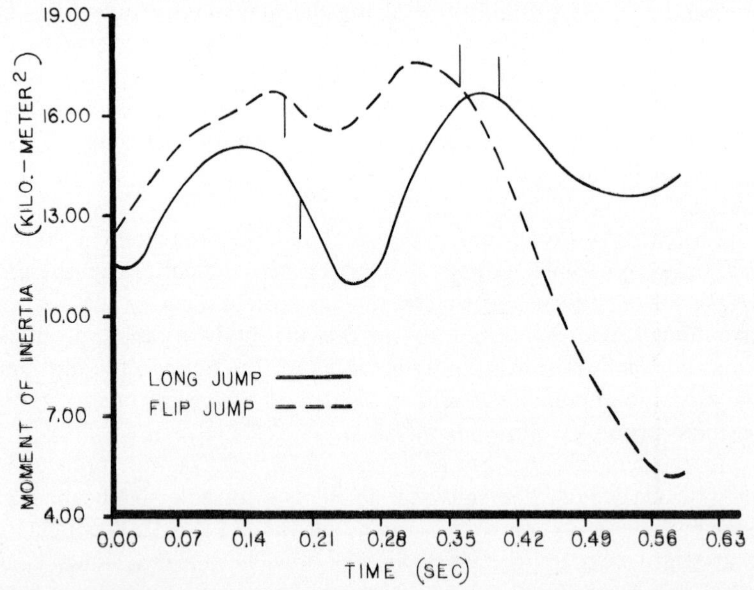

Figure 1. Moment of inertia around the body center of gravity (touchdown and take-off are marked).

styles was in the vertical action of the leading leg. As can be seen in Figure 2, the lower leg velocity during the long jump peaked during the drive phase, and began generating body lift by rapidly decelerating through the take-off phase. The flip jumper's velocity, although peaking in a manner similar to the long jump style, did not radically decrease, and therefore did not generate body lift.

This observation was surprising, but not unexplainable, since to rapidly decelerate the leading leg in the manner utilized in the long jump would generate forces that would make completion of the forward flip very difficult, if not impossible. Therefore, during the take-off phase of the flip jump, the vertical velocity of the leading leg is held relatively constant and the leg is tucked to complete a low rotational inertia situation.

This loss of lower body lift again showed up in the body center of gravity results of the flip jump. The trajectory of the flip jump COG parabola was noticeably lower than that for the long jump. Since the horizontal velocities at take-off were equal in both styles, it is evident that this loss of lift is a definite disadvantage to the flip jump style.

The fact that the horizontal velocity of the flip jumper at take-off did not exceed that of the conventional style was unexpected. However, inspection of the horizontal force platform results indicated additional

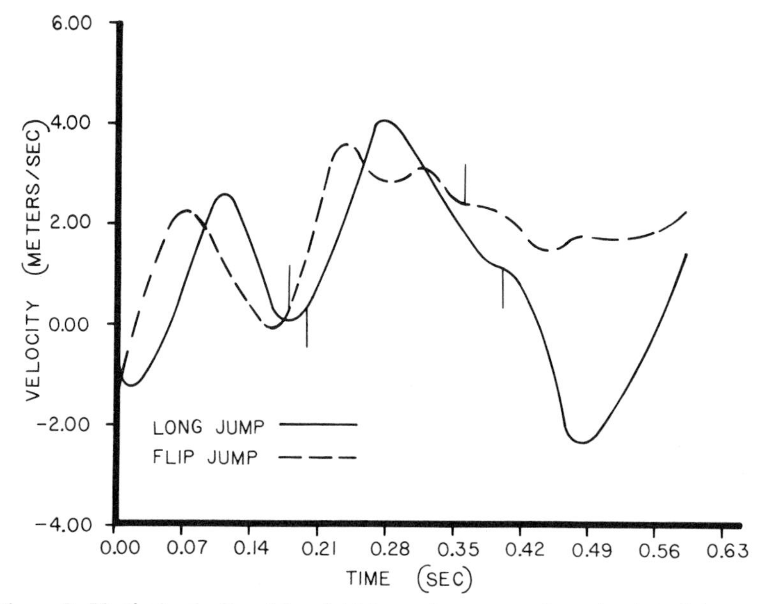

Figure 2. Vertical velocity of the right lower leg center of gravity (touchdown and take-off are marked).

breaking action was initiated to enhance body rotation at the expense of horizontal projection.

Body rotation was further enhanced by an unexpected eccentric force generated in the vertical direction during the latter stages of the ground phase. It is evident that the flip jumper must generate additional rotational forces to complete the turn before the pit is reached.

All other parameters between the two styles did not differ significantly. The fact that the distances in the jumps examined were the same was due only to the advantage gained in the flip jump landing technique. In conclusion, this initial investigation of the flip jump does not support its projected superiority.

Figures 1 and 2 are only two of 64 plots generated to aid in analysis of the techniques of the two styles. Thirty-two point or segment results were plotted for each style to simplify comparisons.

FUTURE IMPLICATIONS

FILMDATA is not a revolutionary development. The ideas, the parts, and the pieces have been available for many years. What FILMDATA does do is bring together the best trade-off possibilities that give the average researcher the greatest capability from limited resources.

As implemented, FILMDATA can be utilized to aid in the analysis of any planar activity. Also, as pointed out previously, FILMDATA was written so that transformation from a planar to three-dimensional program would not be a monumental task for those with the necessary cinematographical resources.

REFERENCES

Dempster, W. T. 1955. Space requirements of the seated operator. WADC Technical Report. Aerospace Medical Research Laboratory, Wright-Patterson Air Development Center, Wright-Patterson Air Force Base, Ohio.

Miller, M. I., and K. L. Petak. 1973. Three-Dimensional Cinematography. *In:* Carol J. Widule (ed.), Kinesiology III, pp. 14–19. AAHPER, Washington, D.C.

Wylie, C. R. 1966. Advanced Engineering Mathematics. McGraw-Hill, New York.

Analysis of the somersault long jump

M. R. Ramey
University of California at Davis, Davis

One of the new techniques being developed in track and field is the use of a forward somersault in the flight phase of the long jump. The technique evolved from a recognition that during the support phase the jumper develops angular momentum that results in a forward rotation of the body if the jumper stays in a fixed position during the flight phase. Athletes react to this effect by executing arm and leg motions that effectively attenuate the rotating effect. Over the years, three long jumping techniques have thus evolved: The Sail, the Hang, and the Hitchkick. Ramey (1974) showed that each of these techniques requires differing amounts of angular momentum to produce satisfactory jumps, the hitchkick generally the highest. The amount of angular momentum produced and the control of the resulting rotation in the flight phase are often the cause of erratic performances by many jumpers. However, Ecker (1974) correctly reasoned that instead of trying to control the unwanted forward rotation in the flight phase, the athlete could use it to turn all of the way over and produce a somersault.

Obviously, since the technique is still in its infant stages of development there have been few reports on the fundamentals involved in its correct execution. In the present paper the somersault jumps of two jumpers are studied. Both were members of the University of California track teams, but the strong event for either subject was not the long jump. Subject L. G. was a high jumper on the men's team and

---

This study is part of a larger study of human motion simulation funded by the National Science Foundation, Grant GK 41273. The opinions, findings, and conclusions expressed herein are solely those of the author.

J. D. was a sprinter on the women's team. Both were learning the somersault technique for purposes of future competition. The analysis presented of the jumps of these individuals uses force plate data, high-speed cinematography, and a computer simulation model of the flight phase.

DATA GATHERING

A special runway was used that had a two-component force plate situated at the take-off area. The force plate was recessed so that its top surface was at the same elevation as the runway. The subjects were asked to execute their jumps in the usual manner, taking off from the force plate. This provided records of the horizontal and vertical forces of the support phase of the jump. A high-speed camera, operating at 100 frames per second, was placed normal to the plane of the jump and approximately 30 m away.

In the subsequent analysis, the force plate measurements were used to show the force histories of the support phase, and to determine the changes in velocity through this phase. To make an estimate of the angular momentum developed during take-off, each frame of the support phase was used to determine the horizontal and vertical distances from the jumper's mass center to the forces on the take-off foot. The mass distribution, and the location of mass centers and moments of inertia of the various body segments were based on the data for the United States Air Force 50th percentile man as presented by Scher and Kane (1969).

A nine-segment hinge-connected system of rigid bodies, previously described by Ramey (1973), was used to simulate the flight phase of the jumps. The computer simulation was used to compare the somersault technique to long jump simulations of more traditional techniques. In this instance the positioning of the body segments during the flight phase was based on an analysis of the films of the jumper. The initial horizontal and vertical velocities used in previous simulations were then applied to the execution of the somersault technique. In this manner, one has the same athlete executing the long jump with several different techniques but the same initial parameters. The primary variable then becomes the angular momentum.

RESULTS AND DISCUSSIONS

Figure 1 shows photographs of typical somersault jumps as executed by the two subjects. Comparison of the support phase shows the differ-

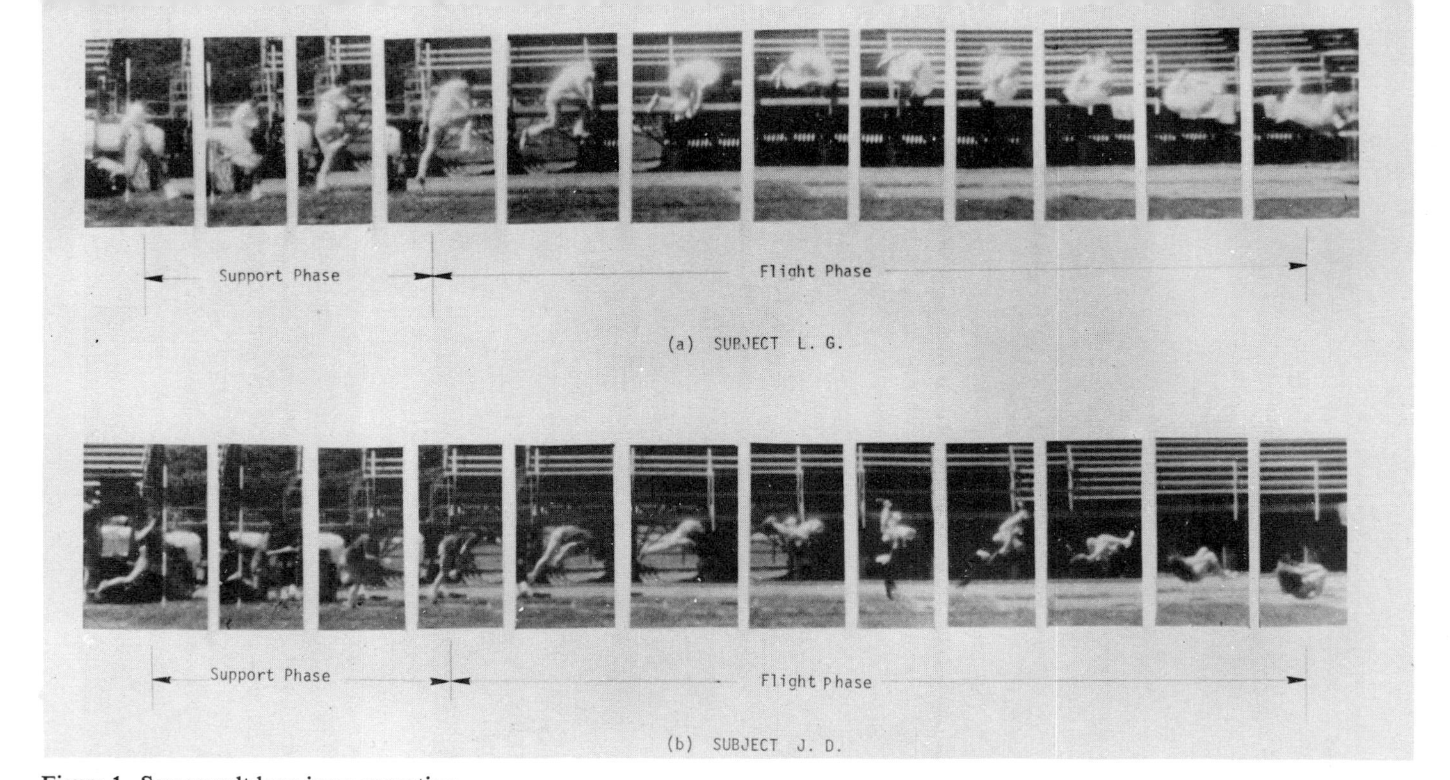

Figure 1. Somersault long jump execution.

ence in the techniques. In Figure 1*a,* Subject L. G. clearly jumps vertically before going into the tucked position, while Subject J. D. (Figure 1*b*) attempts to use the arms to provide additional clockwise rotation prior to take-off. The execution in Figure 1*b* is not recommended, but is sometimes the case for those beginning to learn the technique and having a fear that they may not turn all of the way over in the air.

Figure 2 shows the force histories, the variation of the mass center displacement relative to the supporting foot, and the estimate of the moment of force about the mass center for the support phase of each jump. In common with the results of previous studies (Ramey, 1970) the horizontal force primarily acts against the jumper slowing him/her down (see Figure 2b). (The implication is that the *resultant* force on the jumper throughout most of the support phase is generally upward and backward, not upward and frontal as sometimes shown in texts.) The force histories are known to provide a measure of the take-off velocities and the integral of the moment has been shown to be a measure of the angular momentum that occurs during the flight phase (Ramey, 1974).

Since the angular momentum is a primary factor in obtaining successful somersault jumps, it is of value to investigate the factors that produce it. With reference to Figure 2*f* one finds that the moment is given as:

$$M = F_y x + F_x y \tag{1}$$

Here F_x and F_y are the horizontal and vertical forces acting on the supporting foot, and x and y are the horizontal vertical distances from the jumper's mass center to the point of application of the ground reaction forces.

Since the horizontal force (F_x) acts opposite the direction of motion most of the time, its product with y generally produces a moment in the clockwise sense with reference to Figure 2*f*. The vertical force (F_y) on the other hand, produces counterclockwise rotation when the mass center is behind the supporting foot and clockwise rotation when the mass center is in front of the supporting foot. Equation (1) shows that the sum of the two effects produces a null, clockwise, or counterclockwise rotation at any instant depending upon the magnitudes of the forces and moment arms. Thus, in Figure 2*e* the moments are originally positive as a result of the large vertical force and then negative (clockwise) as the moment arm to the vertical force becomes smaller.

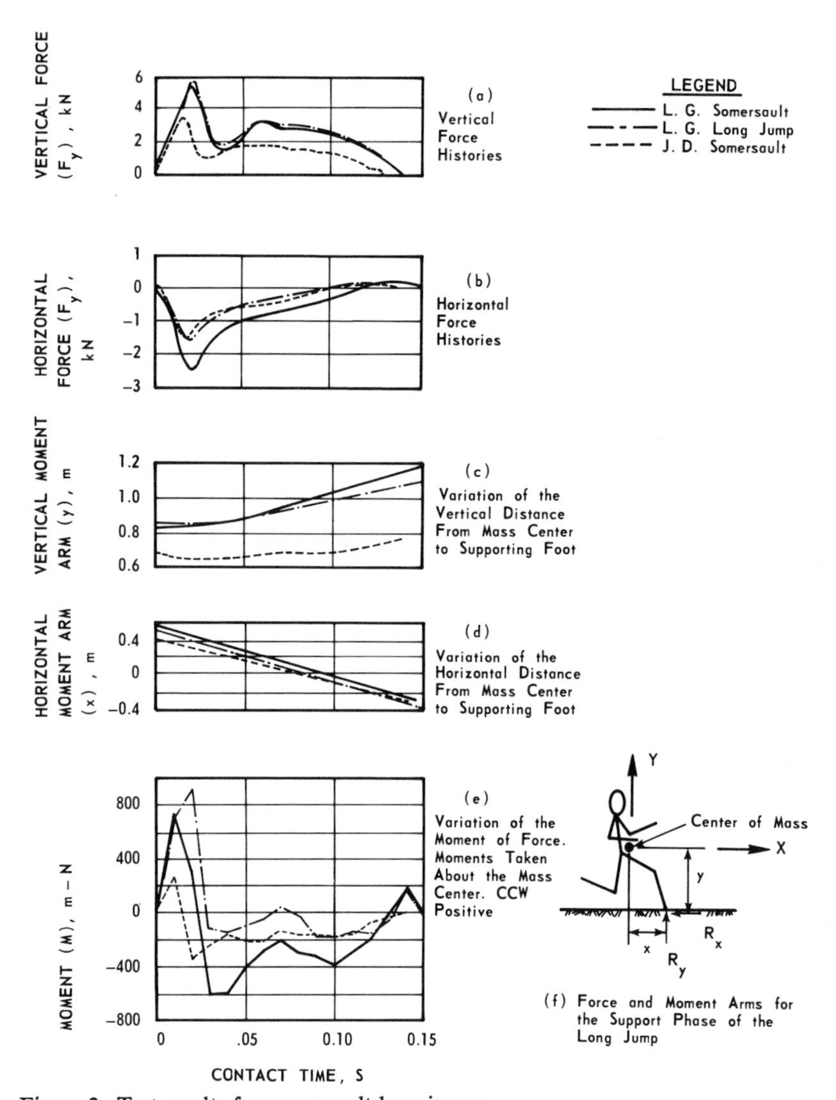

Figure 2. Test results for somersault long jumps.

In addition to the somersault jump shown in Figure 1*a,* Subject L. G. performed a conventional long jump. Comparing the forces, moment arms, and moment between L. G.'s conventional and somersault jumps, one finds that the horizontal force played a significant role in developing the angular momentum (refer to Figure 2). Observe that for these two jumps the vertical and horizontal moment arms (y and x) and the vertical force history (F_y) are nearly the same. However, the horizontal force is considerably larger in the somersault case. Estimation of the angular momentum by integration of the moment curve showed that the somersault jump produced a forward rotation (as it must) while a similar calculation for the conventional jump indicated a slight backward rotation. A backward rotation in the conventional jump was also observed in the films.

In the case of Subject J. D. it is seen in Figure 2 that the forces produced were considerably smaller than those of Subject L. G. Nevertheless, since J. D. did indeed perform a somersault, the angular momentum must have been nearly the same. Calculation of angular momentum by integration of the moment yielded values of 22 Newton meters per second (Nms) and 27 Nms clockwise for Subjects J. D. and L. G., respectively. This serves to illustrate that the support forces and moment arms can combine to produce similar angular momenta for vastly different executions of the jump. Obviously, since L.G. produced larger vertical forces, his jump resulted in considerably more time in the air to execute the maneuver.

In an effort to compare somersault jumps to more conventional techniques, a simulation of the flight phase of L. G.'s jump was made on the computer using the same body parameters and take-off velocities as reported in Ramey (1973). As expected, the angular momentum required to yield an acceptable landing for the somersault jump (41 Nms) was considerably higher than for the conventional jumps (20 Nms).

Since the initial velocities were the same for the two simulated jumps, the distance that the center of mass traveled in flight was the same (air resistance neglected). Ecker (1974) cites an unpublished report indicating that the distance from the mass center to the feet is greater at the landing in somersault cases than in conventional techniques. This also was found to be the case in the simulations. However, in the simulation it was observed that the distance between the mass center and the supporting foot at take-off was correspondingly smaller for the somersault jump. Thus, although it would appear that greater distance is achieved in somersault jumps due to the feet being so much further ahead of the mass center upon landing than in traditional cases, it is possible that some of this extra distance is lost at the take-off.

CONCLUDING REMARKS

There are obviously many factors that combine to make somersault long jumps successful. This article presents data from a small sampling of some of these factors. Here an indication of the importance of the support forces and the jumper's body positioning during this phase are reported. However, it is not clear whether the somersault technique is superior to conventional techniques even though some athletes have obtained better performances using the somersault. Future studies are certainly necessary before one can decide which technique is most suitable for a jumper. For such studies, one must obtain subjects with good proficiency in executing both somersault and conventional long jumps. Further, measures of the forces occurring during the support phase combined with cinematography that permits three-dimensional analyses of the entire jump are necessary. To produce a more complete analysis, simulation models treating three-dimensional motion should be developed.

REFERENCES

Ecker, T. 1974. The somersault long jump. Athletic Journal 54: 12–13, 77–80.

Ramey, M. R. 1970. Force relationships of the running long jump. Med. Sci. in Sports 2: 3, 145–151.

Ramey, M. R. 1973. Significance of angular momentum in long jumping. Res. Quart. 44: 488–497.

Ramey, M. R. 1974. The use of angular momentum in the study of long jump take-offs. *In:* R. C. Nelson and C. A. Morehouse (eds.), Biomechanics IV, pp. 144–148. University Park Press, Baltimore.

Scher, M. P., and Kane, T. R. 1969. Alteration of the state of motion of a human being in free fall. Technical Report No. 198. Division of Applied Mechanics, Stanford University, Stanford, California.

Kinetics and kinematics of the take-off in the long jump

C. Bosco, P. Luhtanen, and P. V. Komi
University of Jyväskylä, Jyväskylä

Take-off in the long jump is a movement in which the athlete tries to develop as much vertical velocity as possible without an appreciable loss of horizontal velocity developed during the approach run. Knowledge of the kinetics and kinematics of the execution of the long jump take-off is, however, still not sufficient for understanding of all the factors that characterize a good take-off movement. This study was therefore undertaken to investigate with elite long jumpers the variations in ground reaction forces and displacements of the body center of gravity at take-off. A special purpose was to identify the possible factors that are significant for obtaining good performance.

METHODS

Four Finnish national level male athletes performed a total of eight long jumps under conditions comparable to actual competition. The ground reaction forces were recorded with a force platform (Komi, Luhtanen, and Viljamaa, 1974) placed on a rubberized asphalt indoor track. The vertical and horizontal components of the ground reaction forces (Figure 1) were measured with four Kyowa LCP 500 KA and two Kyowa LCP 200 KA force transducers, respectively. After amplification the force signals were stored on magnetic tape (Philips Analog 7) for subsequent analysis with HP 2116 C laboratory computer, which gave the output variables shown in Figure 2.

This study was supported in part by Grant No. 11671/79/74 from the Ministry of Education, Finland.

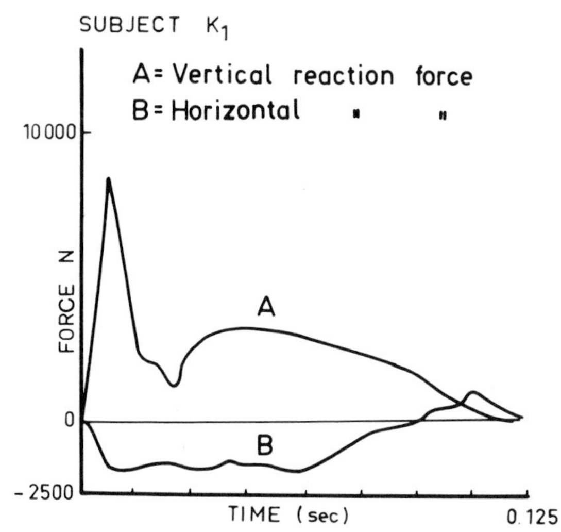

Figure 1. Example of ground reaction forces for one subject (records retouched).

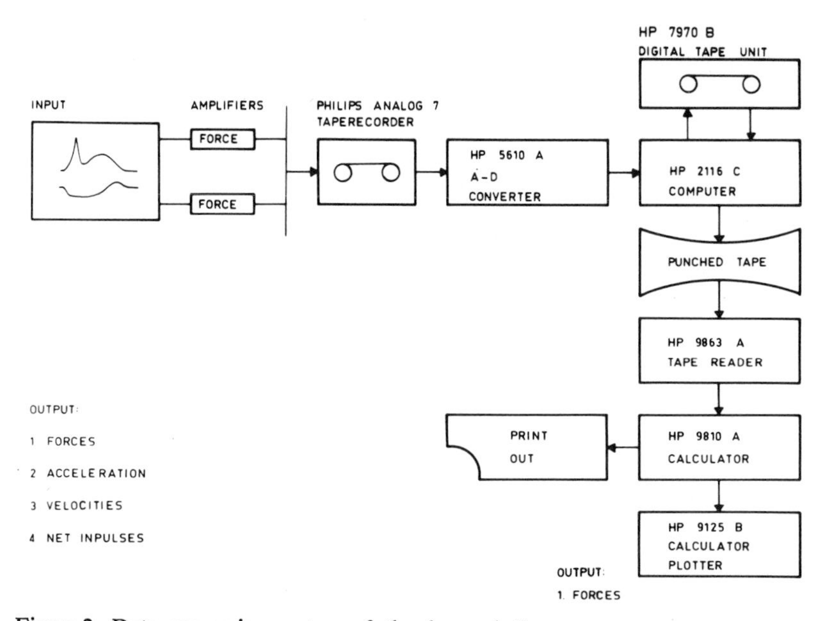

Figure 2. Data processing system of the force-platform measurements.

For comparison, each performance was also filmed with a Locam 16 mm camera set to operate at 100 frames per second and placed perpendicularly to the plane of motion at a distance of 25 m from the force platform. Using Dempster's (1955) references of landmarks for

x–y coordinates of the various body segments, the appropriate cinematographic data were obtained with a semiautomatic Vanguard motion analyzer and further processed with the computer (Figure 3).

RESULTS AND DISCUSSION

Because of the large number of parameters investigated this section is limited to those variables which were considered as most important for the performance in the jumps of the study. Table 1 gives individual and mean data for some of the selected parameters.

The detailed analysis revealed that the contact period[1] seemed to be of importance for successful performance. The contact time correlated negatively (— 0.89) with the jump length. Also the vertical velocity (V_v) at this phase was observed to be related to the final performance and it accounted for 60 percent for the total vertical velocity gained at take-off (Figure 4).

There was a lack of correlation between the final V_v at take-off and the length of jump. However, a clear interdependence was observed between V_v and reduction in the horizontal velocity (V_h) during the

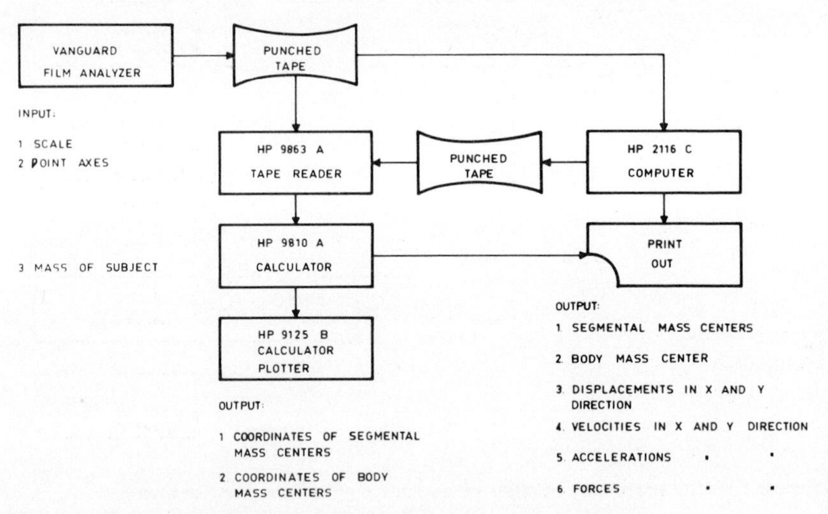

Figure 3. Data processing system of the cinematographic analysis.

[1]The contact time on the force platform was divided into two periods, first and second. From these the first half contact period is equal to the time spent on the platform between the first touch and the moment when the imaginary line, "point of application to center of gravity," reached the vertical position.

Table 1. Individuals and means data from selected variables calculated from the force platform and film analysis

	Horizontal velocity (V_h) at impact (m/sec)	Horizontal velocity (V_h) at take-off (m/sec)	Vertical velocity (V_v) at take-off (m/sec)	Angle of projection (degrees)	Contact time on platform (sec)	Length of jump (m)
K 1	9.78	8.77	2.81	18	0.122	6.85
K 2	9.88	8.70	2.98	19	0.122	6.86
V 1	9.80	8.20	3.35	22	0.109	7.33
V 2	9.80	8.45	3.26	21	0.103	7.39
T 1	8.60	7.23	3.10	23	0.125	6.12
T 2	9.65	8.90	2.70	17	0.109	7.00
S 1	9.70	8.60	2.77	18	0.118	6.92
S 2	9.80	8.36	3.03	20	0.125	7.12
Mean	9.63	8.40	3.00	19.75	0.1166	6.95
Standard Deviation	0.42	0.53	0.23	2.12	0.008	0.39

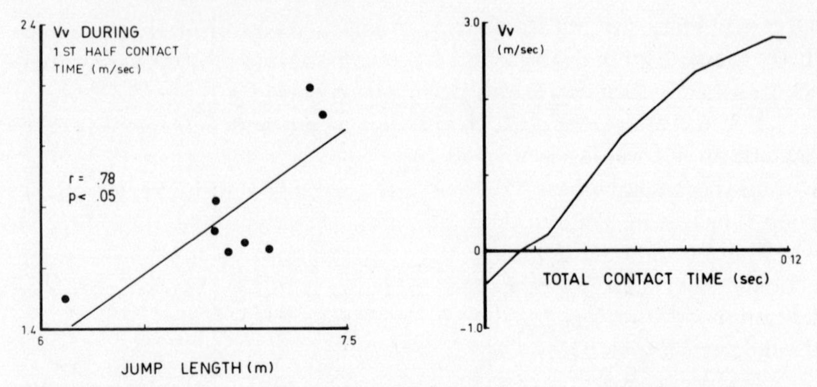

Figure 4. Relationships between the vertical velocity (V_v) of the first half contact period and the length of jump (*left*). Development of V_v during the contact period (*right*).

two contact periods. The correlations were negative (— 0.87) and positive (0.90), respectively, for the first and second period. Good performances were characterized by only minor reductions in V_h during the first contact period. This reduction coincided with the changes in the position of the jumper's center of gravity (CG). In good performance CG began to rise immediately after the first touch on the platform, but in a poorer jump the CG remained at about the same height during the early contact phase (Figure 5).

The obtained results seem to emphasize the importance of the first half contact period—but not the second half—for the execution of a

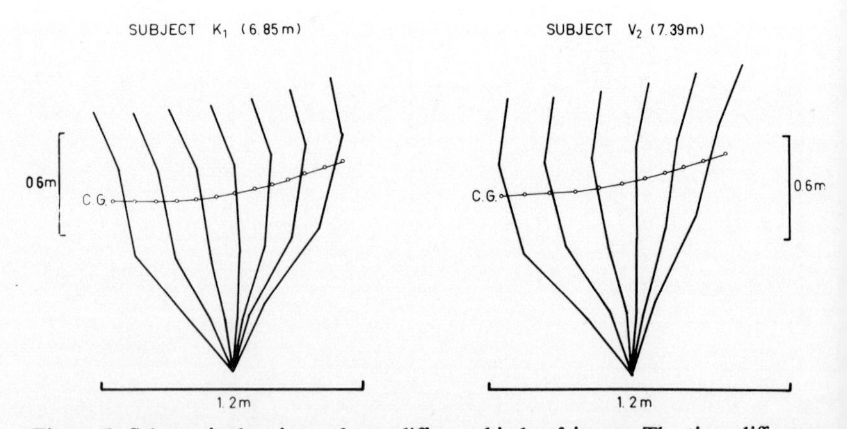

Figure 5. Schematic drawings of two different kinds of jumps. The time difference between the circles of the path of the center of gravity was 0.01 sec.

successful jump. An athlete desiring to improve his performance should direct the vertical and horizontal force components in such a way that the gain in V_v does not decrease V_h too much.

For the entire period and also for the second contact time, the increase in V_v was accompanied by almost parallel reduction in V_h. During the second phase, V_v had low correlation with the length of jump, which is opposite to that observed during the first period (Figure 4). The fact that greater increases in V_v during the first phase were obtained with smaller reductions in V_h (Figure 4) implies that the utilization of elastic energy during the take-off has been different in the different jumps.

A closer inspection of the displacements of CG revealed that the poor jumper in addition to failing to raise CG during the early contact phase also needed longer time to reach the maximum $V_v \cdot V_h$ values at impact were almost equal (range 9.65 to 9.88 m/sec) in all jumps analyzed with the exception of T_1, whose V_h was only 8.60 (Table 1). Thus among seven of eight jumps investigated the V_h at impact cannot be regarded as a limiting factor for a successful performance. The proper technical execution of the take-off should therefore rely on the utilization of the elastic energy that is stored in the muscles during the eccentric (stretching) contraction at impact. If the stretching phase is too long, it should naturally lead to reduction in the stored elastic energy (Cavagna, Dusman, and Margaria, 1968) and subsequently

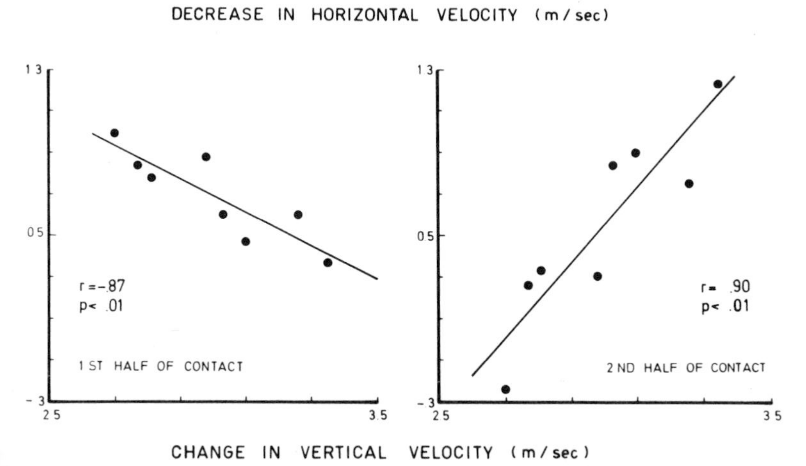

Figure 6. Relationship between the change in horizontal velocity and change in vertical velocity during the first (*left*) and second (*right*) half of the contact period.

cause lowering of the CG, just as was the tendency for poorer performances in the present study. Thus it would seem logical to emphasize the importance of the gain in V_v during the early contact period. How an athlete is able to benefit from this take-off period depends on the ability of his leg extensor muscles to utilize the coupling of the eccentric–concentric contractions at and immediately following the impact.

REFERENCES

Cavagna, G. A., B. Dusman, and R. Margaria. 1968. Positive work done by a previously stretched muscle. J. Appl. Physiol. 24(1): 21–32.

Dempster, W. T. 1955. Space requirements of the seated operator: geometrical, kinematic and mechanical aspects of the body with special reference to the limbs. WADC Technical Report 55–159, Wright-Patterson Air Force Base, Ohio.

Komi, P. V., P. Luhtanen, and K. Viljamaa. 1974. Measurement of instantaneous contact forces on the force-platform. Research report No. 5, Department of Biology of Physical Activity, University of Jyväskylä, Finland.

Biomechanical analysis of intermediate and steeplechase hurdling technique

D. A. Kaufmann and G. Piotrowski
University of Florida, Gainesville

A number of biomechanical investigations of the hurdling technique have been reported by Abbot (1948), Farmer (1948), French (1948), Larcher (1971), Mitchell and Hopper (1971), and Riddle (1971). However, only one study (French, 1948) investigated the effects of hurdling on a curve.

PROCEDURES

Six subjects, three intermediate hurdlers and three steeplechasers, were filmed by a 16 mm Beauleiu high speed (69.7 fps) camera in the sagittal plane of action and by a 16 mm Bolex high speed (64.0 fps) camera in the frontal plane as they passed over the hurdle or steeplechase barrier at the center of the curve of a 440-yard track. Table 1 presents the general data on the six subjects. All subjects were instructed to run all-out toward the hurdle or barrier simulating a race-type performance. The steeplechasers not only hurdled the barrier but in some trials utilized the step-on-the-barrier technique.

The following bodily landmarks were used in the analysis: (a) top of head, (b) center of forehead, (c) lobe of left ear, (d) middle of neck, (e) top of manubrium, (f) acromion process of both scapulae, (g) lateral epicondyle of left humerus, (h) medial epicondyle of right humerus, (i) entire surface of both wrists, (j) navel, (k) greater trochanter of both femurs, (l) lateral condyle of left femur, (m) medial condyle of right femur, (n) lateral malleolus of left fibula, (o) medial malleolus of right tibia, and (p) toes of both shoes.

Table 1. Subject data

Subject	Event	Number of trials*	Age	Height (cm)	Weight (kg)	Years of competition	Best time	Lead leg	Honors
Novice	440 intermediate hurdles	2	23	178	68.3	0	—	Right	None
Varsity	440 intermediate hurdles	3	21	175	76.5	3	52.9	Left	Best Indoor High School Hurdler, N.Y. 1971
Champion	440 intermediate hurdles	4	22	182	67.3	3	51.6	Left	National Junior College Champion, 1973
Novice	3,000 meter steeplechase	2 2	21	172	69.9	1	9.27	Right	None
Varsity	3,000 meter steeplechase	2 3	24	175	67.3	3	8.42	Right	Mid-American Conference Champion
Champion	3,000 meter steeplechase	2 3	29	180	63.6	6	8.27	Left	U.S.A.-Russian Meet, 1973

*Where two numbers are recorded, the first number denotes the hurdle technique and the second number denotes the step-on-the-barrier technique.

The cameras were calibrated by the stroboscopic method using a Strobotac Stroboscope. In the frontal plane, contourograms were determined from the film by use of a Recordak Film analyzer. Kodak Plus X Reversal film (A.S.A. 50) was used with an f-stop setting of 22 because of the constant brightness of the sun.

In the sagittal analysis the film was projected one frame at a time onto a sheet of paper to about 1/15 life size. Two distinct reference points were located in the background, and their locations carefully determined relative to the hurdle. These two points, which appeared in every frame of the entire film, were used as an absolute size and scale reference, and in effect made the accuracy of the input data invariant with respect to the projector location.

Eighteen points were marked onto a sheet of paper for every frame analyzed. The first two were the reference points. Then, 16 landmarks of the body were marked, thus defining the locations of both toes, ankles, knees, hips, wrists, elbows, and shoulders, as well as the top and bottom of the head.

The coordinates of each of these 18 points were then encoded by means of a digitizer. This device was connected to a card punch that automatically punched the coordinates on any series of points on computer cards. A computer program was written to aid in the analysis of the large amount of data generated. The anthropometric model developed by Contini (1972) was used to assess the relative weights of individual body segments, and locate their centers of mass. The computer program located the center of mass for each extremity, and for the entire body. Angles of the knee and hip of the lead leg were computed, as well as the forward trunk lean. The mass moment of inertia of the lead leg about the hip was computed for each frame, and the rate of extension of the knee joint estimated. All of the information acquired and calculated by the program was printed in tabular form, and selected information was graphically displayed using a printer–plotter subroutine.

To ascertain the scatter in the data resulting from the uncertainty in the location of the data points, the length of the lead leg (ankle to knee to hip) was routinely calculated for each frame. Within a given trial, the distribution of leg lengths averaged within 2 percent of the value given by Drillis, Contini, and Bluestein (1964) with a standard deviation of about 3 percent of the leg length. As a second check on the validity of the data, the elevation of the center of mass of one hurdler was plotted against time for the period that he was completely airborne. This curve matched within 1 cm the path predicted by ballistic theory.

RESULTS AND DISCUSSION

Figure 1 depicts the three hurdle subjects clearing the hurdle during a typical performance. Observations from the contourograms indicated:

1. All intermediate hurdlers leaned their trunks and heads toward the inside (left) of the curve. The novice hurdler had a great inward head lean ($X = 1.65$ degrees) but little inward trunk lean ($X = 1.5$ degrees).
2. In contrast, the champion hurdler had greater inward head lean ($X = 10$ degrees versus $X = 3.3$) and greater inward trunk lean ($X = 10.3$ degrees versus $X = 3.8$) when compared to the varsity hurdler.

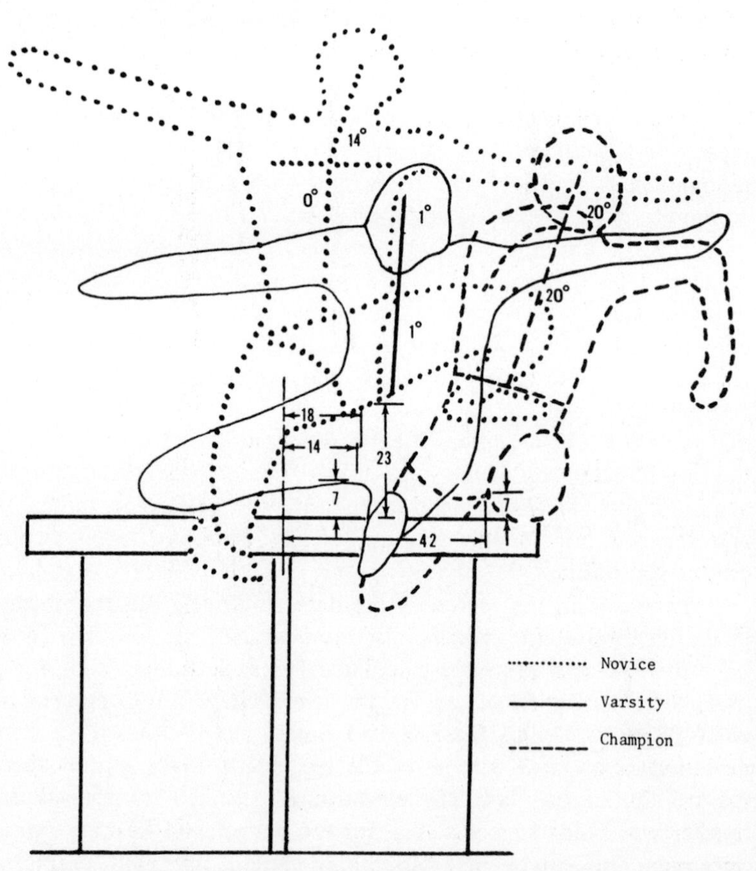

Figure 1. Contourograms of hurdlers clearing hurdle in frontal plane. All linear measurement in centimeters.

3. The champion hurdler cleared the hurdle an average of 43.8 cm away from the center of the hurdle compared to 16.0 cm for the varsity hurdler and for the novice hurdler.

Observations from the contourograms of the steeplechasers indicated that the linear height of the crotch above the barrier for all three steeplechasers was relatively high compared to the intermediate hurdlers. Using only data from hurdling the barrier and not the step-on-the-barrier technique, the steeplechasers carried their crotch 13.6 cm (25.7 versus 12.1 cm) higher over the barrier than the intermediate hurdlers. This is probably due to the fear of injury from hitting the heavy barrier. The crotch of the steeplechasers was carried 2.5–20 cm higher over the barrier in the step-on-the-barrier technique than in the hurdling technique.

Figure 2*A* summarizes the locations of the take-off and landing points for all the runners. The champion hurdler was seen to take off much farther from the obstacle than any of the other runners. In fact, study of the path of his lead leg and the center of mass showed that both attained a peak elevation about 10 cm in front of the hurdle. By contrast, the novice hurdler's lead leg was still rising as it passed the hurdle. The landing points were nearly identical for the three hurdlers. The champion steeplechaser hurdled the barrier with a fairly short stride, indicating that an entirely different style was employed in clearing a fixed barrier. In the step-on-the-barrier technique, the novice took off from farther away than the champion, but all three were in the same area.

The clearance between center of mass and the top of the hurdle did not vary substantially between the champion and varsity hurdlers, while the novice elevated his center of gravity much higher (Figure 2*B*). Similar results were found for the steeplechasers hurdling the barrier. These findings were in agreement with the observations on crotch clearance made from the frontal contourograms described above. Qualitatively it was noted that the better the runner, the smoother the transition from purely horizontal to vertical motion. As shown in Figure 2*C*, the clearance between the toe of the lead leg and the obstacle generally followed the same pattern, with the lower clearance corresponding to the better runner.

One important aspect of both events involves the rapid change of the lead leg from a fully flexed to fully extended configuration as the obstacle was approached. This maneuver was characterized by (a) the minimum mass moment of inertia of the lead leg in the fully flexed configuration, (b) the degree of flexion of the knee as the leg passes

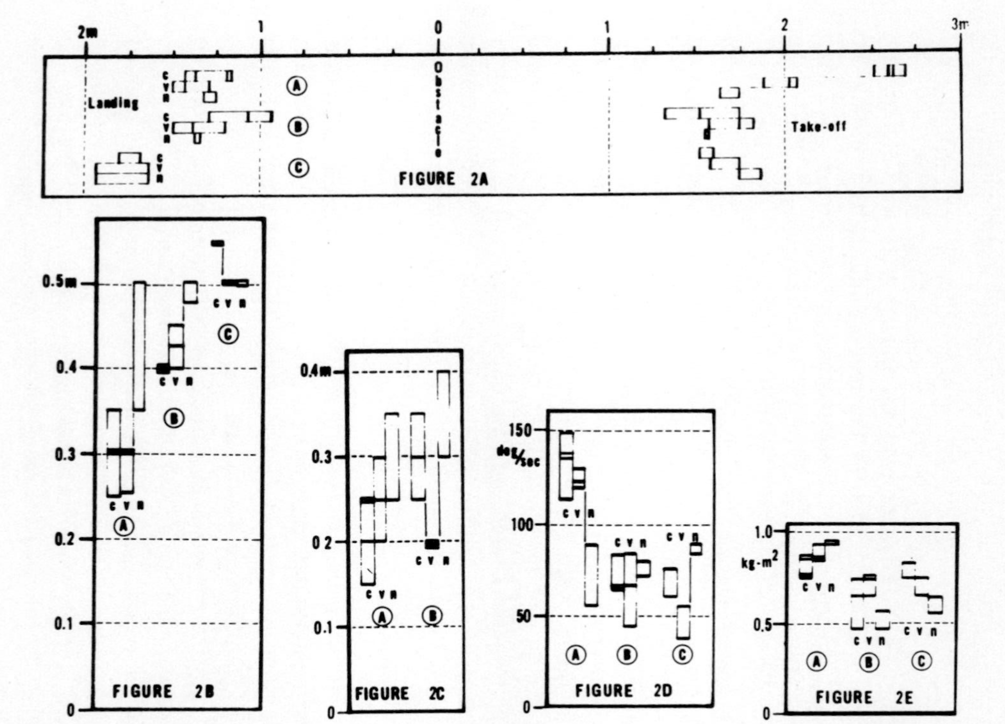

Figure 2. Several characteristics of the performances by champion (c), varsity (v), and novice (n) athletes are compared. A refers to intermediate hurdles, B to hurdling the steeplechase barrier, and C to stepping onto the steeplechase barrier. *A* compares the take-off and landing points. *B* depicts the clearance between the runner's center of mass and the top of the obstacle. *C* shows the clearance between the tip of the lead leg shoe and the top of the obstacle. *D* contrasts the peak rates of extension of the lead leg knees during the initial phases of the leap. *E* illustrates the differences in the minimum moment of inertia of the lead leg about the hip at the beginning of the leap.

over the hurdle, and (c) the rate of extension of the knee. The champion hurdler demonstrated the lowest value for the mass moment of inertia, the greatest amount of extension of the knee, and the most rapid rate of extension of the knee, as summarized in Figures 2*D* and 2*E*. These differences were not all evident in the data for the steeplechasers who as a group extended their knees much slower than the hurdlers. The steeplechasers did, however, achieve lower mass moments of inertia prior to take-off, thereby reducing the torque required to produce a given amount of angular acceleration about the hip.

ACKNOWLEDGMENT

The digitizer used in this study was made available through the courtesy of the Department of Radiation Therapy, College of Medicine, University of Florida.

REFERENCES

Abbot, R. R. 1948. A cinematographical analysis of the technique of hurdling. M. S. thesis, University of Illinois, Urbana.

Contini, R. 1972. Body segment parameters. Artificial Limbs 16: 1–19.

Drillis, R., R. Contini, and M. Bluestein. 1964. Body segment parameters. Artificial Limbs 8: 329–351.

Farmer, L. W. 1948. A mechanical analysis of the hurdles. M. A. thesis, University of Iowa, Iowa City.

French, R. L. 1948. Factors related to low hurdling on a curved track. M. A. thesis, University of California, Los Angeles.

Larcher, C. P. 1971. Relationship of various factors with the performance of male students on the hurdles. Ph.D. dissertation, University of Oregon, Eugene.

Mitchell, L. J., and B. J. Hopper. 1971. Analysis of high hurdles clearance. Track Tech. 44: 1407–1408.

Riddle, P. E. 1971. Cinematographical analysis of women hurdlers. M. S. thesis, University of Illinois, Urbana.

Trends in speed of alternated movement during development and among elite sprinters

P. F. Radford and A. R. M. Upton
McMaster University, Hamilton

One-hundred-meter sprinters in competition typically have faster stride rates than runners in longer races. Although there are considerable changes in the ratio of the ground-contact time to the airborne time of the running strides as the athlete moves towards top speed, the stride rate of a sprinter in a 100-meter race is remarkably constant at between 4.5 and 5.0 strides per second.

In all human locomotion, stride rates cannot legitimately be examined alone. They are affected by velocity (Hill, 1950), the capacity of the muscles to deliver energy during contraction (Cavagna, Komarek, and Mazzoleni, 1971), stride length, and the length, mass (Hoffman, 1971, 1972), and structure of the limbs (Tanner, 1964), and even, perhaps, the individual character of the performer (May and Davis, 1974).

Hill's work established the basic relationships between stride length, stride rate, and limb length in animals. In simple terms, shorter limbs produced shorter strides and faster stride rates. Despite important anatomical differences between animals, this principle is still essentially true. Relationships between stride rate and animal size for animals in the size range from mice to horses were recently reinvestigated and Hill's hypothesis confirmed (Heglund, Taylor, and McMahon, 1974).

Hoffman (1971, 1972) has demonstrated that within the relatively small range of adult human differences, and for the special case of sprinting, stride rate is closely related to leg length, an observation that any competent coach would confirm. Tanner (1964) has also emphasized the importance of limb structure of successful male sprinters.

The relationship between limb length and stride rate might lead one to speculate that young children would use extremely fast stride rates in their running, and in some instances they do. The toppling gait of the very young child may be performed at about two steps per second (May and Davis, 1974), and the running gait of the older child is often much faster than this, but at no time do they show excessively fast stride rates that would be predicted from leg-length alone. The complex interplay of factors that produce stride rates makes interpretation of such locomotor data difficult. The present study was designed to measure one possible contributing factor to stride rate, the rate at which subjects could perform simple, rapid alternated movements.

METHODS

The subjects were 139 males and females ranging in age from 5 to 75 yr, who were scored on a simple tapping test that had a fixed epoch of 20 sec, and was considered suitable for measuring the rates of alternated movement for each of a subject's four limbs. No knowledge of results was given and tests were performed under standardized conditions without an audience. The test was also administered to elite sprinters with proven reputations in international sprinting who were extensively studied by other techniques (Upton and Radford, 1976).

RESULTS

Subjects were grouped into eight age groups, and the age-trends in these data are plotted in Figure 1. These data show a trend only, and additional research is necessary to establish greater accuracy of these curves. Male and female subjects have been combined to make the age trends clearer, but Radford and Upton (1973) reported limb-specific differences between sexes in tapping performance and in learning. Figure 2 shows the ranges, means, and standard deviations of male and female physical education students (20–25 yr) on the tapping test.

In Figure 3 the tapping scores of elite male and female sprinters are plotted against the means and 95 percent confidence intervals of age-matched controls.

DISCUSSION

The age differences in performances on the tapping test show a marked increase in alternation ability during the developmental years. This

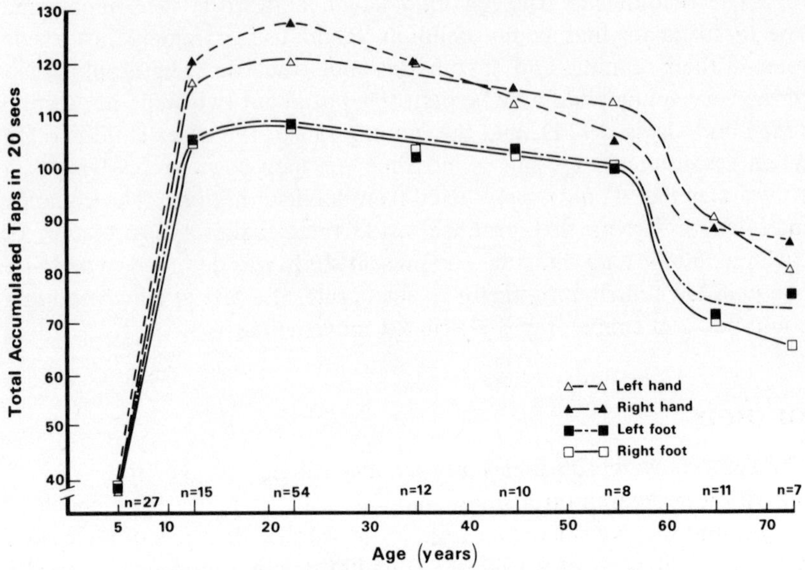

Figure 1. Age trends in scores on a 20-sec tapping test: 5–75 yr of age.

increase may reflect increased facilitation and selective inhibition by the developing nervous system; with the 5-yr-old subjects (and also with other younger subjects not included in this study), there were often synchronous, uncontrolled movements of the contralateral limb, particularly in foot tapping, but this was not seen in older subjects. Accuracy of tapping tended to increase with increased age, as subjects refined alternated movements to involve specific muscle groups rather than the whole limb. It seems likely that locomotor stride rates in childhood are particularly affected by the subject's ability to make alternated movements.

The population of elite sprinters presents considerable problems because of the small size of the sample but a small sample is inevitable if the study is truly confined to elite sprinters. The results of the sprinters can be assessed by comparison with a control population, and it can be seen (Figure 3) that the elite sprinters studied show very fast alternation well outside the 95 percent confidence intervals. Male sprinters' right hand and right foot scores were particularly superior, which is consistent with the report (Radford, 1973) of a strong lateral preference in this group. The highest values in all limbs belonged to the sprinters with the best performances. Levels of present training seemed to have no effect on scores.

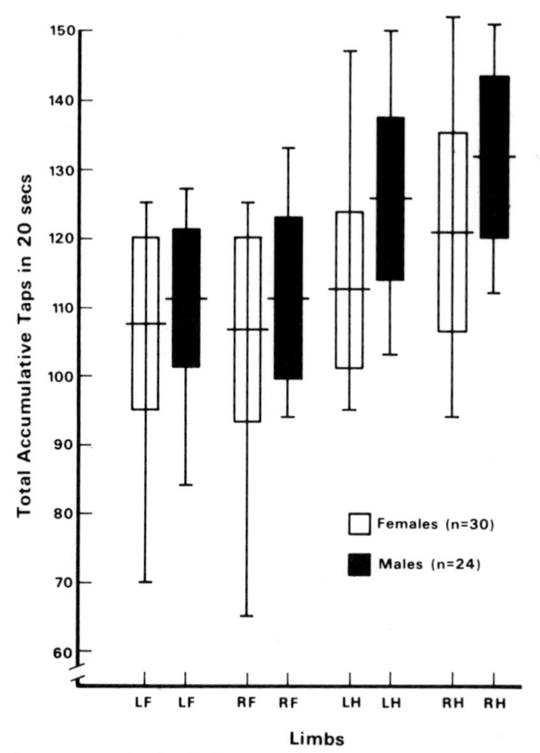

Figure 2. Tapping rates of physical education students showing means, standard deviations, and ranges.

Since the peripheral nervous system and muscular parameters of the sprinters were within normal limits (Upton and Radford, 1976) it would seem likely that the results of this tapping test reflect differences in coordination between the sprinters and the control population, and further studies will be required to assess if this difference is a result of training methods or whether the difference has determined the success of the sprinters.

This simple tapping test has indicated a major difference between elite sprinters and controls. The recent emphasis on developing the feeling for speed and rapid limb alternation in sprinters (towing, fast treadmills, rapid leg-circling, etc.) may well be logical if it can be shown that the major difference between sprinters and other athletes lies in the training of the nervous system rather than in the improvement of the peripheral neuromuscular mechanisms.

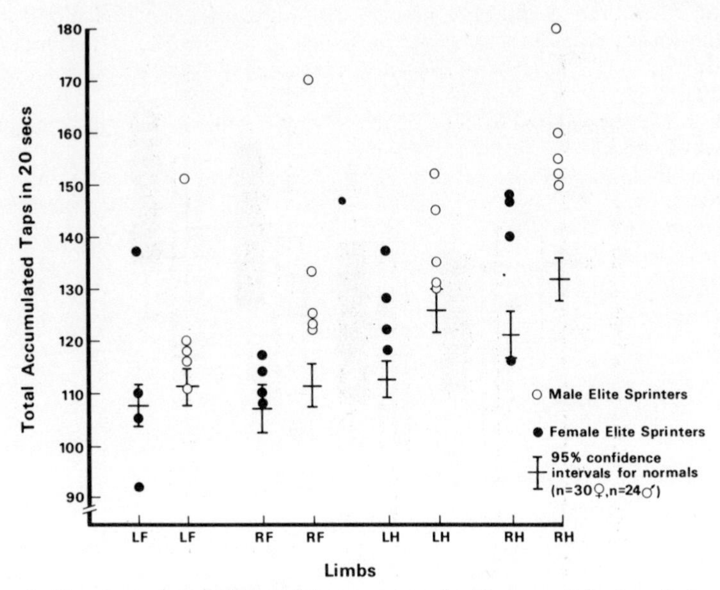

Figure 3. Tapping rates of elite sprinters compared with age-matched controls.

This simple test of alternated movement can be used in the field and may be a practical way of discovering essentially the same information that can be obtained from using sophisticated neurophysiological tests (Upton and Radford, 1976).

REFERENCES

Cavagna, G. A., L. Komarek, and S. Mazzoleni. 1971. The mechanisms of sprint running. J. Physiol. 217: 709–721.

Heglund, N. C., C. R. Taylor, and T. A. McMahon. 1974. Scaling stride frequency and gait to animal size: mice to horses. Science 186: 1112–1113.

Hill, A. V. 1950. The dimensions of animals and their muscular dynamics. Sci. Prog. 38: 209–230.

Hoffman, K. 1971. Stature, leg length, and stride frequency. Track Technique 46: 1463–1469.

Hoffman, K. 1972. Stride length and frequency of female sprinters. Track Technique 48: 1522–1524.

May, D. R. W., and B. Davis. 1974. Gait and the lower-limb amputee. Physiotherapy. 60: 166–171.

Radford, P. F. 1973. Initiating locomotor patterns: a study of lateral preference in elite male sprinters. Paper presented at the 3rd World Congress of the International Society of Sports Psychologists, Madrid, Spain. June 25–30.

Radford, P. F., and A. R. M. Upton. 1973. Speed of alternated movement in elite sprinters and others. Paper presented at the 3rd World Congress of the International Society of Sports Psychologists. Madrid, Spain. June 25–30.

Tanner, J. M. 1964. The Physique of the Olympic Athlete. George Allen and Unwin, London.

Upton, A. R. M., and P. F. Radford. 1976. Motoneurone excitability in elite sprinters. *In:* P. V. Komi (ed.), Biomechanics V-A, University Park Press, Baltimore, pp. 82–87.

Kinematic and dynamic characteristics of the sprint start

W. Baumann
Westfälische Wilhelms-Universität, Münster

Although a number of investigations of the sprint start has been published, there is relatively little information about the quantitative relationships between the sprint start characteristics and the 100 m performance of the sprinter. The purpose of this study was to quantify and to compare some kinematic and kinetic parameters of sprint starts performed over 100 m, by athletes with different levels of ability.

METHODS

The subject sample consisted of 30 experienced male sprinters, classified into three groups according to their best 100 m times in the current or preceding season. The sprinters' qualifications can be noted in Table 1. Group G_1 included nearly all German top class sprinters.

Each athlete performed at least five starts and ran a 25 m distance. The 20 m time was measured by a light barrier and served as a "premium display": In each group the three best individual runs and the three best averages of the five trials were awarded differently. Since

Table 1. Classification of subjects

Group	Number of subjects	100 m time (sec)		
		Range	Mean	S.D.
G_1	12	10.2–10.6	10.35	0.12
G_2	8	10.9–11.4	11.11	0.16
G_3	10	11.6–12.4	11.85	0.24

all subjects knew their mutual rank in their respective groups, they were additionally motivated to confirm or to question this order in terms of the 20 m times.

The force–time functions of each foot were measured with three-component dynamometric starting blocks fed to a magnetic tape recorder. The start and the run up to 6 m were recorded with a 16 mm cinecamera at a frame rate of 200 fps. The Tartan track was prepared with thin polyurethane foil in order to determine width and length of the strides and angular positions of the feet for five successive strides from the spike marks. All recording devices were synchronized by a trigger pulse from the gun. A set of 27 anthropometric parameters was taken from each subject.

For every fifth frame, beginning with the gun signal up to the 5 m distance, the location of the center of mass of the athlete was determined from coordinate measurements on the cinefilm pictures and application of a slightly modified model from that described by Clauser, McConville, and Young (1969). Adequate smoothing and interpolation of the displacement-time data rendered possible the calculation of the velocity-time and acceleration-time functions, respectively. The latter ones were subjected to validity checks (e.g., the vertical acceleration of the center of mass must equal gravitational acceleration in the flight phases). The mean error for the instantaneous values of velocity and acceleration could be estimated as follows: $m_{vel} = \pm 0.1$ msec^{-1}, $m_{acc} = \pm 0.5$ msec^{2}). Note that the evaluation of the impulse, which was assumed to be the most important kinetic parameter in this problem, was not critical to those errors. This procedure was carried out for the best trial of each subject and in addition for the second best trial of the sprinters belonging to G_1. The criterion chosen was the 20 m time.

RESULTS AND DISCUSSION

This study was concerned with a large number of parameters, but not all of them can be presented in this article. In Table 2 some anthropometric data are shown. Obviously the three groups were similar relative to these quantities and therefore possible variations of the "Set" position or of the stride length characteristics could not be attributed to anthropometric differences.

An example of the kinematics, calculated from the point-by-point evaluation of the center of mass, is given in Figure 1. Although the rough course of the curves can be discovered in all individual trials, it must be emphasized that these curves, especially the acceleration-time function, by no means were typical for one of the groups, neither

Table 2. Means and standard deviations of group anthropometrical measurements*

Body parameter	G_1 ($N = 12$) $\bar{X}$	G_1 S.D.	G_2 ($N = 8$) $\bar{X}$	G_2 S.D.	G_3 ($N = 10$) $\bar{X}$	G_3 S.D.
Weight (kg)	73.6	4	72.7	8	72.5	5
Height (cm)	178	5	179	5	180	5
Leg length (cm)	94	4	94	14	94	10
Arm length (cm)	59	3	61	3	62	3
Trunk length (cm)	62	2	66	3	65	3
Thigh circumference (cm)	57	2	57	3	56	2

*The body parameters were determined in accordance with the methods described by Hanavan (1964).

concerning the block action nor the consecutive strides. But they are characteristic for the individual subject like a fingerprint, as has also been found by Payne and Blader (1971). The times t_1 and t_4 in Figure 1 coincide with the corresponding readings from the direct force–time measurements for first reaction rear leg and take-off front leg within $\pm$ 3 msec.

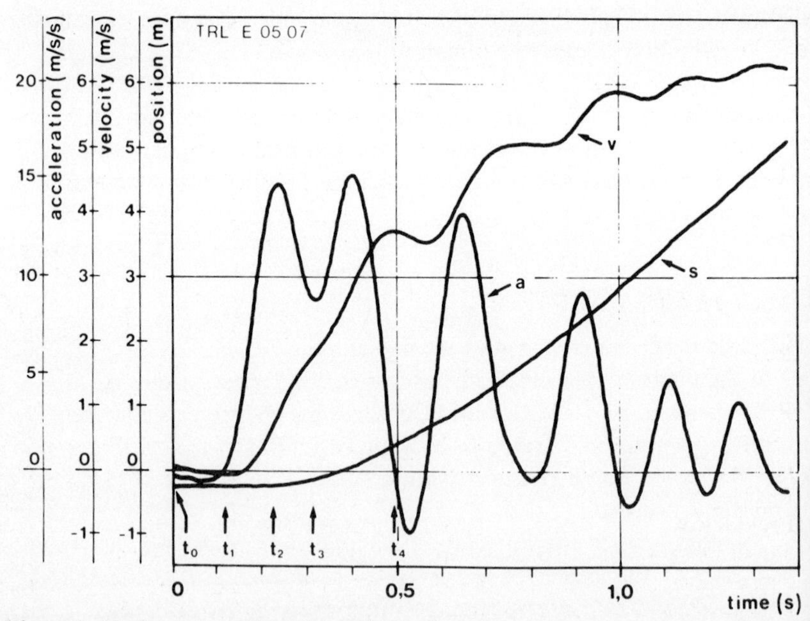

Figure 1. Horizontal position (s), velocity (v), and acceleration (a) of center of mass versus time. t_o, gun signal; t_1, first reaction; t_2, hands off; t_3, rear leg off; t_4, front leg off.

The results of the time measurements are summarized in Table 3. Contrary to the results of Payne and Blader (1971) but in agreement with the findings of Henry (1952), there was a significant difference ($p \leq 0.01$) between the reaction time of the rear foot and that of the front foot. The front foot reacts about 0.02 sec later with the greatest difference between G_1 and G_3.

Except for the lower-level sprinters of G_3, who showed slightly slowed down reactions, the time values with reference to the block phase were nearly identical. Only the time on blocks of the rear leg was somewhat longer for G_1 than for G_2.

The time required to cover such a short distance as 5 m was significantly different for the three groups: 0.08 sec between G_1 and G_2, 0.09 sec between G_2 and G_3 ($p \leq 0.05$).

The corresponding values for the 20 m times were 0.09 sec between G_1 and G_2, 0.18 sec between G_2 and G_3 ($p \leq 0.05$).

Assuming that the 100 m times ($t100$) would have been arranged in the same order if measured in these trials, correlation coefficients between the 5 m time (t_5), the 20 m time (t_{20}) and t_{100} were calculated: $r_{t5/t20} = 0.87$, $r_{t5/t100} = 0.78$, and $r_{t20/t100} = 0.90$. These coefficients clearly indicate the fact that in general high performance levels over 100 m are related to correspondingly high performance levels in the starting event and the main acceleration phase up to 20 m.

Table 3. Means and standard deviations of group time measurement

	Times from gun to event (sec)					
	G_1 ($N = 23$)		G_2 ($N = 8$)		G_3 ($N = 10$)	
Event	$\bar{X}$	S.D.	$\bar{X}$	S.D.	$\bar{X}$	S.D.
Reaction of rear leg	0.101	0.018	0.099	0.015	0.113	0.014
Reaction of front leg	0.117	0.024	0.118	0.028	0.140	0.029
Hands off	0.214	0.040	0.211	0.026	0.211	0.022
Rear leg off	0.302	0.027	0.279	0.046	0.323	0.027
Front leg off	0.470	0.036	0.468	0.020	0.504	0.032
5 m distance*	1.308	0.044	1.387	0.089	1.460	0.039
5 m distance	1.334	0.048	1.418	0.088	1.506	0.043
20 m distance	3.118	0.046	3.210	0.040	3.398	0.074

*This distance is measured from position of center of mass in "Set" position; the other distances are measured from starting line.

A summary of the kinematic and kinetic parameters is given in Table 4. Although the subjects used the block spacings they were accustomed to, there were no marked differences between the groups.

As can be seen from Table 4, the horizontal distance of the center of mass of the athlete to the starting line increased with decreasing performance level, i.e., from G_1 to G_3. Because of these differences and the modified "Set" positions, different proportions of the body weight fell on the hands with group values as follows: G_1, 73–82 percent; G_2, 62–75 percent; G_3, 52–67 percent. These values clearly point to stronger arm and shoulder muscles of the top-level sprinters. Since they are related to higher values of opposite horizontal forces in the hands and feet, the whole movement apparatus has a greater "spring tension" in the "Set" position.

Starting from a mean height of 0.65 m above ground, the center of mass is lifted up to 1.0 m when it reaches the 5-m distance. Although this is nearly the maximum height, it does not mean that the athlete is already in normal running position.

Of course, the kinetic sprint start parameters were expected to be the most important factors, especially since the time-related measurements did not show any appreciable differences. Regarding only the horizontal components in the direction of movement, there were highly

Table 4. Means and standard deviations of group kinematic and kinetic measurements.*

Parameter	G_1 ($N = 23$)		G_2 ($N = 8$)		G_3 ($N = 10$)	
	$\bar{X}$	S.D.	$\bar{X}$	S.D.	$\bar{X}$	S.D.
Distances to starting line:						
Front block (cm)	−60	5	−53	7	−56	6
C.M., horiz. (cm)	−16	4	−20	14	−27	13
C.M., vert. (cm)	66	3	60	7	63	10
Block spacing (cm)	28	5	26	5	25	3
Max. block acceleration (m/sec²)	15.4	2.0	13.2	1.7	12.2	2.4
Average block acceleration (m/sec²)	10.0	0.8	8.6	0.7	7.8	0.7
Block velocity (m/sec)	3.6	0.2	3.1	0.15	2.9	0.2
Block impulse (N • sec)	263	22	223	20	214	20
Total impulse of 1st three strides (N • sec)	200	16	214	49	204	26

*C.M., Center of Mass, determined in the "Set" position. All parameters, except for the vertical C.M. distance to starting line, relative to the horizontal axis, positive in direction of movement, the starting line being the origin of the reference frame.

significant differences between the three sprint groups for the maximum as well as the average block acceleration. The results for horizontal block velocity, i.e., velocity of the center of mass when front foot leaves the ground, of 3.6 m/sec for G_1, exceeded all previously reported values.

CONCLUSION

Summarizing these results it is beyond question that a good start is characterized by great forces exerted in the horizontal direction, which means that the successful start depends predominantly on the strength of the hip, knee, and foot extensor muscles. Differences in block time values have very little to do with 100 m performance.

REFERENCES

Clauser, C. E., J. T. McConville, and J. W. Young. 1969. Weight, volume and center of mass of segments of the human body. AMRL Technical Report 69–70, Wright-Patterson Air Force Base, Ohio.

Hanavan, E. P. 1964. A mathematical model of the human body. AMRL Technical Report 64–102, Wright-Patterson Air Force Base, Ohio.

Henry, F. M. 1952. Force-time characteristics of the sprint start. Res. Quart. 23: 301–318.

Payne, A. H., and F. B. Blader. 1971. The Mechanics of the Sprint Start. *In:* J. Vredenbregt and J. Wartenweiler (eds.), Biomechanics II, pp. 225–231. S. Karger, Basel.

Analysis of the changes in progressive speed during 100-meter dash

Y. Murase
Nagoya Gakuin University, Seto

T. Hoshikawa
Aichi Prefectural University, Nagoya

N. Yasuda
Chukyo University, Toyota

Y. Ikegami and H. Matsui
University of Nagoya, Nagoya

The present study was concerned with the variations in the running speed, the running forms, and the anaerobic energy output in the 100 m sprint and the relationships of these variables in each subject during the three phases (acceleration, maximum velocity, and deceleration) of the 100 m sprint.

PROCEDURES

The subjects consisted of one excellent runner (A), one poor runner (B) who was experienced in running training, one good runner (C) with little experience in running training, and one ordinary student (D). Their physical characteristics are shown in Table 1. The speed during 100 m running was measured electrically at every 10 m utilizing a phototransistor apparatus. Each running form (a complete cycle) was filmed from the side at a rate of 30 frames per second with a 35 mm camera (Cameflex CM-3). In order to compare the running forms at the acceleration phase, the maximum-speed phase, and the deceleration

Table 1. Physical characteristics of the subjects, and the data obtained from the films

Subject	Age (years)	Height (cm) Weight (kg)	Vertical movement of C.G. Whole body (cm) Right leg (cm) Stride length (cm)			Speed of leg's C.G maximum speed (m/sec) minimum speed (m/sec)			Angles of knee joint maximum angles (degree) minimum angles (degree)		
			acce.	max.	dece.	acce.	max.	dece.	acce.	max.	dece.
A	22	181 76	6.6 18.7 216	5.6 18.8 222	6.5 17.2 222	13.2 5.5	13.3 6.1	12.2 5.3	143 21	142 25	145 22
B	20	180 67	6.4 19.3 212	8.1 20.9 220	6.1 18.9 216	12.6 5.0	12.6 4.5	11.5 4.4	148 25	150 26	150 25
C	20	171 70	5.2 18.4 190	4.0 18.3 199	4.8 18.9 198	13.4 5.3	14.7 6.3	12.0 5.7	145 40	145 30	146 35
D	21	170 56	7.0 16.5 185	6.9 16.1 188	5.8 18.0 181	12.5 4.3	12.8 5.9	11.1 4.0	149 34	148 27	149 27

phase, the subjects were asked to run three times. A 35 mm NAC motion analyzer was used to obtain horizontal and vertical coordinates for the reference marks on the whole body. The body's center of gravity (C.G.) was determined by Matsui's method (1958) in which the human body was divided into 15 segments and the total C.G. was obtained by measuring the C.G. in each of 15 segments. The stride length was obtained by measuring the track of spiked shoes with a tape measure. The anaerobic power was determined from the mechanical work done by pedaling a Monark bicycle ergometer with maximum effort for 10 sec (Ikuta and Ikai, 1972). The rotational frequency of the wheel was measured using a photocell with the output recorded by an oscillograph. The work load used throughout the experiments was 5–7 kg. Power was calculated by the following equation: Power (kgm/sec) = rotational frequency of the wheel/sec $\times$ 1.62 (m) $\times$ work load (kg) (where 1.62 is the circumference of the wheel.

RESULTS

Running Speed

All subjects attained almost maximum speed at the 30 m point from the start (Figure 1). The maximum speed of Subject A was 9.9 m/sec and he could maintain that speed for the distance of 30–70 m from the start. The maximum speed of Subjects B, C, and D was 9.4, 9.3, and 9.1 m/sec, respectively. The speeds of all subjects were decreased at the deceleration phase. The speed at the 90 m point from the start was as follows: Subject A, 9.2 m/sec; Subjects B and C, 8.6 m/sec; Subject D, 7.9 m/sec. These speeds were equivalent to the following: Subject A, 92.9 percent of his maximum speed; Subject B, 91.5 percent; Subject C, 92.5 percent; Subject D, 86.8 percent. The running time in seconds of each subject was as follows: Subject A, 11.4; Subject B, 12.3; Subject C, 12.4; Subject D, 12.8.

Running Form

The running form was examined for the four variables analyzed, which were as follows. These results are shown in Table 1 and Figure 2. The vertical movement of the C.G. of the whole body for each subject was as follows: Subject A, 5.6–6.6 cm; Subject B, 6.1–8.1 cm; Subject C, 4.0–5.2 cm; Subject D, 5.8–7.0 cm. No significant differences were observed among the subjects in the acceleration phase, the maximum-speed phase, and the deceleration phase.

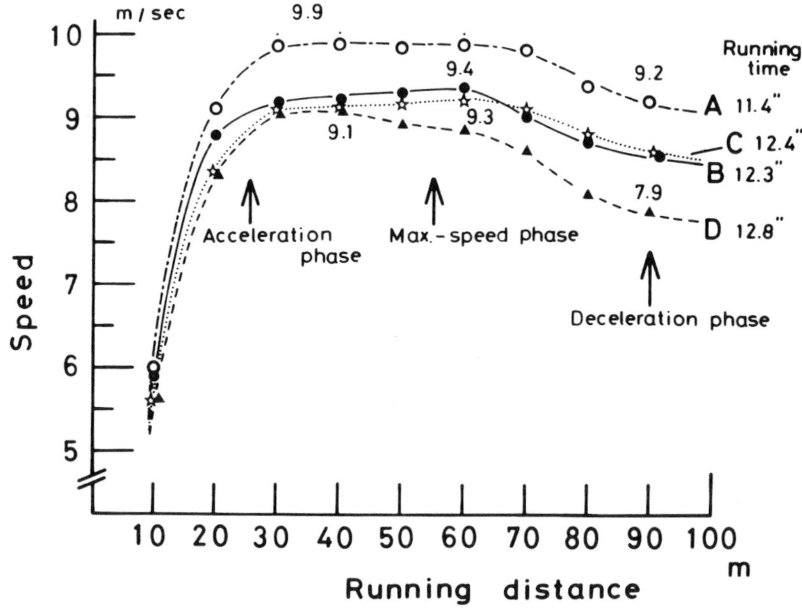

Figure 1. The speed curves during the 100 m running.

No differences were found in the vertical movements of the C.G. of subjects' right legs. The horizontal velocities of the C.G. of the subjects' right legs were less during the support phase. Among four subjects, the slowest speed ranged from 4.0 for Subject D to 6.3 m/sec for Subject C. On the other hand, the speed was increased to the maximum degree during the swing phase. The maximum speeds of four subjects ranged from 11.1 to 14.7 m/sec. When the speeds of the C.G. of the right legs of two subjects, A and B, who had almost the same stride length, were examined, it was observed that Subject A was 1.0–2.0 m/sec faster than Subject B through one cycle, as shown in Figure 2. The C.G. of the right leg of each subject attained its greatest velocity during the maximum-speed phase and its slowest speed occurred in the deceleration phase.

The minimum angles of the knee joints of the four subjects ranged from 21° to 40°, while the maximum ones were 142°–150°. Comparing the results of each subject, the knee joints of the trained Subjects A and B were well-flexed and the duration of the flexed form was lengthened in comparison with the untrained Subjects C and D. The running time in this experiment was close to the best time for each subject. For instance, the times for three trials for Subject A were 10.8, 10.8, and 10.7 sec, respectively.

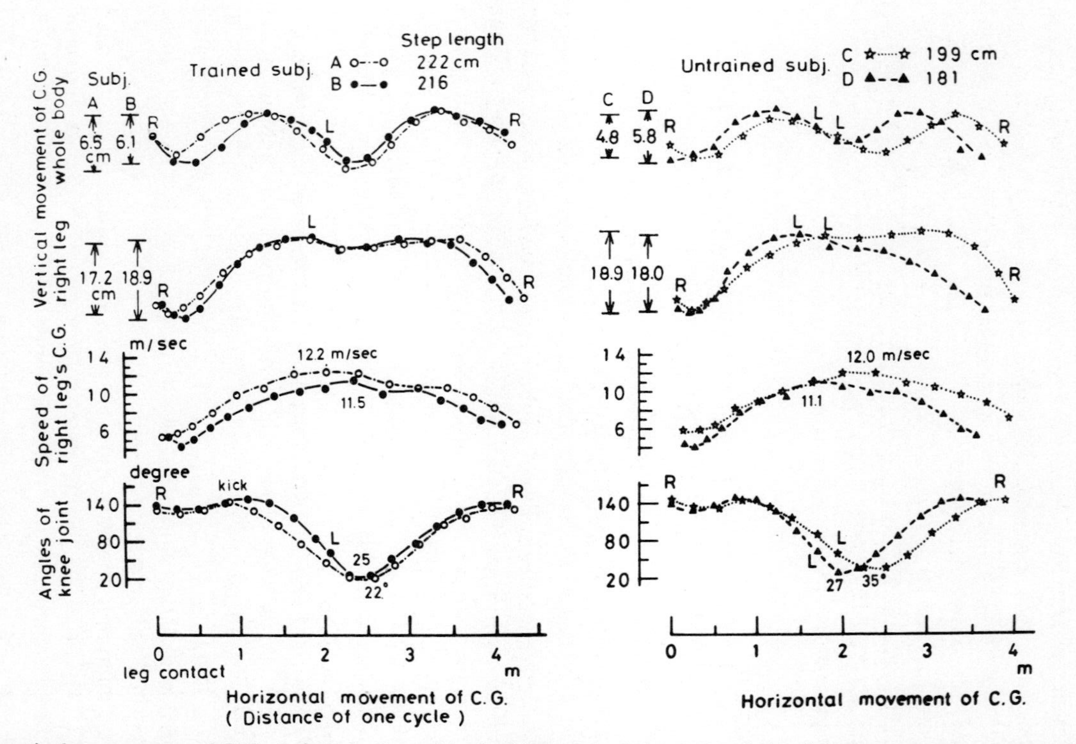

Figure 2. The vertical movements of C.G. (whole body and leg), the speed of the C.G. of the right legs and the angles of the knee joint for the excellent runner, Subject (*open circle*) and the poor runner, Subject B (*closed circle*) and untrained Subjects C (*star*) and D (*closed triangle*) at the deceleration phase (90 m-point from the start). R, right leg make a contact to the ground; L, left leg contact.

Anaerobic Power

The maximum power obtained by Subjects A, B, and C was 7 kg with the work load and by Subject D was 6 kg. The trained Subjects A and B showed the highest increase in power just after the initial movement (Figure 3). The maximum power of Subject A, the fastest runner, was 104.4 kgm/sec. The maximum values for Subjects B, C, and D were 95.1, 89.9, and 78.9 kgm/sec, respectively. As the run progressed, the power decreased for all subjects. The power exerted for each subject after 10 sec was as follows: Subject A, 89.8 kgm/sec (86.0 percent of his maximum value); Subject B, 79.8 kgm/sec (83.9 percent); Subject C, 82.3 kgm/sec (91.2 percent); Subject D, 69.0 kgm/sec (87.5 percent).

DISCUSSION

Some differences in horizontal movement of C.G. were found between trained Subjects A and B, and untrained Subjects C and D. This horizontal movement of C.G. resulted in an increased stride length. It was indicated by Högberg (1952), Groh (1973), and Hoshikawa, Matsui, and Miyashita (1973) that running speed and the stride length were closely related to each other. On the other hand, no difference was found in the vertical movement of C.G. as a result of training and run-

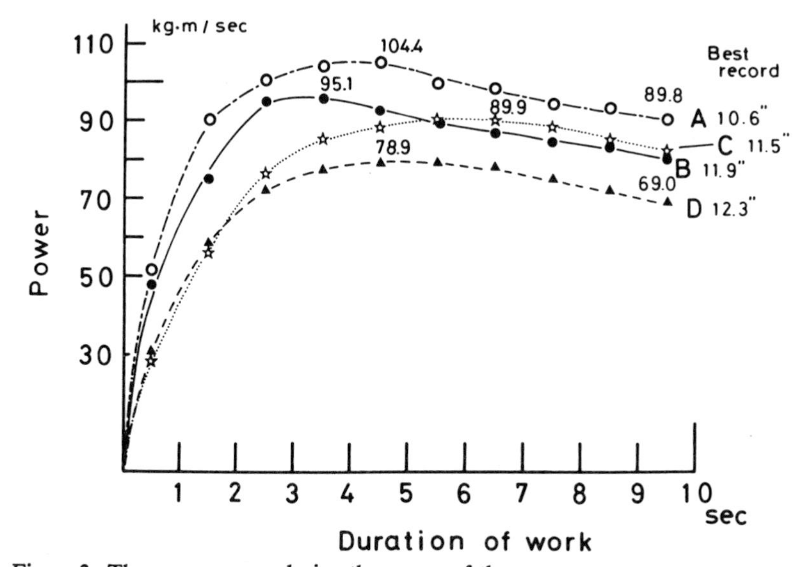

Figure 3. The power curves during the course of the run.

ning ability. This is in agreement with the report by Cavagna, Saibene, and Margaria (1964) that the vertical work is constant and independent of speed. The values obtained in this experiment agree with the ones reported by Fenn (1930) and Cavagna, Saibene, and Margaria (1964). When we examined the leg movement, Subject A was characterized as follows: the knee was flexed acutely after take-off, and then the leg was brought rapidly forward. Such movement of the leg is described as mechanically rational movement by Fenn (1930) and Slocum and James (1968). That is, the acute flexion of the knee joint reduces the moment of inertia of the leg, thus facilitating the movement of the leg forward. Subject B had 1.0–2.0 m/sec slower speed of his leg's C.G. than Subject A, although he had a well-flexed knee joint. The difference in running speed between Subject A and Subject B was due to the slower speed of leg's C.G. described above. In the case of Subject D, who had the shortest stride length, the C.G. of his leg was lowered when his leg was brought forward. This movement means that his knee is not flexed enough during the swing phase or the position of his knee joint is relatively low.

With respect to anaerobic power, Subject A, the fastest runner, displayed higher ability than others, whereas Subject D, the slower runner, showed the lowest power among the subjects. These results were also in agreement with the reports by Margaria, Aghemo, and Rovelli (1966) and Ikuta and Ikai (1972). Judging from the results in this investigation, the different running speeds of the subjects are due to both the running form, especially in the leg movement and anaerobic power. In order to explain the decrease of running speed in the deceleration phase, the following reasons could be considered: (a) deterioration of running form by fatigue of neuromuscular coordination; or (b) the decrease of phosphagen in the muscle, which is the direct energy source for mechanical work.

Comparing the running form from the film and the stride lengths in the acceleration phase, maximum-speed phase, and the deceleration phase, no significant difference was found except the decrease of speed of the C.G. of the subjects' legs and the stride length in the deceleration phase.

Anaerobic power decreased after it reached a maximum value. The decrease of speed in the C.G. of the subjects' legs and decrease in stride length in the deceleration phase would be related to the power decrease during the course of the run. It may be concluded that the decrease in the running speed in the 100 m dash is due more to the decrease in the high-energy phosphate compounds (phosphagen), which are the most direct energy source, as pointed out by Margaria, Aghemo, and Rovelli (1966), than to the variation of running forms. From the present study, it should be emphasized that anaerobic power is the most important factor in sprint running.

CONCLUSIONS

1. No differences in the vertical movements of C.G. of the whole body and right leg during the 100 m run were observed among the subjects with different abilities.
2. For an excellent runner, the speed of his leg's C.G. was greater than for a poorer runner with the same stride length.
3. With respect to anaerobic power, as determined by pedaling the Monark's bicycle ergometer, the excellent runner showed the highest power among all the subjects.
4. The decrease in running speed during the deceleration phase is related positively to the decrease in anaerobic power during the course of the run.

REFERENCES

Cavagna, G. A., F. P. Saibene, and R. Margaria. 1964. Mechanical work in running. J. Appl. Physiol. 19: 249–256.

Fenn, W. O. 1930. Work against gravity and work due to velocity changes in running. Amer. J. Physiol. 93: 433–462.

Groh, H. 1973. Leistungsbestimmende kinematische und dynamische bewegungsmerkmale beim 100-m-lauf. (The determination of kinetic and dynamic characteristics during the 100-m running.) Sportarzt und Sportmedizin 9: 207–210.

Högberg, P. 1952. Length of stride, stride frequency, "flight" period and maximum distance between the feet during running with different speeds. Arbeitphysiologie 14: 431–436.

Hoshikawa, T., H. Matsui, and M. Miyashita. 1973. An analysis of running pattern in relation to speed. *In:* S. Cerquigilini, A. Venerando, and J. Wartenweiler (eds.), Biomechanics III, pp. 342–348. University Park Press, Baltimore.

Ikuta, K., and M. Ikai. 1972. Study on the development of maximum anaerobic power in man with bicycle ergometer. Res. J. Phys. Ed. (Japan) 17: 151–157.

Margaria, R., P. Aghemo, and E. Rovelli. 1966. Measurement of muscular power (anaerobic) in man. J. Appl. Physiol. 21: 1662–1664.

Matsui, H. 1958. The movement and the center of gravity in man. Science of Physical Education Ltd., Tokyo.

Slocum, D. B., and S. L. James. 1968. Biomechanics of running. JAMA. 205: 721–728.

Model for estimating the influence of stride length and stride frequency on the time in sprinting events

R. Ballreich
Johann Wolfgang Goethe-Universität, Frankfurt/Main

From the general definition of speed, $v = \frac{s}{t}$, the following relationship can be derived: $v = l \cdot f$ (v, mean speed of running; l, mean stride length; f, mean stride frequency). On the basis of this definition, five logical possibilities exist for improving performance in sprinting events [increase of mean speed (v), thereby decrease the running time (l)]:

1. $v + \Delta v = (l + \Delta l) \cdot f \qquad (f \simeq \text{constant})$
2. $v + \Delta v = l \cdot (f + \Delta f) \qquad (l \simeq \text{constant})$
3. $v + \Delta v = (l + \Delta l) \cdot (f + \Delta f)$
4. $v + \Delta v = (l + \Delta l) \cdot (f - \Delta f); [(l + \Delta l) \cdot (f - \Delta f) > l \cdot f]$
5. $v + \Delta v = (l - \Delta l) \cdot (f + \Delta f); [(l - \Delta l) \cdot (f + \Delta f) > l \cdot f]$

Since there are contradictory findings in literature (e.g. Gundlach, 1963; Hoffman, 1965; Ballreich, 1972; Groh, 1972; Letzelter, 1972) it was the purpose of this study to solve this problem by the use of a different methodology.

METHODS

The available data were appropriate for the question: What is the influence of l and f on the variation of v? To answer this question it was necessary to vary v and to estimate the proportionate effects of l

and f on v by measuring both variables and computing their influence by adequate statistics. This variation can be done intra- and interindividually: (a) by the intraindividual variation of v within the interval of submaximal speed (procedure 1), or (b) by selecting samples composed of subjects (interindividually) who can be classified on the criterion of different mean speeds (procedure 2).

The present study was confined, because of insufficient data, to procedure 2. For this procedure we had five samples, three composed of male ($S_{1\text{-}3}$) and two composed of female subjects ($S_{4,5}$), respectively (Table 1).

Mean stride length, mean stride frequency, and running time were measured by video recorder and by special short-time stride length recordings. As can be seen from other findings each of the five variables, namely (1) length of lower extremities, (2) running time, (3) sex, (4) running distance (100 or 200 m), (5) running interval (interval of positive and negative acceleration, interval of relatively constant speed), has a different influence and therefore was standardized by the following criteria: (1) stride length was defined relative to leg length (body height) by calculating the stride-length index [stride length/leg length (body height)]; stride frequency was defined relative to leg length (body height) by calculating the stride-frequency index [stride frequency · leg length (body height)]; (2) selection of the sample resulted in a relatively homogeneous level of performance; (3) selection of the sample was based on sex; (4) running distance was specified in samples, and (5) the running intervals were specified as far as was possible ($S_{11\text{-}13}$).

Table 1. Mean sprinting times for the five samples and three subgroups of Sample 1

Group	n	x (sec)	S.D. (sec)
S_1	30	11.80	0.81
S_{11}	9	10.91	0.34
S_{12}	11	11.65	0.24
S_{13}	10	12.76	0.34
S_2*	15	10.41	0.13
S_3	47	10.40	0.19
S_4†	32	11.57	0.22
S_5‡	32	23.64	0.64

*100 m semifinalists of the 1972 Olympic Games.
†100 m sprinters of the 1972 Olympic Games.
‡200 m sprinters of the 1972 Olympic Games.

The statistical method of regression analysis ascertained the following statement: If l and f change by Δl and Δf, respectively, there will be mean speed changes within a definite limit of percentage ($p \geqslant 95$ percent) of all cases by Δv. By changing l and f a comparable amount, the induced change in v can be looked upon as a criterion of influential weight. This model presupposes that l and f are independent variables. Since this is not the case the model had to be modified: the covariation of l and f caused by a change in l and f, respectively, was taken into account while computing their influence:

$$
\begin{aligned}
v_1(x) &= G(f_x, F_1(f_x)) && \text{(1a)}\\
v_1(x+s) &= G(f_{x+s}, F_1(f_{x+s})) && \text{(1b)}\\
\Delta v_1 &= v_1(x+s) - v_1(x) && \text{(1)}\\
v_2(x) &= G(F_2(l_x), l_x) && \text{(2a)}\\
v_2(x+s) &= G(F_2(l_{x+s}), l_{x+s}) && \text{(2b)}\\
\Delta v_2 &= v_2(x+s) - v_2(x) && \text{(2)}
\end{aligned}
$$

Equation 1 describes the variation of running speed, v, induced by the variation of stride frequency; f takes into account the covariation of stride length, l. Equation 2 describes the variation of running speed, v, induced by the variation of stride length, l, taking into account the covariation of stride frequency, f. Indicators for the relative influential weight of f on v relative to the effect of l on v are therefore Δv_1 and Δv_2, respectively.

RESULTS

Table 2 shows the results obtained by applying the modified regression-analytic criterion, equations 1 and 2, respectively. Δv_1 (column 2) and Δv_2 (column 3), respectively, indicate the minimal change of mean running speed occurring with a probability of 95 percent as far as the arithmetic mean of stride frequency f (column 4) and mean stride length l (column 6), of the respective samples, S_i (column 1) increases by one standard deviation. Relative increases in the standard deviations are given in $s\ (f)$ (column 5) and $s\ (l)$ (column 7), respectively.

Except samples $S_{11\text{-}13}$, an increase of stride frequency induces a higher mean speed and consequently a greater improvement in running time (about 0.1–0.2 sec) than does a comparable increase in stride length. The samples S_2–S_5 consist of male and female world champion sprinters. Thus, sprinting for top level sprinters can probably be

Table 2. Variations in mean running speed as a result of changes in stride frequencies and stride lengths.

1	2	3	4	5	6	7
S_i	Δv_1 (m/sec)	Δv_2 (m/sec)	f (strides/sec)	$s(f)$ (strides/sec)	l (m)	$s(l)$ (m)
S_1	0.31	0.37	4.41	0.26	1.70	0.11
S_{11}	0.04	0.15	4.54	0.20	1.81	0.10
S_{12}	−0.06	0.06	4.48	0.22	1.68	0.09
S_{13}	0.07	0.04	4.21	0.23	1.63	0.09
S_2	0.20	0.00	4.67	0.33	2.10	0.10
S_3	0.03	0.00	4.58	0.19	2.10	0.09
S_4	0.17	−0.01	4.44	0.21	1.95	0.08
S_5	0.11	0.00	4.18	0.21	2.03	0.09

improved to a greater degree by developing the technical (coordination) rather than the conditioning (power) component. According to this result, the improvement of the running time for top level sprinters by 0.1 up to 0.2 sec is the result of the second logical possibility mentioned; i.e., the increase of mean speed depends on the increase of mean stride frequency, assuming that the mean stride length is relatively constant.

Since this statement is in a way an average result that refers to the total distance (100 or 200 m), one must check whether and to what degree this result may also be valid for other running intervals (intervals of submaximal running speed, interval of relatively constant running speed). A criterion for the appropriateness of this verification of the results can be obtained from the findings of the samples S_{11}–S_{13}. For S_{11-13} the influence of stride frequency and stride length on running speed within a special running interval (the interval of positive acceleration) was examined. Table 2 shows that the total of sample S_1 as well as the upper (S_{11}) and middle groups (S_{12}) have a higher influence of stride length on running speed compared with stride frequency while the influence of stride frequency dominates in the lower group (S_{13}).

REFERENCES

Ballreich, R. 1972. Ausgewählte biomechanische Probleme des 100 m Sprints (Selected biomechancial problems of the 100 m sprinting event). Biomedizin und Training, pp. 16–24.

Groh, H. 1972. Leistungsbestimmende kinematische und dynamische Bewegungsmerkmale beim 100 m Lauf (Kinematic and dynamic factors in the 100 m sprinting event) Biomedizine und Training, pp. 24–29.

Gundlach, H. 1963. Laufgestaltung und Schrittgestaltung im 100-m-Lauf (Running pattern and stride pattern in 100 m running event). Theorie und Praxis der Körperkultur 3: 254–267; 4: 346–359; 5: 418–425.
Hoffman, K. 1965. (Unpublished study.) Die Abhängigkeit der Schrittlänge und Schrittfrequenz von der Körpergröße bzw. Beinlänge (Relations of stride length, stride frequency, and body height relative to leg length).
Letzelter, M. 1972. Auswirkungen biomechanischer Bewegungsanalysen auf das Training des Sprinters (Biomechanical movement analysis and training process of sprinting). Biomedzin und Training, pp. 10–16.

Water sports

Electromyographic study of swimming in infants and children

H. Tokuyama
Osaka Kyoiku University, Osaka

T. Okamoto
Kansai Medical School, Osaka

M. Kumamoto
Kyoto University, Kyoto

Adult swimming movements have been studied electromyographically by Ikai, Ishii, and Miyashita (1964), Barthels and Adrian (1971), and Lewillie (1971). However, no electromyographic study of swimming in infants and children has been reported in the literature. In order to examine the characteristics of swimming in infants and children, we recorded EMGs, goniograms, and forms on events which they were able to swim, and analyzed in detail the functional mechanisms of the muscles.

PROCEDURES

In this experiment, 37 subjects, from a 10-month-old infant to unskilled and skilled adults, were employed. EMGs were recorded from 14 muscles of the upper and lower extremities and the trunk, utilizing an 18-channel multipurpose electroencephalograph with water-proofed surface electrodes, 10 mm in diameter. The subjects swam using their best form, with the electrodes attached to their body, in an indoor heated swimming pool equipped with underwater windows. The movements of the swimmers were filmed with 16 mm movie cameras, from the side, front, and back through the underwater windows. A signal in each frame and the goniogram of the knee joint were recorded simultaneously with the electromyograms during the swimming.

RESULTS

Figure 1 shows the electromyograms of the crawlstroke of a 6-yr-old child in the initial period of learning to swim by imitating adults, and those of a specially trained adult swimmer. The results showed that there were some differences in the movements and the discharge patterns.

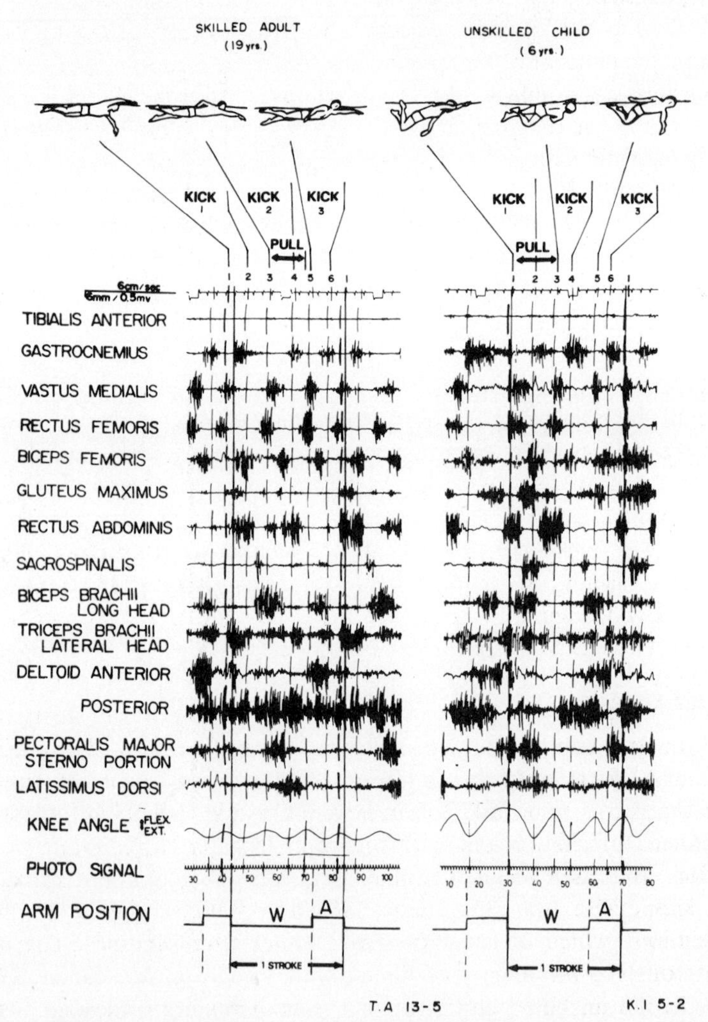

Figure 1. Electromyograms of the crawlstroke of the 6-yr-old child in the initial period of learning to swim by imitating adults and a specially trained adult swimmer.

As to the forms, the body of the skilled adult was maintained almost parallel to the surface of water, while the hip joint of the unskilled child was permitted to drop down. The knee flexion of the unskilled child was greater than that of the skilled adult. The goniograms of the knee joint showed a rhythmical pattern in the skilled adult, while in the unskilled child it was irregular.

In catching the water, the skilled adult caught water by the hand and finger tips with the elbow being extended, but the unskilled child caught water by the hand and the forearm with elbow flexed.

During the kick movements, the EMG discharge patterns of the vastus medialis and the biceps femoris that produce knee and hip extension were almost the same in both subjects. Discharges were observed in the biceps brachii, the triceps brachii, the posterior deltoid, the pectoralis major, and the latissimus dorsi, all of which act during the arm pull. But the periods during which the discharges appeared were remarkably different. In the unskilled child, these five muscles participated immediately after catching water, but in the skilled adults, they did not act until the middle part of the stroke. The different period of discharge seemed to result from the different ways of catching water.

In the backstroke, the same tendency was observed in the forms and electromyograms as in the case of the crawlstroke.

In the breaststroke, however, notable differences in the movements and the discharge patterns appeared between the skilled adults and the infants and children who learned the style of the breaststroke earlier by imitating adults. These differences can be seen in Figure 2.

The horizontal velocity of the trochanter major was measured from the motion picture filmed of the same subjects. As shown in the upper part of Figure 3, the skilled adult increased the horizontal velocity during the kick and arm pull phases, and decreased in the glide and leg recovery phases. However, in the case of the unskilled child the horizontal velocity of the kick and the arm pull phases hardly increased.

In the early part of the kick phase of the skilled adult, remarkable discharges were observed in the tibialis anterior, the vastus medialis, and the rectus femoris. This indicates that the skilled adult extended the knees, catching water with the entire sole while maintaining dorsiflexion, and, as a result, obtained large propulsive forces. Throughout the kick phase, the pronounced discharges were seen in the rectus abdominis, which is considered to be important in fixing the pelvis for strong leg kicks.

In the unskilled child, on the other hand, the discharge was not noticeable in the tibialis anterior, and marked discharge was not seen in the rectus femoris and the vastus medialis, which are both knee

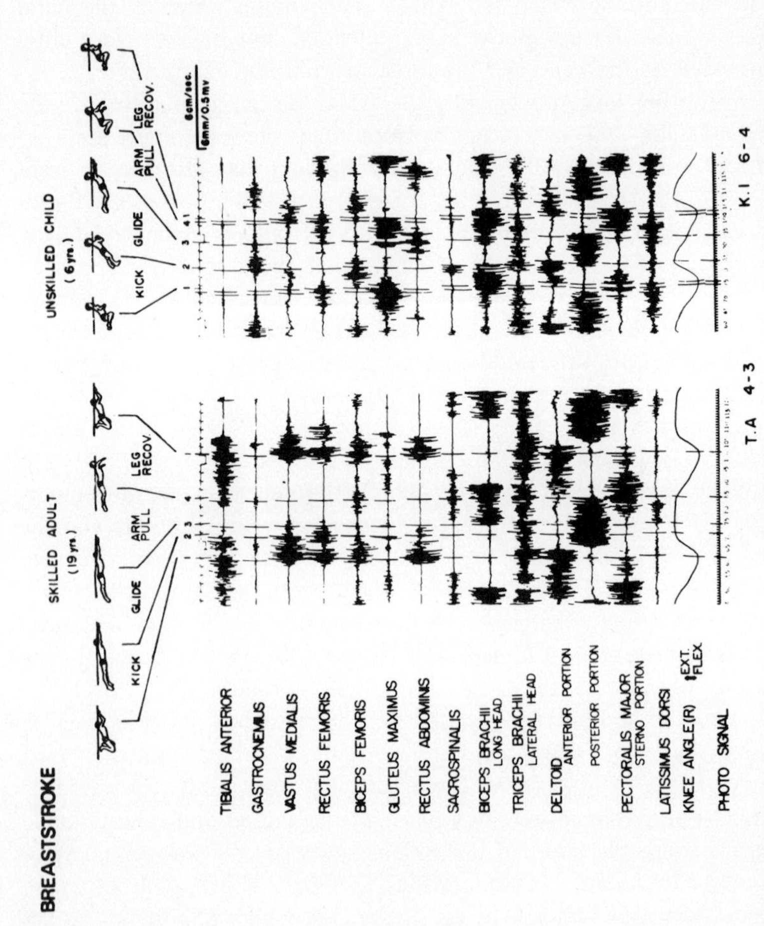

Figure 2. Electromyograms of the breaststroke of a 6-yr-old child in the initial period of learning to swim imitating adults and the specially trained adult swimmer.

BREASTSTROKE

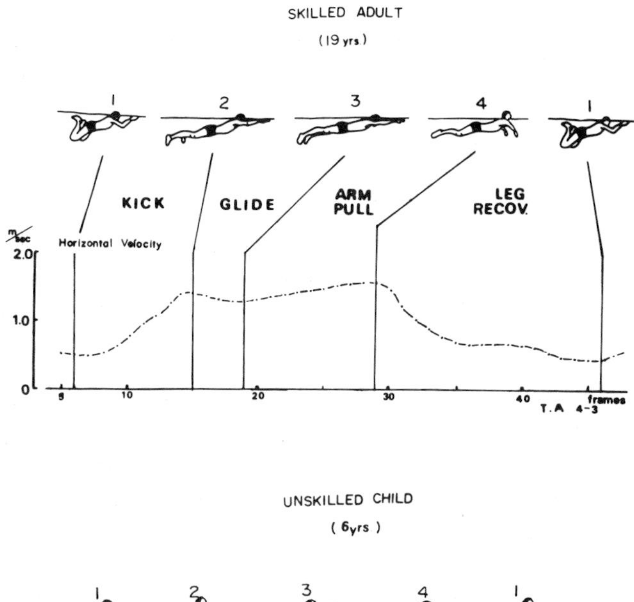

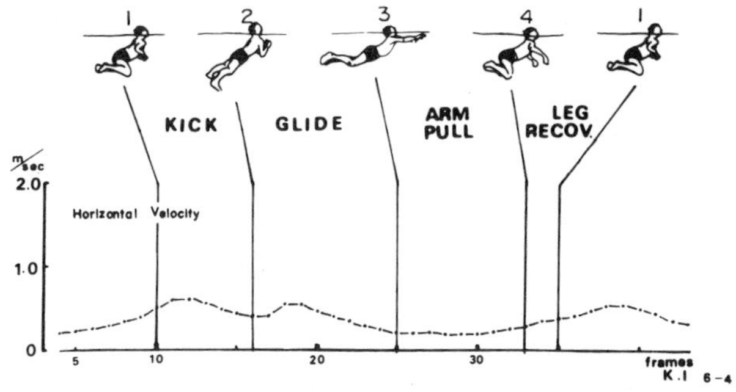

Figure 3. The curves of the horizontal velocity of breakststroke of the trochanter major measured from motion pictures of the same subjects shown in Figure 2; the skilled adult (*upper*); the 6-yr-old child (*lower*).

extension muscles. These discharge patterns indicate that the unskilled child did not perform the dorsiflexion in order to catch water with the entire soles, and that heavy load was not put on the knees, even if the knees were extended. It follows that the extensive discharges were not present in the rectus femoris and the vastus medialis, which were actively involved with knee extension. That is, the unskilled child performed an "ineffective kick" and was not able to acquire enough propulsive forces. This fact makes it clear that the unskilled child had difficulty in using his ankles under water. In the unskilled child, no discharge appeared in the rectus abdominis, and the pelvis was not well fixed in the kick phase.

As to the pull phase of the skilled adult, the arm pull was performed while the body and lower limbs were maintained almost parallel to the surface of water. For the skilled adult, notable discharges were seen in the EMGs of the biceps brachii, the posterior portion of deltoideus, and the latissimus dorsi during the arm pull phase, while there was almost no discharge in the lower limb muscles, except in the biceps femoris at the latter part of this phase. The discharge of the biceps femoris indicated that it was acting in order to maintain the hyperextension of the hip joint. This shows that the skilled adult got large propulsive forces by flexing the elbows and pulling the upper arms toward the chest.

In the unskilled child, however, noticeable discharges were observed in the rectus femoris throughout the arm pull phase. This was probably because the muscles involved in flexing the thigh acted strongly, and thus the backward forces caused by pulling of the arms neutralized the forward forces created by flexing the thigh until the propulsive forces were hardly present. This indicated that during the initial period of learning to swim, it was very difficult for the infants and children to keep the trunk and lower limbs nearly parallel to the surface of water during the arm pull phase.

DISCUSSION

In general, the arms were flexed with the legs extended and the legs were flexed with arms extended from the body. These were natural movements, which can be seen in the extension and flexion of the upper and lower limbs in the standing long jump, vertical jump, and free standing exercises. In the breaststroke of the infants and children, when the arms were pulled, the legs also were mostly flexed as shown in Figure 2. This appeared to be a natural reflex movement. Such movement is not effective for acquisition of speed in swimming, however natural it may seem to be. It is rather necessary to keep the body and lower limbs parallel to the surface of water during the arm pull phase as the skilled adults did, and to correct the posture in order to decrease water resistance. This fact shows that it is very difficult to learn the technique of the breaststroke as compared to the crawlstroke and the backstroke.

We considered at what age it would be possible to correct the form of the infants and children, and to have it approximate the form of the skilled adults. As far as our experiments are concerned, a 5-yr-and-10-month-old child who has been trained for about 2 yr at a swimming school and has already learned to swim the crawlstroke and backstroke, demonstrated that both the main discharges and the coordinated movements of the upper and lower limbs were similar to the skilled adult pattern.

CONCLUSION

The conclusions of this investigation can be summarized as follows:

1. In the crawlstroke and backstroke, both the main discharges of the hip and knee extensors and the coordinated movements of the upper and lower limbs in the infants and children in the initial period of learning to swim were about the same as the pattern of the skilled adults. In the breaststroke, however, the child's patterns were quite different from the skilled adult patterns.
2. In the kick phase of the breaststroke, the skilled adults showed discharges that indicated that the knee was actively extended with dorsiflexion maintained, but the infants and children in the initial period of learning exhibited knee extension without active dorsiflexion.
3. During the arm pull phase in the breaststroke, the skilled adults displayed discharge patterns that indicated that a hyperextension of the hip joint occurred with the knee joint extended. In the infants and children in the initial period of learning, however, the discharge patterns indicated that flexion of the hip joint occurred with the knee flexed.
4. In the 5-yr-and-10-month-old child who had been trained for about 2 yr at a swimming school, both the main discharges and the coordinated movements of the upper and lower limbs were already similar to the skilled adult pattern.

REFERENCES

Ikai, M., K. Ishii, and M. Miyashita. 1964. An electromyographic study of swimming. Jap. Res. J. Phys. Educ. 7: 55–87.

Barthels, K. M., and M. J. Adrian. 1971. Variability in the dolphin kick under four conditions. *In:* L. Lewillie and J. P. Clarys (eds.), First International Symposium on Biomechanics in Swimming, pp. 105–118. Universite Libre de Bruxelles, Brussels.

Lewillie, L. 1971. Graphic and electromyographic analysis of various styles of swimming. *In:* J. Vredenbregt and J. Wartenweiler (eds.), Biomechanics II, pp. 253–257. S. Karger, Basel.

Electromyographic study of the breaststroke

M. Yoshizawa
Fukui University, Fukui

H. Tokuyama
Osaka Kyoiku University, Osaka

T. Okamoto
Kansai Medical School, Osaka

M. Kumamoto
Kyoto University, Kyoto

Adult swimming has been studied cinematographically by many investigators. However, only few electromyographic studies of the breaststroke have been carried out (Ikai, Ishii, and Miyashita, 1964; Lewillie, 1971, 1974). In our experiment, a more detailed electromyographic and cinematographic analysis was attempted.

PROCEDURES

The subjects ($n = 21$) employed in this experiment were well skilled in their coordination of the upper and the lower limbs in the breaststroke, and they were ranked in the highest class (Olympians), a high class (members of a university swimming club), and a poor class (average adults) according to their swimming records.

Electromyograms were recorded by an 18-channel multipurpose electroencephalograph (manufactured by Sanei Sokki Co., 1A 53 type) using waterproofed surface electrodes, 10 mm in diameter with 50-m shielded wires. The shielded wires were arranged at the back of the subjects and were tied to a pole at a distance of about 1 m from the body and were carried along with the subjects by an assistant at pool side.

The 14 muscles of the upper and lower extremities and the trunk that participate in the breaststroke were selected for reference in this electromyographic study of swimming (see Figure 3 below). This decision was based on previous work by Ikai, Ishii, and Miyashita (1964) and on the studies of the fundamental movements of the upper and lower extremities by Okamoto et al. (1966a, b, 1967, 1968a, b, 1972). The recording of the electromyograms was limited to the right side of the body. In the case of some subjects tested, electrogoniograms of the ankle, the knee, the hip, and the elbow joints were recorded simultaneously with the EMGs; in the case of all other subjects, the recording of the electrogoniograms was limited to the knee joint.

The swimming pool for this study was 25 m in length, 15 m in width, 1.4 m in depth, and equipped with underwater windows. A 16 mm motion picture filmed the subjects from the side, the front, and the back through these windows. The signal pulse of each frame of the motion picture was also recorded simultaneously with the EMGs.

Each subject swam at full, moderate, and slow speeds according to his swimming ability.

RESULTS

As shown in Figure 1, on the basis of the electrogoniograms of the ankle, the knee, the hip, and the elbow joints, one cycle of the breaststroke was divided into four phases. These four phases consisted of the kick phase during which the knee joint extended; the glide phase in which the upper and lower extremities were in almost a straight line after the kick; the arm pull phase in which the elbow joint became flexed, the upper arms came near the trunk, and all joints of the lower extremities were maintained in the same state as the glide phase; and the leg recovery phase during which the knee joint began to flex up to maximum flexion.

As shown in Figure 2, each of these four phases depicted in Figure 1 had a tendency to increase or decrease the speed.

In general, the kick and the arm pull phases showed a tendency to increase the speed, and in the glide and in the leg recovery phases there was a tendency to decelerate. The kick phase showed a greater effect on increasing the speed than the pull phase, and the recovery phase had a greater influence on decreasing the speed than the glide phase.

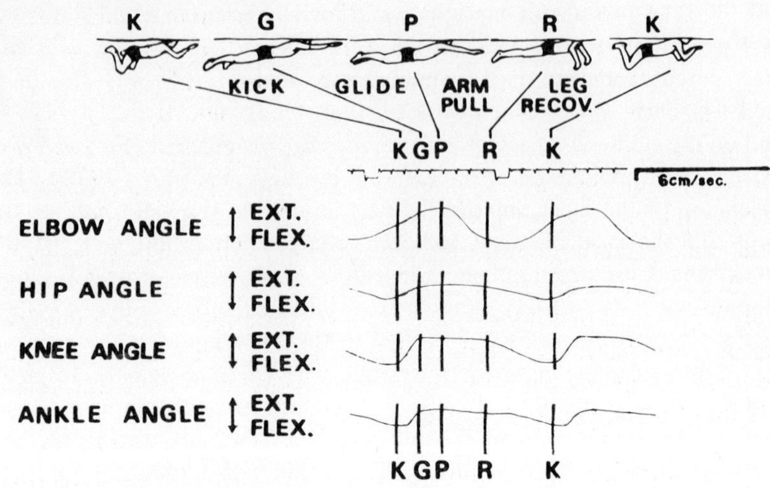

Figure 1. Electrogoniograms of the elbow, hip, knee, and ankle in the breaststroke of an adult; the kick phase (K-G), the glide phase (G-P), the arm pull phase (P-R), the leg recovery phase (R-K).

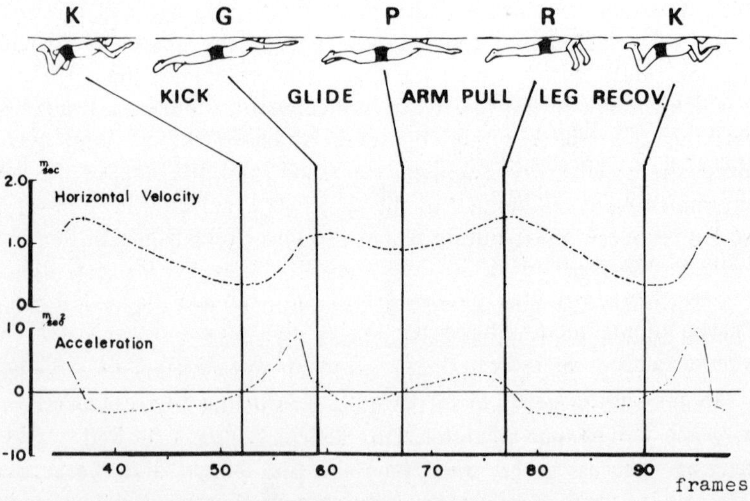

Figure 2. Recordings of the horizontal velocity and acceleration of the trochanter major as calculated from the motion picture films.

The muscular functions in each of the four phases were analyzed using the electromyograms and the electrogoniograms of the knee joint for the trials at full speed in the three cases of the top, the good, and the poor subjects (see Figure 3).

The Kick Phase

All subjects had in common large action potentials generated by the tibialis anterior, the vastus medialis, and the rectus femoris which indicates the active extension of the knee joint with the flexion of the ankle joint. These were observed during the first half of the kick phase.

In the case of the best swimmer, the action potential of the tibialis anterior indicated that dorsiflexion of the foot continued during the first part of the leg recovery phase and remained constant until after the middle of this phase; but in the cases of the good and the poorer swimmers these action potentials disappeared in the middle phase.

As the studies of Carpenter (1938) and Lindeburg (1964) have already pointed out, the best angle of the knee joint for extension with optimum muscular efficiency is about 120 degrees. In order to get a large propulsion from the legs, the dorsiflexion of the foot should occur during the latter half of the kick phase of the stroke, that is, when the knee joint is extended at about 120 degrees.

In the case of the best swimmer, the large action potentials of the gastrocnemius appeared during the latter half of the kick phase; but in the case of the poorer subjects, during the initial half of this phase. The activity observed in the poorer subjects indicates that they performed plantar flexion much earlier than the best subject, which resulted in a less effective kick.

At average and slow speeds, however, the EMGs of the tibialis anterior in the high-level, good, and poor swimmers were similar. It seems that the good and the poor subjects performed an effective kick when they did not swim at full speed.

In the case of the top and the good subjects, high levels of activity of the rectus abdominis were observed during the first half of the kick phase; but in the case of the poor subjects, no activity was observed in the rectus abdominis during this phase. This pattern for the rectus abdominis indicated that the pelvis was firmly stabilized.

In all subjects, the EMG of the anterior deltoid indicated that flexion of the shoulder joint occurred during the first half of kick phase, and the activity of the triceps brachii indicated that extension of the elbow joint also occurred throughout this phase. The action potentials of the posterior deltoid indicated that active elevation of

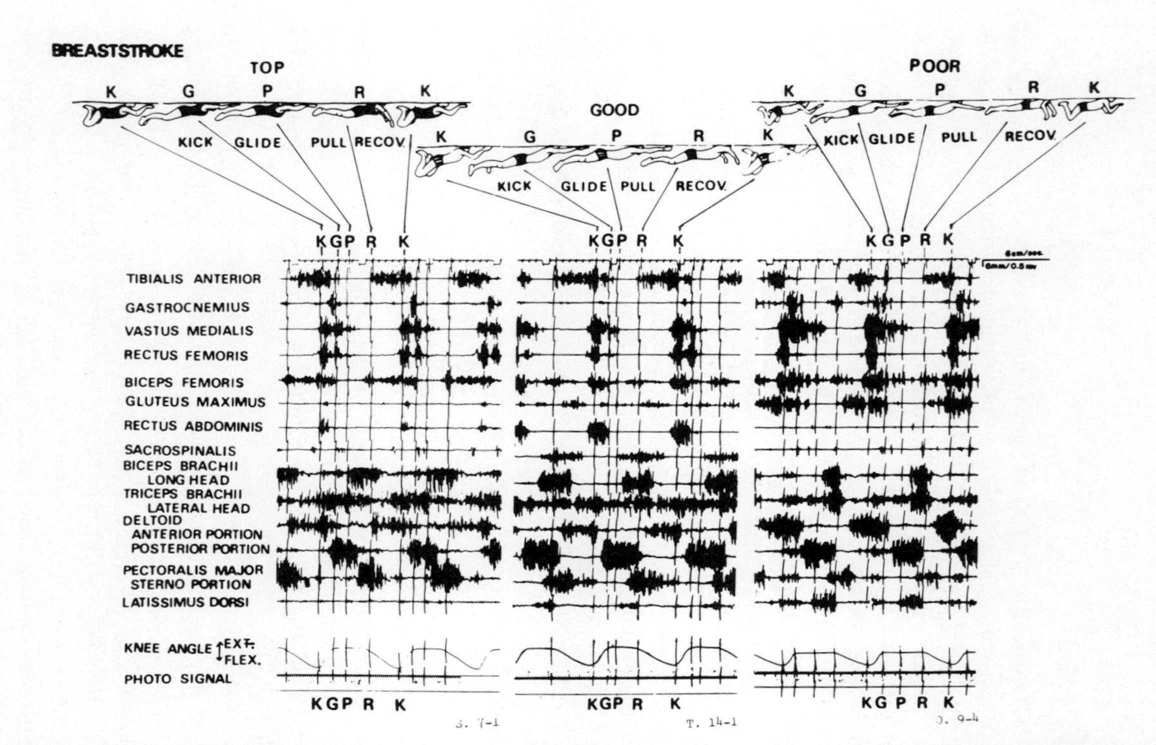

Fgure 3. Electromyograms and electrogoniograms of the knee joint, and positions of the breaststroke at full speed in the top (the Olympian), the good (members of the university swimming club), and the poor subjects (average adults).

the shoulder girdle was accomplished between the latter half of the kick phase and the next glide phase. Almost all of the good subjects had an EMG pattern of the posterior deltoid somewhat similar to the top swimmer; but the poor swimmers did not show any activity in this muscle at this stage in the stroke cycle.

The Glide Phase

In the first half of the glide phase, the high levels of activity in the vastus medialis and the rectus femoris which decreased during the latter half of the kick phase were observed in the case of the best subject. It seems that this pattern meant that full extension of the knee joint occurred after the feet were almost together. But the activity in the rectus femoris was only slight in the case of the poor subjects. The low level activity indicated that these subjects did not execute full extension of the knee joint as the top swimmer did.

Slight activity was observed in the other leg muscles except in the vastus medialis and the rectus femoris in the case of the best subject. But in the cases of a few of the good swimmers and all the poor subjects, the activity in the biceps femoris was observed. This appeared to be unnecessary if one is to accept the EMG pattern of the best swimmer as the ideal model.

The continuous activity of the posterior deltoid which continued from the latter half of the kick phase was observed throughout the glide phase in the cases of the best and the good subjects; but a high level of activity appeared during the latter half of the glide phase in the case of the poor swimmers. In the cases of the top and the good subjects, this EMG pattern of the posterior deltoid meant that the shoulder elevation not only minimized the water resistance, but also assisted in positioning the arm at a more forward position for the next pull phase.

The Arm Pull Phase

In all subjects, co-contraction of the biceps brachii and the triceps brachii was observed during the arm pull phase. The analysis of the film indicated that co-contraction provided active flexion of the forearm with the elbow joint stabilized. In the case of the top subject, the high level of activity in the biceps brachii started at the beginning of the pull phase; but in the cases of the good and the poor subjects, it started fairly late in this phase. This EMG pattern indicated that the top subject performed flexion of the elbow joint earlier than others. From the front view film, it was recognizable that the top subject performed the elbow-up pull with the elbow joint flexed, with the upper arm at shoulder level, and the forearm angled to the right of the water surface. This elbow-up pull is an effective motion in order to get large propulsion.

The high level of activity of the posterior deltoid was observed during the initial half of the arm pull phase, and that of the pectoralis major was evident throughout the latter half. It is evident from the results of previous electromyographic studies of the fundamental movements of the upper extremities (Okamoto et al., 1966a, 1972) that the upper arm moved diagonally outward and downward throughout the first half, and inward throughout the second half of the arm pull phase.

In the case of the best subject, the slight activity of the latissimus dorsi participating in shoulder extension was observed, but in the cases of the good and poor subjects, the activity levels of the latissimus dorsi was high.

According to the films, the arms of the best subject were never extended beyond the shoulder level during the pull; but the good and the poor subjects went beyond the shoulder level, their elbows dropped down.

The slight activity observed in the biceps femoris worked to keep the hip joint extended during the latter half of the pull phase in all subjects, but the EMGs of other muscles showed very little activity throughout this phase.

The Leg Recovery Phase

In the case of the top subject, slight, continuous activity of the biceps femoris was observed throughout the leg recovery phase, but in the cases of the good and the poor subjects, there was a tendency for the activity to decrease or even disappear during this phase.

In the case of the best swimmer, it was easily recognized in the motion picture film that the knee joint was flexed with hyperextension of the hip joint; but in the cases of the good and the poor subjects, such hip hyperextension was not observed throughout the initial half of the leg recovery phase. The best subject minimized the water resistance by a minimum forward flexion of the thigh.

CONCLUSIONS

From these data of the breaststroke of the top level swimmer who swam most effectively, the following conclusions seem to be warranted.

In the kick phase, the EMG patterns of the rectus abdominis, the tibialis anterior, the vastus medialis, and the rectus femoris showed a stabilization of the pelvis and an active extension of the knee joint with the ankle joint in dorsiflexion.

In the arm pull phase, the activity patterns of the biceps brachii, the triceps brachii, and the latissimus dorsi showed the elbow-up pull with the elbow joint flexed, the upper arm kept at the shoulder level, and the forearm angled to the right of the water surface in the pull phase.

In the glide phase, the activity patterns of the vastus medialis and the rectus femoris showed full extension of the knee joint. The EMG pattern of the posterior deltoid showed that the shoulder elevation kept the water resistance minimal and worked to perform an arm pull at a more forward position.

In the recovery phase, the high-level, continuous activity pattern of the biceps femoris showed that a minimum forward flexion of the thigh also kept the water resistance at a mimimum.

REFERENCES

Carpenter, A. 1938. A study of angles in the measurement of the leg lift. Res. Quart. 9: 370–372.

Ikai, M., K. Ishii, and M. Miyashita. 1964. An electromyographic study of swimming. Jap. Res. J. Phys. Educ. 7: 55–87.

Lewillie, L. 1971. Graphic and electromyographic analysis of various styles of swimming. *In:* Vredenbregt and J. Wartenweiler (eds.), Biomechanics II, pp. 253–257. S. Karger, Basel.

Lewillie, L. 1974. Telemetry of electromyographic and electrogoniometric signals in swimming. *In:* R. C. Nelson and C. A. Morehouse (eds.) Biomechanics IV, pp. 203–207. University Park Press, Baltimore.

Lindeburg, F. A. 1964. Leg angle and muscular efficiency in the inverted leg press. Res. Quart. 35: 179–183.

Okamoto, T., K. Takagi, and M. Kumamoto. 1966a. Electromyographic study of extension of the upper extremity. Japan. J. Phys. Fitness 15: 37–42.

Okamoto, T., M. Kumamoto, and K. Takagi. 1966b. Electromyographic study of the function of M. adductor longus and M. adductor magnus. Japan. J. Phys. Fitness 15: 43–48.

Okamoto, T., K. Takagi, and M. Kumamoto. 1967. Electromyographic study of elevation of the arm. Jap. Res. J. Phys. Ed. 11: 127–136.

Okamoto, T. 1968a. Electromyographic study of the function of M. rectus femoris. Jap. Res. J. Phys. Ed. 12: 175–182.

Okamoto, T. 1968b. A study of the variation of discharge pattern during flexion of the upper extremity. J. Dept. Lib. Arts Kansai Med. School 12: 111–122.

Okamoto, T., M. Kumamoto, and N. Yamashita. 1972. Electromyographic study of fundamental movements of the upper extremity. Sixth International Congress of Physical Medicine, (Barcelona, Spain), Vol. 2, pp. 288–291.

Variability of myoelectric signals during swimming

L. A. Lewillie
Université Libre de Bruxelles, Bruxelles

The first study of myoelectric signals during swimming was published by Ikai, Ishii, and Miyashita (1964).

Since that time, the use of telemetry, however, has allowed a greater freedom of movement for the swimmer, and EMGs have been recorded and quantified with reference to the maximal isometric contractions (Lewillie, 1968a; 1968b; 1971a; 1971b). A first approach to joint angle measurement and its implications on the muscular activity was presented at the 4th International Seminar on Biomechanics (Lewillie, 1974).

The repeatability of the movements of qualified swimmers is well known and seems so high that Pugh et al. (1960) used them to estimate the speed at the surface of the water of Channel swimmers. The fluctuations of intracycle speed have been studied by Miyashita (1971). Therefore, it appears important to evaluate the variability of myoelectric signals before one can present a complete description of muscular contractions during swimming.

METHODS

Ten swimmers of different levels of ability were tested repeatedly. The method has been used previously; surface electrodes at 15-mm intervals, a two-channel waterproofed transmitter, receivers, and recordings on magnetic tapes. Recordings on paper are then accomplished, by means of a 700-cm/sec polygraph, and the direct record is completed through a digital and analog analyzer.

The calibration was established based on 12–20 isometric contractions, ranging from 10 to 100 percent maximum; the force was exerted against a strain-gauge dynamometer.

A record of the movements of the swimmer was obtained by means of a motion picture camera. The synchronization between the camera and the EMG was accomplished by means of a flashing light placed in the visual field of the camera and connected to the magnetic tape recorder (Figure 1).

The total duration of each stroke cycle and various myoelectric parameters (maximal intensity, total electrical activity, and total electrical activity per unit of time) were studied.

RESULTS AND DISCUSSION

Using the number of movements of Channel swimmers to estimate their speed appears reliable, since the duration of the cycles of swimming movements performed by good swimmers is almost constant. For a series of 200 crawl stroke cycles performed in succession by a swimmer of international caliber we found:

mean duration	182/100 sec
range	178–186
variability coefficient	$\frac{100\ \sigma}{X}$ 1.09 percent

This consistency is maintained, even when for technical reasons, the swimmer must break the rhythm after each length of the pool. The variability appears greater for swimmers of a lower level of ability while swimming a length of the pool. Moreover, the mean duration of cycles fluctuates from one pool lap to another. For a total mean duration of 175/100 sec (based on 224 consecutive cycles), the means per length fluctuated from 160 to 191/100 sec. The variability coefficient for the 224 cycles was 11 percent; the same coefficient per length ranged from 3.4 to 6.1.

The variability of quantitative statements about surface electromyogram has been studied recently by Grieve and Cavanagh (1974) during locomotion. Duration repeatability from one cycle to the next seems very close to the requirements ($\pm$ 20msec) as proposed by these authors. Using the maximal isometric contraction as a reference level, EMG offers the same repeatability. The maximum amplitude of the EMG of the triceps brachii, recorded during 20 consecutive lengths

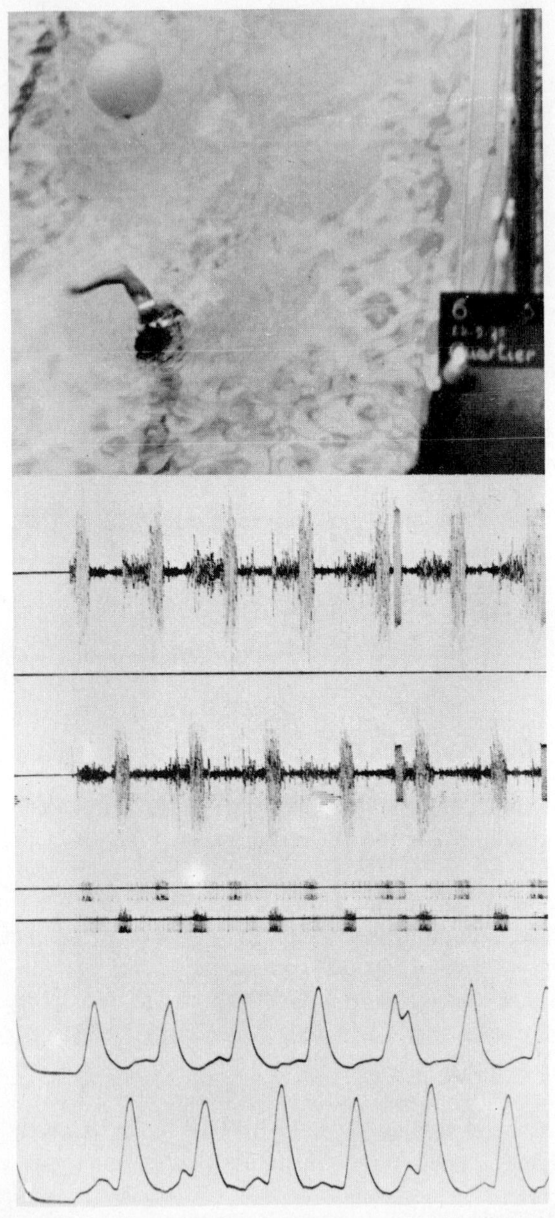

Figure 1. Photograph of a swimmer showing the flashing light and the corresponding signal on the right side. For the electromyogram, the different channels are, respectively, EMG of m. triceps brachii (left and right arm) and their digital and analog conversions.

of the pool (194 consecutive cycles), shows a mean of 86 percent of the isometric maximum. In fact, that mean is composed of 16 partial means of 86 percent; the four others were 82, 87, 89, 89 percent, respectively. The different intralength variability ranged from 81 to 90 percent.

The duration of the contraction varied slightly, in much the same way for the various arm muscles tested, but the total electrical activity versus time remained almost constant. For the same subject, we found a mean duration of contraction of 78/100 sec for the triceps brachii (range 70–88) but the mean for total electrical activity (time for the 194 lengths), which was 69 percent, was in fact composed of identical values: 172 contractions were at 69 or 70 percent; the remaining 22 had only a slight difference, the lowest being 64 percent, the highest 72 percent of the maximum. These values were largely within the limits of the precision of the method.

This was true for each arm tested. It is important that the muscular activity does not appear very different between left and right arm for symmetric styles like the breast stroke. In the crawl stroke, the traction exerted by the arm is not disturbed by respiration, but the rotation of the body shows clearly in the EMG of the quadriceps femoris.

CONCLUSION

The repeatability of swimming movements by highly skilled swimmers appears exceptionally high, as measured by both duration and quantified electromyography. A limited number of cycles may therefore be accepted as valuable information for movement analysis and for studies of fatigue.

Further studies are needed on speed, pressure, and different biomechanical parameters of the hand and on the activity of the muscles of the lower leg. The precision of the quantified EMG based on an isometric contraction calibration is adequate for such studies, provided the 100 percent actually corresponds to the maximal contraction and the position of the limb is correctly chosen.

ACKNOWLEDGMENT

We are gratefully indebted for the technical assistance given by Mr. Robeaux with respect to the various apparatus employed.

REFERENCES

Grieve, D. W., and P. R. Cavanagh. 1974. The validity of quantitative statements about surface electromyograms recorded during locomotion, Scand. J. Rehab. Med. Suppl. 3: 19–25.

Ikai, M., K. Ishii, and M. Miyashita. 1964. An electromyographic study of swimming, Japan. Res. J. Phys. Ed. 7: 55–67.

Lewillie, L. A. 1968a. Telemetrical analysis of the electromyogram. *In:* J. Wartenweiler, E. Jokl, and M. Hebbelinck (eds.), Biomechanics, pp. 147–149. S. Karger, Basel.

Lewillie, L. A. 1968b. Analyse télémétrique de l'électromyogramme du nageur, Travaux Soc. Méd. Belge Educ. Phys. 20: 174–177.

Lewillie, L. A. 1971a. Quantitative comparison of the electromyogram of the swimmer. *In:* L. Lewillie and J. P. Clarys (eds.), First International Symposium on Biomechanics in Swimming, pp. 155–159. Université Libre de Bruxelles, Brussels.

Lewillie, L. A. 1971b. Graphic and electromyographic analysis of various styles of swimming. *In:* J. Vredenbregt and J. Wartenweiler (eds.), Biomechanics II, pp. 253–257. S. Karger, Basel.

Lewillie, L. A. 1973. Muscular activity in swimming. *In:* S. Cerquilini, A. Venerando, and J. Wartenweiler (eds.), Biomechanics III, pp. 440–445. S. Karger, Basel.

Lewillie, L. A. 1974. Telemetry of electromyographic and electrogoniometric signals in swimming. *In:* R. C. Nelson and C. A. Morehouse (eds.), Biomechanics IV, pp. 202–207. University Park Press, Baltimore.

Miyashita, M. 1971. An analysis of fluctuations of swimming speed. *In:* L. Lewillie and J. P. Clarys (eds.), First International Symposium on Biomechanics in Swimming, pp. 53–57. Universite Libre de Bruxelles, Brussels.

Pugh, L. G. C. E., O. G. Edolm, R. H. Fox, H. S. Wolff, G. R. Hervey, W. H. Hammond, J. M. Tanner, R. H. Whitehouse. 1960. A physiological study of Channel swimming. Clin. Sci. 19: 257–273.

Analysis of dynamic forces in crawlstroke swimming

V. V. Belokovsky and V. V. Kuznetsov
Institute of Physical Culture, Moscow

In order to define nonutilized biomechanical reserves for increasing speed, it is necessary to determine the dynamic pattern of the swimmer's movements.

One of the main tasks for improvement of swimming technique is to find which patterns of movements of arms and legs will result in maximum speed for the swimmer. The aim of this paper is to provide a description of the dynamics of freestyle patterns.

The general equation for describing the motion of the body in water can be written as:

$$\frac{d\vec{P}}{dt} = \vec{\mathcal{F}} = \Sigma_i F_i$$

$$\frac{d\vec{L}}{dt} = \vec{\mathcal{M}} = \Sigma_i \, [\vec{r_i}\vec{F_i}] \qquad (1)$$

where P is the impulse of forces, F is the resulting force, M is the resulting force moment, and L is the moment arm of movement's quantity.

In the following equation, axis z is vertical while axis x shows the swimmer's direction of movement. Let us assume that our inertial system of calculation moves against a still water surface with the average speed of a swimmer, V_0. The beginning of the coordinate is placed at the point of the center of mass.

Such an inertial coordinate system is equivalent to the following physical model: a horizontally placed swimmer is being propelled around by water with speed, $V_\infty = -V_0$, influencing him with force

$$/\vec{F}_b/ \simeq C(t)V^n(t).$$

Here ^{n} is an actual number. The function $C(t)$ depends on the individual properties of the swimmer's body.

To eliminate the propulsion part of the speed in this coordinate system, the swimmer must periodically "push" against water with his legs F_{jH} and arms F_{jp}

$$\vec{F}_{jp}(t) = \vec{F}_{jp}(t + T_p) \tag{2}$$

$$\vec{F}_{jH}(t) = \vec{F}_{jH}(t + T_H), \qquad T_p = KT_H = T \tag{3}$$

where K is a whole number.

Since in the chosen system the mass center deviates relative to the point of coordinates, then:

$$\langle\vec{P}\rangle = \frac{1}{T}\int_{t_1 - \frac{T}{2}}^{t_1 + \frac{T}{2}} \vec{P}(t)\,dt = 0 \tag{4}$$

The force impulse $P(t)$ is a constant periodic function of time, hence

$$\int_{t_1 - \frac{T}{2}}^{t_1 + \frac{T}{2}} \frac{d\vec{P}(t)}{dt}\,dt = 0 \tag{5}$$

So from Equation (1) of the system and also from Equation (5) we obtain

$$\langle \vec{\mathfrak{F}} \rangle = \frac{1}{T} \int_{t_1 + \frac{T}{2}}^{t_1 - \frac{T}{2}} \vec{\mathfrak{F}}(t)\, dt = 0 \tag{6}$$

There are different points of view about the part played by swimmer's legs in freestyle swimming. Actually, all of these views could be categorized into two main groups. The first says that leg movements are necessary only for body stabilization. In this case the legs are viewed as "brakes" although necessary "brakes." In such a situation, leg movements should be used which would offer minimum resistance but would effectively stabilize the body.

The second group views leg movements as creating propulsive power and at the same time stabilizing the body. In this case, it is necessary to determine movements which would create maximum propulsive power.

But it is accepted by everybody that the main propulsive power in freestyle swimming is created by the arm strokes:

$$\mathfrak{F}_x(t) \simeq \sum_{j=1}^{2} \mathrm{F}_{jp}{}^{(x)}(t) - C(t)V^n(t) \tag{7}$$

Combining (6) and (7),

$$2 \int_{t_1 - \frac{T}{2}}^{t_1 + \frac{T}{2}} F_p(t)\, dt = \int_{t_1 - \frac{T}{2}}^{t_1 + \frac{T}{2}} C(t)V^n(t)dt \tag{8}$$

Here $V_n(t)$ equals the swimmer's speed in relation to the stationary coordinate system. The relationship to the average resistance force appears as

$$\frac{1}{T} \int_{t_1 - \frac{T}{2}}^{t_1 + \frac{T}{2}} C(t)V^n(t)\, dt \simeq C_0 V_0{}^n$$

$$+ K_{CV} \sqrt{D_v D_c}\, n V_0{}^{n-1} + \frac{C_0 n(n-1)}{2} V_0{}^{n-2} D_V \tag{9}$$

where

$$V(t) = V_0 + V_1(t), \qquad C(t) = C_0 + C_1(t)$$

$$V_0 = \frac{1}{T}\int_{t_1 - \frac{T}{2}}^{t_1 + \frac{T}{2}} V(t)\, dt;$$

$$C_0 = \frac{1}{T}\int_{t_1 - \frac{T}{2}}^{t_1 + \frac{T}{2}} C(t)\, dt$$

$$K_{cv} = \frac{1}{T\sqrt{D_c D_v}}\int_{t_1 - \frac{T}{2}}^{t_1 + \frac{T}{2}} V_1(t) C_1(t)\, dt \tag{10}$$

$$D_V = \frac{1}{T}\int_{t_1 - \frac{T}{2}}^{t_1 + \frac{T}{2}} V_1^2(t)\, dt$$

This represents dispersion of speed fluctuation.

$$D_C = \frac{1}{T}\int_{t_1 - \frac{T}{2}}^{t_1 + \frac{T}{2}} C_1^2(t)\, dt$$

This represents dispersion of fluctuation, $C_1(t)$. Assuming that

$$/\, V_1(t)\, / << V_0$$
$$/\, C_1(t)\, / << C_0 \tag{11}$$

Then equation (1) for component Px will be

$$M\frac{dV_1(t)}{dt} \simeq \sum_{j=1}^{2} F_{jp}{}^{(x)}(t) + [C_0V_0{}^n]$$
$$+ C_0nV_0{}^{n-1}V_1(t) + \tfrac{1}{2}C_0n(n-1)V_0{}^{n-2}V_1{}^2(t)$$
$$- nC_1V_0{}^{n-1}V_1(t) - C_1(t)V_0{}^n \qquad (12)$$

In Equation (12) we omit decreasing members of the second and the next order.

Since the functions $V(t)$ and $Fp(t)$ are uninterrupted periodical functions they can be shown as Fouriers rows.

$$\sum_{j=1}^{2} F_{jp}{}^{(x)}(t) = \sum_{m=-\infty}^{+\infty} S_m e^{i\omega_0 mt}$$

$$V_1(t) = \sum_{m=-\infty}^{+\infty} a_m e^{i\omega_0 mt}$$

$$C_1(t) = \sum_{m=-\infty}^{+\infty} b_m e^{i\omega_0 mt}$$

$$S_m = \frac{1}{T}\int_{t_1-\frac{T}{2}}^{t_1+\frac{T}{2}} \sum_{j=1}^{2} F_{jp}{}^{(x)}(t)e^{-im\omega_0 t}\,dt$$

$$a_m = \frac{1}{T}\int_{t_1-\frac{T}{2}}^{t_1+\frac{T}{2}} V_1(t)e^{-im\omega_0 t}\,dt$$

$$b_m = \frac{1}{T}\int_{t_1-\frac{T}{2}}^{t_1+\frac{T}{2}} C_1(t)e^{-im\omega_0 t}\,dt$$

From equation (8) we get

$$S_0 = C_0V_0{}^n + K_{CV}\sqrt{D_vD_c}\; nV_0{}^{n-1}$$
$$+ \tfrac{1}{2}C_0n(n-1)V_0{}^{n-2}D_v \qquad (14)$$

Including equation (13) in equation (12) we get a correlation connecting coefficients of Fourier's rows:

$$iMm\omega_0 a_m = S_m - nC_0 V_0^{n-1} a_m - V_0^n b_m. \quad (15)$$

From equation (15) we find a_m:

$$a_m = \frac{S_m - V_0^n b_m}{iMm\omega_0 + nc_0 V_0^{n-1}}, \qquad m \neq 0 \quad (16)$$

So the final equation for the average speed of the swimmer along the X axis is

$$V_1(t) = \sum_{/m/=1}^{\infty} \frac{S_m - V_0^n b_m}{nc_0 V_0^{n-1} + iMm\omega_0} e^{im\omega_0 t} \quad (17)$$

where $\omega_0 = \frac{2\pi}{T}$ is the frequency of arm movements.

Equation (17) describes the relationship between the fluctuation speed $V_1(t)$ and such components as the spectrum of force, S_m, and drag force (resistance), b_m.

To determine the coefficients S_m and b_m, it is necessary to examine two hydrodynamic models of the swimmer's movements.

It is notable that when ω_0 is increasing, the fluctuation speed is decreasing and hence, with increased frequency of strokes (ω_0) the swimmer loses less energy in relation to average propulsion speed.

EXPERIMENTAL RESULTS

The recording of speed and the recording of forces were obtained with the help of the above mentioned technique, which is discussed in our earlier works and is also similar to previously published analogous methods.

Wire strain gauges used for measuring the forces were membrane systems that converted (transformed) pressure into electrical measured signal.

The intracycle speed was recorded by measuring the tension vibration (oscillation) of the unwinding cord on the 10-m distance.

When we tested highly experienced swimmers, we found that the speed fluctuations and the arm forces were time periodic functions and therefore could be expressed as Fourier's rows (Equation 13).

A periodic function of resistance force relative to time also can be expressed as Fourier's series.

The spectrum speed component, spectrum stress component, and spectrum function components $c(t)$ are connected with the following relationship:

$$b_m = \frac{2C_0}{V_0} a_m + \frac{iM\omega}{V_0^2} a_m - \frac{1}{V_0^2} S_m \tag{18}$$

where $m \neq 0$

b_m is the spectrum function component $c(t)$
a_m is the spectrum component of speed fluctuation
S_m is the spectrum stress component

As seen from the above formula, the spectrum component of the function $c(t)$ can be calculated according to the known spectrum components, a_m and S_m. When $m = 0$:

$$\overline{F} = C_0(V_0^2 + \Omega_V) + K_{CV} V_0 \sqrt{\Omega_C \Omega_V} \tag{19}$$

When the correlation coefficient is equal to zero:

$$K_{CV} = 0 \tag{20}$$

We then get:

$$\overline{F} = S_0 = C_0 V_0 + C_0 \Omega_V \tag{21}$$

where D_V is the dispersion of swimmer speed fluctuations.

This relation shows that the greater the amplitude of speed fluctuations relative to the mean value, the greater the force consumption.

The calculations of our experimental results follow:

$m = 5 \quad f = 2.08\text{Hz}$ | $m = 11 \quad f = 4.58\text{Hz}$

$b_5 = 2.60 \dfrac{Kg \cdot cek^2}{m^2}$ | $b_{11} = 2.17 \dfrac{Kg \cdot cek^2}{m^2}$

$\varphi b_5 = 4.92\ rad$ | $\varphi b_{11} = 0.75\ rad$

$\dfrac{b_5}{C_0} = 0.35$ | $\dfrac{b_{11}}{C_0} = 0.29$

CONCLUSION

The results enabled us to conclude that such a theoretical approach is suitable for estimating the resistance forces during swimming and hence for the mathematical modeling of swimmer movement dynamics.

Thus, if the spectrum stress component, the spectrum component of resistance forces, and the spectrum component of speed fluctuations are known, then the mathematical modeling of swimmer movements can be used with the help of a computer.

A practical aspect of this work may include the development of recommendations on speed stabilization inside the cycle by means of force redistribution between the actions of both arms or the actions of the arms and legs of the swimmer.

REFERENCES

Jenkins, G. M., and D. G. Watts. 1969. Spectral Analysis. Holden Day, San Francisco.

Jensen, R. K., and B. Blanksby. 1975. A model for upper extremity forces during the underwater phase of the front crawl. *In:* J. P. Clarys and L. Lewillie (eds.), Swimming II, pp. 145–153. University Park Press, Baltimore.

Miyashita, M. 1975. Arm action in the crawl stroke. *In:* J. P. Clarys and L. Lewillie (eds.), Swimming II, pp. 167–173. University Park Press, Baltimore.

Efficiency in swimming the front crawl

H. C. G. Kemper
Universiteit van Amsterdam, Amsterdam

R. Verschuur
Universiteit van Amsterdam, Amsterdam

J. P. Clarys
Vrije Universiteit Brussel, Brussels

J. Jiskoot
Akademie voor Lichamelijke Opvoeding, Amsterdam

H. Rijken
Nederlands Scheepsbouwkundig Proefstation, Wageningen

Since Karpovich (1933) mentioned the frictional wave-making and eddy resistance as applied in ship-building research, attempts have also been made to determine these partial drag components in relation to the human body (Alley, 1952; Clarys, Jiskoot, and Lewillie, 1973). Efficiency, drag, and velocity measurements are not complete without an energy expenditure investigation.

The purpose of this study was to establish the mechanical efficiency (*E*) of subjects swimming the front crawl at a constant velocity by means of measurement of the oxygen consumption ($\dot{V}O_2$) and mechanical work (W).

This study was a part of a larger research project on water resistance of the human body supported by the Instituut voor Morfologie (Prof. P. Brouwer) Vrije Universiteit Brussel, Brussels, Belgie.

METHODS

Subjects

The experiments were conducted using 63 male Caucasian physical education students (Amsterdam) as subjects. Thirteen students were excluded for reasons of incomplete results. The subjects had different degrees of swimming abilities, varying from poor to Olympic levels of performances. All subjects were volunteers and were familiarized with the test procedures since they participated in previous experiments of a similar nature (Clarys et al., 1973, 1974, 1975; Jiskoot and Clarys, 1975). The mean physical characteristics of the investigated population are shown in Table 1.

Procedures

All tests were carried out in a ship-model towing tank of 200 × 4 × 4 m. During the experiments water temperature was 24–28°C and air temperature 19–23°C. The test apparatus for the measurement of water resistance and propulsion have previously been described and applied by Jiskoot and Clarys (1972, 1975), Clarys et al. (1973, 1974), Clarys and Jiskoot (1975), and Van Manen and Rijken (1975). $\dot{V}O_2$ was measured continuously and automatically using an Ergo-analyzer (Mijnhardt B.V.) installed on the towing carriage. The swimmer inhaled through a snorkel and exhaled through a breathing valve (Dräger) with a dead space (V_d) of 100 cc, mounted on a diver's full mask. The expired air ($\dot{V}_E$) was collected via a connecting tube of 3.5 m in a dry spirometer (Dordrecht) and analyzed for oxygen content (F_EO_2) by a paramagnetic analyzer (Servomex) and for carbon dioxide (F_ECO_2) by an infrared analyzer (Mijnhardt) (Kemper et al., 1975).

Table 1. Mean ($\bar{x}$), standard deviation (S.D.) and range of the physical characteristics of all subjects; ($n = 50$, male, 18–24 yr old)

Measurement	Height (cm)	Weight (kg)	Volume (dm³)	Greatest body cross-section area (cm²)	Surface area (m²)	Max. free swimming speed (V_{max}) (m/sec)
$\bar{x}$	179.1	75.7	73.8	787.30	1.69	1.46
S.D.	16.2	9.4	12.3	128.96	0.17	0.22
min.	164.3	58.5	49.5	596.5	1.39	1.00
max.	195.9	115.0	141.5	1251.5	2.09	1.85

Heart rate (f_h) was recorded with a cardiotachometer (Rood) from ECG breast electrodes and registered by a recorder (Servogor).

The following tests and measurements were carried out:

1. Measurement of the $\dot{V}O_2$ at rest ($\dot{V}O_2$ rest) was measured with the subject hanging on the towing device in a vertical position in the water.

2. Measurements of the $\dot{V}O_2$ were made while swimming the front crawl at a constant velocity of 0.75 m/sec ($\dot{V}O_2$ swim). This velocity was chosen below the lowest maximum free swimming speed (V_{max}) assuming that all subjects would be able to keep it up continuously during a 4–5 min swim. All subjects adapted themselves to this carriage velocity through a pacing device attached to the towing carriage in front of the swimmer. $\dot{V}O_2$ during the fourth minute was taken as the measure for the energy expenditure.

3. Measurement of the passive drag (D_p) was obtained by towing the body in a streamlined prone position with the arms extended forward at a speed of 0.75 m/sec. The subjects wore the same diver's full mask and ECG electrodes as during the $\dot{V}O_2$ measurements.

4. Mechanical efficiency (E) was calculated by using the ratio between mechanical work (W) and energy metabolism (M):

$$E = \frac{W}{M} \times 100 \qquad (1)$$

W was expressed as the product of D_p and distance swum at a velocity of 0.75 m/sec. M was calculated from the oxygen consumption above resting level ($\dot{V}O_2$net):

$$\dot{V}O_2\text{net} = \dot{V}O_2\text{swim} = \dot{V}O_2\text{rest} \qquad (2)$$

In equation (1) both W and M were expressed in watts, assuming 1 kg/m·min = 0.163 watt and 1 liter O_2 = 348.583 watt.

RESULTS AND DISCUSSION

In Table 2 the results of $\dot{V}O_2$ and respiratory exchange ratio (R) during the third and fourth minutes of swimming are listed. $\dot{V}O_2$ is expressed relative to body weight and body volume of the subjects, which is the usual practice in exercise physiology. Body volume was chosen on the bases of previous experiments (Clarys et al.,

Table 2. Mean ($\bar{x}$), standard deviation (S.D.) and range of oxygen consumption ($\dot{V}O_2$) per dm^3 body volume and per kg body weight, respiratory exchange ratio (R) of 50 subjects in the third and fourth minutes of swimming at a constant speed of 0.75 m/sec

	$\dot{V}O_2$ swim				R	
Measurement	(3 min) [ml/(dm^3 · min)]	(4 min) [ml/(dm^3 · min)]	(3 min) [ml/(kg · min)]	(4 min) [ml/(kg · min)]	(3 min) —	(4 min) —
$\bar{x}$	41.73	43.37	40.34	41.89	1.03	1.07
S.D.	7.03	8.01	6.49	7.33	0.13	0.13
min.	21.1	26.2	25.91	26.08	0.68	0.82
max.	57.6	55.5	54.85	57.94	1.36	1.42

1974) in which body volume turned out to be an important parameter in relation to the total resistance during swimming. The most significant increase in $\dot{V}O_2$ swim during the fourth minute with 1.64 (ml/dm^3·min) or 1.55 (ml/kg·min) indicates that the majority of the swimmers reached a steady state in their $\dot{V}O_2$. This statement was confirmed by R, which had a mean value of 1.03 and 1.07 in the third and fourth minutes. $\dot{V}O_2$ swim as well as R show considerable interindividual differences as illustrated by the large range of the results. The f_h shows an identical trend with values varying from 102 to 188 beats/min.

The values for D_p and E are listed in Table 3. The D_p in our subjects were on the average higher than those reported in the literature (Holmer, 1974; Di Prampero et al., 1974). The mean of E of 3.6 percent was on the average lower than those reported in elite swimmers by Holmer (1974) (6–7 percent) and Di Prampero et al. (1974) (about 4.2 percent). However, W in these studies was calculated (1) from drag during swimming (active drag, D_a), which is presumably

Table 3. Mean ($\bar{x}$), standard deviation (S.D.), and range of passive drag (D_p) during towing with a speed of 0.75 m/sec and mechanical efficiency (E) in the fourth minute of swimming the front crawl at a constant submaximal speed of 0.75 m/sec

	D_p (kg)	E (%)
$\bar{x}$	3.41	3.64
S.D.	0.77	1.43
min.	2.05	1.99
max.	6.10	8.33

50–100 percent higher than in passive towing (Holmer, 1974) and (2) from results of experiments with top class swimmers. Both of these factors would tend to yield a higher E.

If body form is an important factor in D_p (Clarys et al., 1974), a high relationship between these factors should exist. Pearson product-moment correlation coefficients (r) between body height, weight, volume, surface area, greatest body cross section area, body slenderness (height/volume$^{1/3}$) and W were generally low (< 0.25), except for height ($r = -.56$), indicating that body height is the best predictor of W. Since D_p is an important factor in determining W, expressing M of subjects with different body types of $\dot{V}O_2$ relative to weight or volume seems rather useless in swimming. A better method would be to relate the $\dot{V}O_2$ to body height.

In general it has been held that expressing the $\dot{V}O_2$ independent of weight, volume, or height affords the theoretically correct method However, convincing evidence has been presented by Katch (1972) showing that these ratio scores may not express $\dot{V}O_2$ independent of the above-mentioned physical characteristics and that use of these scores can result in spurious correlations with other variables. We calculated a weight-adjusted (W-A $\dot{V}O_2$), volume adjusted (V-A $\dot{V}O_2$), and a height-adjusted oxygen uptake (H-A $\dot{V}O_2$) (Henry, 1956). The computed Pearson correlation coefficient between H-A $\dot{V}O_2$ and height was essentially zero, thus satisfying the major conditions for its use. In addition we correlated the H-A $\dot{V}O_2$ with the E of swimming. The r was $-.47$, which was significant at the 1 percent level of probability indicating a certain but not a high negative relationship.

From the 50 subjects we selected two groups: a group of nine experienced top-class swimmers (excellent swimmers) and a group of nine swimmers with the lowest V_{max} (poor swimmers).

The energy metabolisms (M) of these two contrasting groups are compared in Table 4. One can conclude: (1) the excellent swimmers showed a lower $\dot{V}O_2$ swim in relation to body height; (2) the poor swimmers increased their $\dot{V}O_2$ during the fourth minute with respect to the third one. The excellent swimmers, however, on the average showed the same $\dot{V}O_2$. In some individuals there was even a slight decrease. (3) As a consequence, R in the poor swimmers was considerably higher compared with the excellent group. It was an indication of the fact that in the former during the fourth minute a considerable part of the total M was due to anaerobic metabolism. (4) All these features are reflected in the efficiency of these groups: in the poor swimmers E was 3.33 percent in the third minute and this ratio

Table 4. Height-adjusted oxygen uptake (H-A $\dot{V}O_2$), respiratory exchange ratio (R), and mechanical efficiency (E) of two contrasting groups in the third and fourth minute of swimming the front crawl with a speed of 0.75 m/sec

Measurements	Unit	Excellent swimmers ($n = 9$) $V_{max} = 1.76$ ($\pm$ 0.07)		Poor swimmers ($n = 9$) $V_{max} = 1.18$ ($\pm$ 0.13)		Mean differences between groups
		$\overline{X}$	S.D.	$\overline{X}$	S.D.	
H-A $\dot{V}O_2$ (3 min)	liter/min	2.53	0.30	3.22	0.49	−0.69
H-A $\dot{V}O_2$ (4 min)	liter/min	2.56	0.50	3.46	0.55	−0.90
diff.	liter/min	+0.03	0.34	+0.24	0.22	
R (3 min)		0.88	0.12	1.06	0.11	−0.18
R (4 min)		0.94	0.10	1.09	0.08	−0.15
diff.		+0.06	0.05	+0.03	0.05	
E (3 min)	%	4.76	1.61	3.33	0.80	+1.43
E (4 min)	%	4.87	1.84	2.95	0.61	+1.92
diff.	%	+0.11	0.91	−0.38	0.36	

decreased to 2.95 percent, compared with 4.76 and 4.87 percent, respectively, in the excellent group. These data are in agreement with the results of Di Prampero et al. (1974). (5) The *r* between H-A $\dot{V}O_2$ and *E* of the total group of —.47 increased to —.65 when only the data of the 18 subjects of the contrasting groups were used.

REFERENCES

Alley, L. E. 1952. An analysis of water resistance and propulsion in swimming the crawl stroke. Res. Quart. 33: 253–270.

Clarys, J. P., J. Jiskoot, and L. Lewillie. 1973. A kinematographical, electromyographical, and resistance study of waterpolo and competition front crawl. *In:* S. Cerquilini, A. Venerando, and J. Wartenweiler (eds.), Biomechanics III, pp. 446–452. S. Karger, Basel.

Clarys, J. P., J. Jiskoot, H. Rijken, and P. J. Brouwer. 1974. Total resistance in water and in relation to body form. *In:* R. C. Nelson and C. A. Morehouse (eds.), Biomechanics IV, pp. 187–196. University Park Press, Baltimore.

Clarys, J. P., and J. Jiskoot. 1975. Total resistance of selected body positions in the front crawl. *In:* L. Lewillie and J. P. Clarys (eds.), Swimming II pp. 110–119. University Park Press, Baltimore.

Di Prampero, P. E., D. R. Pendergast, D. W. Wilson, and D. W. Rennie. 1974. Energetics of swimming in man. J. Appl. Physiol. 37: 1–5.

Henry, F. M. 1956. Evaluation of motor learning when performance levels are heterogeneous. Res. Quart. 27: 176–181.

Holmer, I. 1974. Physiology of swimming man. Acta Physiol. Scand. 407 (suppl.).

Jiskoot, J., and J. P. Clarys. 1975. Body resistance on and under the water surface. *In:* J. P. Clarys and L. Lewillie (eds.), Swimming II, pp. 105–109. University Park Press, Baltimore.

Karpovich, P. V. 1933. Water resistance in swimming. Res. Quart. 4: 21–28.

Katch, F., E. D. Michael, and S. M. Horvath. 1967. Estimation of body volume by underwater weighting: Description of a method. J. Appl. Physiol. 23: 811–813.

Katch, V. 1972. Correlational v. ratio adjustments of body weight in exercise-oxygen studies. Ergonomics 15: 671–680.

Kemper, H. C. G., R. A. Binkhorst, R. Verschuur, and A. C. A. Vissers. 1975. Vergelijkend onderzoek naar de betrouwbaarheid van de Ergo-analyser (Reliability of the Ergo-analyzer). Gen. en Sport 4: 6–10.

Van Manen, J. D., and H. Rijken. 1975. Dynamic measurement techniques on swimming bodies of the Netherlands Ship Model Basin. *In:* J. P. Clarys and L. Lewillie (eds.), Swimming II, pp. 70–79. University Park Press, Baltimore.

Determination of man's drag coefficients and effective propelling forces in swimming by means of chronocyclography

J. Klauck and K. Daniel
Deutsche Sporthochschule, Köln

The water resistance of the human body and the propulsive forces developed by the movements of arms and legs have been determined from various points of view by many authors (Karpovich, 1935; Alley, 1952; Counsilman, 1955; Schramm, 1958/59; Kent and Atha, 1971; Zaciorskij and Safarjan, 1972). Generally the towing method was used to obtain the data. In this study, another method is presented. Based on solutions of differential equations for a gliding person and for an individual swimming by leg kicking only, formulas are obtained to determine the "drag" coefficient of the gliding body and the propulsive force together with a "loss" drag factor in swimming using the leg kick. The results of these methods are compared to those obtained from the conventional method.

DRAG COEFFICIENT OF THE HUMAN BODY MOVING PASSIVELY IN WATER

The drag coefficient normally represents an expression of the quantitative factor obtained from measured towing force at a constant velocity. A functional dependence can be given by the following expression according to the general laws of hydrodynamics:

$$k_r = \frac{F_t}{v^2} \tag{1}$$

where k_r is the drag coefficient, F_t is the measured resistance force, and v is the towing velocity. On the other hand, a differential equation for a gliding body in the water can be established by

$$mv = -k_r v^2 \tag{2}$$

where m is the mass of the body, v is the acceleration of the body, k_r is the drag (or resistance) coefficient, and v is the velocity of the body. The solution of equation (2) gives information about the time-velocity curve of the body, and can be expressed by

$$v[t] = \frac{v_0}{v_0 \frac{k_r}{m} t + 1} \tag{3}$$

where v_0 is the initial velocity at $t = 0$.

Application of equation 3 consists of determining the velocity–time curve (horizontal velocity), fitting the experimental curve into the theoretical model, and computing k_r. A comparison between the results obtained from the towing method and from those using the method described above can be made, since both methods refer to the same force–velocity relation.

MEASURING PROCEDURE

Towing Method

The towing apparatus was a motor-driven platform moving above the water surface of a 50-m pool. A tube with strain gauges was attached to the platform. A grip was fixed at the lower end of this tube, so the swimmer could hold on with his hands. The strain gauges measured the force in the towing direction either backward or forward. Towing velocity varied from 0.8 to 2.0 m/sec. The towing force was measured continuously and printed on recorder paper. Mean values of the towing forces were then computed.

Optical-Computational Method

A lighted lamp was attached to the swimmer's head, and the gliding process was filmed by a chronocyclographical camera from a vertical direction to the moving swimmer producing a picture of an interrupted light trace (Figure 1). The time interval between two successive gliding phases on the picture was 50 msec. A reference light point system located in the moving plane of the gliding swimmer permitted the evaluation of the local coordinates of each light trace point with a film analyzing system. By differentiation of the obtained local coordinates in a moving direction, the velocity–time curve of the measured body point was established. Before differentiation, the local coordinates were smoothed slightly. An example of a measured and computed time–velocity curve is shown in Figure 2.

RESULTS

Towing and gliding measurements were made with five male test subjects. Typical results of both measuring methods are given in Table 1 for three test subjects. The drag coefficients were computed using equations (1) and (3), respectively. The drag coefficient obtained by applying the towing method shows a noticeable decrease in the velocity range from 0.8 to 1.2 m/sec, according to the findings of Alley (1952). This effect results from the influence of the fluid stream acting during the towing process at different towing velocities when a lift occurs to the swimmer's body. The magnitude of this effect depends on the location of the point the towing device is attached to the swimmer's body. The drag coefficients obtained from gliding motion measurements remained constant within the measuring accuracy over a wide range of moving velocities; their values reached those obtained at higher velocities from towing measurements.

Figure 1. Chronocyclogram of a gliding swimmer.

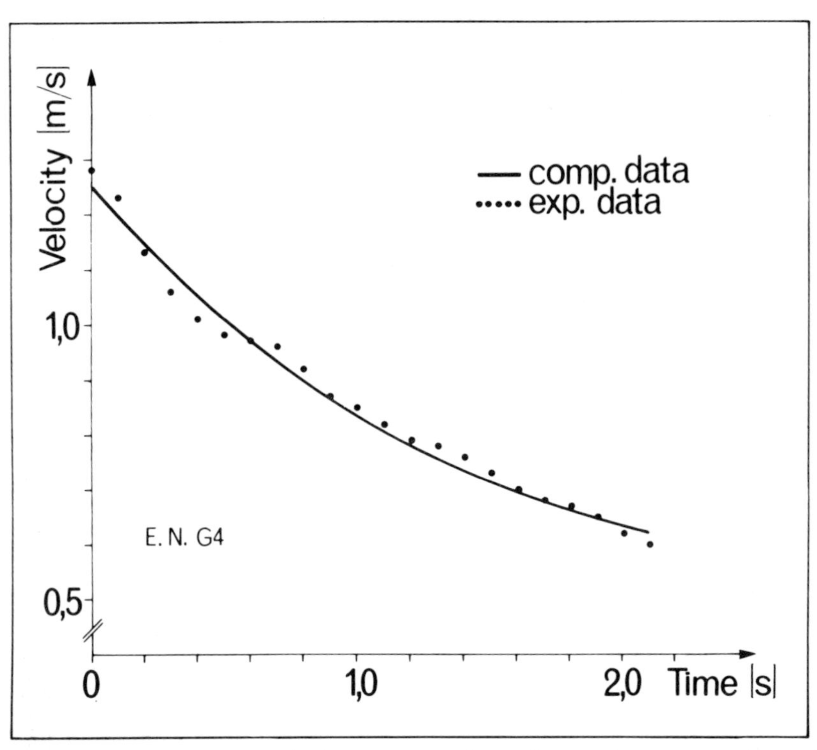

Figure 2. Time-dependent velocity decrease of a gliding swimmer, experimental and computed curve.

Table 1. Water resistance (drag) coefficients of three individuals

Test person	Method	Velocity (m/sec)	k_r (kg/m) (mean ± S.E.)
01 A. N.	Towing	0.8	41.6 (±4)
$m = 80$ kg		1.0	32.8 (±4)
		1.2	28.0 (±4)
	Opt./comp.	0.7–1.2	20.8 (±2)
02 E. N.	Towing	0.8	34.6 (±3)
$m = 72$ kg		1.0	28.1 (±3)
		1.2	26.6 (±3)
		1.4	23.0 (±3)
	Opt./comp.	0.7–1.4	26.6 (±2)
05 R. K.	Towing	0.8	38.4 (±4)
$m = 80$ kg		1.0	40.0 (±4)
		1.2	33.6 (±4)
	Opt./comp.	0.6–1.0	35.2 (±2)

RELATIONSHIP BETWEEN WATER RESISTANCE AND PROPELLING FORCE DURING LEG KICKING

Applying leg kicking during towing measurements showed a decrease of towing force, depending on the towing velocity. This effect resulted from the generation of a propulsive force by the swimmer's leg kick. The measured towing force F_{tl} can be written

$$F_{tl} = F_p - F_{rl} \tag{4}$$

where F_p is the propulsive force and F_{rl} is the resistance force.

Assuming again the law of squares for F_{rl} from towing velocity, equation (4) becomes

$$F_{tl} = F_p - k_{rl}v^2 \tag{5}$$

where k_{rl} is the drag coefficient of the swimmer's body executing the leg kick. The factor k_{rl} is not necessarily identical to the drag coefficient defined in equation (1), since the swimmer's body varies its form during the leg kicking motion. Additionally from hydromechanical considerations, the results show that the propulsive force F_p decreases with increasing towing velocity. The decrease of F_p can be written in a functional form:

$$F_p = F_{p0} - dv^2 \tag{6}$$

where F_{p0} is the maximum propulsive force exerted in still water and d may be called the decrease factor. The combination of equations (5) and (6) yields

$$F_{tl} = F_{p0} - Av^2 \tag{7}$$

where $A = d + k_{rl}$ is the "loss" drag factor.

The measured towing force F_{tl} depends to a high degree on the performance of the leg kick. A complete separation of both factors d and k_{rl} is not possible. Equation (7) describes an overall effect of the leg kicking motion.

A differential equation can be established for a swimmer executing the leg kick under the same assumptions as cited above:

$$mv = F_{p0} - Av^2 \tag{8}$$

with the same constants F_{p0} and A. It should be noticed that the agreement between equations (7) and (8) is only formal since in equation (8), the accelerative effect of propulsion is dealt with, and equation (7) is based on a test condition without acceleration occurring.

In the initial condition, with the swimmer's velocity at v_0 at a certain time t_0, a solution of equation (8) is given by

$$v[t] = \frac{v_l \operatorname{tgh} \{v_l B[t - t_0]\} + v_0}{1 + \frac{v_0}{v_l} \operatorname{tgh} \{v_l B[t - t_0]\}} \tag{9}$$

with the abbreviations

$$v_l = \sqrt{\frac{F_{p0}}{A}}, \qquad B = \frac{A}{m}$$

The limiting velocity v_l occurs if F_{p0} equals Av^2 and the swimmer's acceleration becomes zero.

The physical situation described by equation (8) with a velocity–time curve of equation (9) may be more realistic by letting a swimmer begin to swim by leg kick from a resting position. The evaluation of the velocity curve permits the determination of F_{p0} and A.

MEASURING PROCEDURES

Measurements of the leg kick propulsive force were made by the towing apparatus already described. The test subjects were ordered to kick with maximum effort in each towing test. Using the optical-computational method the swimmers started from a resting position, also exerting a full speed leg kick. Optical instrumentation was the same as that used in the gliding tests. Velocity–time curves of the swimmer's motion were computed as for the gliding situation. An example is given in Figure 3. Additionally, a separate measurement was taken of the limiting velocity of each swimmer. These measurements served as control data of the corresponding values obtained from the chronocyclogram.

For the determination of "loss" drag factor A in both measuring methods refer to equations (7) and (9), respectively.

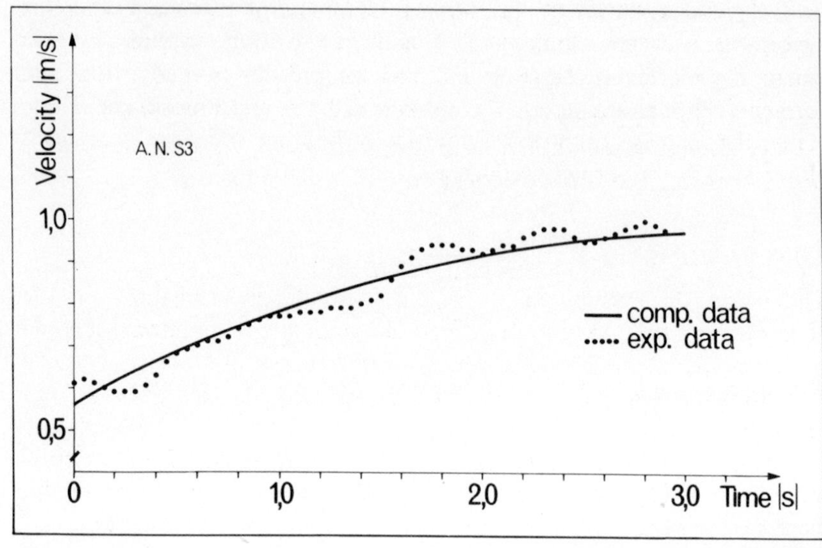

Figure 3. Velocity curve of a swimmer starting from a resting position, executing crawl leg kick.

RESULTS

The values v_l, A, and F_{po} from both towing and optical measurements are presented in Table 2. Looking at Table 2, a tendency can be seen. If the measured value of v_l differs only slightly from that obtained from towing and optical-computational method, "loss" drag factor A and

Table 2. Limiting velocity, combined resistance and decrease factors in crawl leg kicking motion. t, towing method, o-c, optical-computational method

Test person	Measured v_1 (m/sec)	Computed v_1 (m/sec)	A (± S.E.) (kg/m)	F_{po} (± S.E.) (N)
0.1 A. N.	0.96	t 0.95	29.6 (±2)	27.0 (±5)
		o/c 0.96	54.2 (±3)	51.0 (±2)
		o/c 1.03	34.4 (±2)	37.0 (±2)
		o/c 1.06	25.6 (±2)	29.0 (±2)
02 E. N.	1.09	t 1.04	25.9 (±2)	28.0 (±6)
		o/c 0.96	72.0 (±2)	66.0 (±2)
		o/c 1.08	20.2 (±2)	23.6 (±2)
05 R. K.	0.90	t 1.00	37.6 (±2)	38.0 (±5)
		o/c 0.87	27.2 (±2)	21.0 (±3)
		o/c 0.88*	36.8 (±2)	29.0 (±3)

*Stretched arms supported by a kicking board.

force F_{p0} also show no great differences. This means that the towing and optical methods are equivalent to each other. With respect to tests where the measured value of the limiting velocity was not reached, further investigations about the reasons for this must be made. Especially, the increase of A and F_{p0} at lower limiting velocities requires a closer look at the leg action during kicking.

CONCLUSIONS

The drag measured by the towing method depends strongly on the construction of the towing device. By the optical-computational method presented here, the results show not only the validity of the gliding test cited by Cureton (1971), but in addition, a resistance (drag) factor can be defined.

Natural body position in the water is not changed by external influences. Therefore, a more reliable value of water resistance coefficient can be expected using the gliding process analyzed optically.

Crawl leg kicking force and "loss" drag factor can also be determined by the optical-computational method. The same advantages as cited above are valid here, but further investigations must be made to explain the results presented in Table 2.

REFERENCES

Alley, L. E. 1952. An analysis of water resistance and propulsion in swimming the crawl stroke. Res. Quart. 23: 253–69.

Counsilman, J. E. 1955. Forces in swimming two types of crawl stroke. Res. Quart. 26: 127–38.

Cureton, Th. K. 1971. Biomechanics of swimming with interrelationships to fitness and performance. *In:* L. Lewillie and J. P. Clarys (eds.), Proc 1st Int. Symp., Biomechanics in Swimming, pp. 31–52. Université Libre de Bruxelles, Brussels.

Karpovich, P. V. 1935. Analysis of the propelling force in the crawl stroke. Res. Quart. 5: 49–58.

Kent, M. R., and J. Atha. 1971. Selected critical transient body positions in breast stroke and their influence upon water resistance. *In:* L. Lewillie and J. P. Clarys (eds.), Proc. 1st Int. Symp., Biomechanics of Swimming, pp. 119–125. Université Libre de Bruxelles, Brussels.

Schramm, E. 1958/59. Untersuchungsmethoden zur Bestimmung des Widerstandes, der Kraft und der Ausdauer bei Schwimmsportlern (Procedures for Determination of a Swimmer's Resistance, Force, and Endurance). Wiss. Zt. der DHfK 2: 161–80.

Zaciorskij, V. M., and I. G. Safarjan. 1972. Untersuchungen von Faktoren zur Bestimmung der maximalen Geschwindigkeit im Freistilschwimmen (Investigation of Factors for the Determination of Maximum Speed in Free Style Swimming). Theorie und Praxis der Körperkultur 21: 695–709.

Functional evaluation of kayak paddlers from biomechanical and physiological viewpoints

A. Dal Monte and L. M. Leonardi
C.O.N.I.—Sports Medicine Institute, Rome

The functional evaluation of high-performance athletes is normally assessed by using the treadmill or the bicycle ergometer. While these methods are satisfactory when applied to athletes whose activity concerns the lower body musculature, they are inadequate when the sport they practice involves the upper limbs and the trunk, like in canoeing (kayaking).

In order to reduce this disadvantage, some scientists have developed specific types of ergometers capable of producing in the laboratory the characteristic movements of the sport activity being studied.

The purpose of this work was to analyze the metabolic and cardiocirculatory responses and the energy cost of a group of athletes engaged in kayaking by using a particular type of simulated ergometer developed at the Sports Medicine Institute of Rome. The data obtained were compared with those obtained on the same subjects during a classic bicycle ergometer test.

PROCEDURES

The test was carried out on 11 male athletes of the Italian National Team, whose anthropometric characteristics are given in Table 1.

The simulated kayak ergometer built at the Sports Medicine Institute was based on a load-measuring apparatus of a bicycle ergometer (variable couple and constant power type, Elema-Schönander, Stockholm). Handles which reproduced the slope of the kayak paddle were

Table 1. Physical characteristics of subjects

Mean age (yr)	Mean height (cm)	Mean weight (kg)	Mean ponderal index
22.78 ±2.52	180.36 ±4.90	79.76 ±5.74	23.85 ±0.55

applied to this apparatus. A special frame on which a kayak seat and a footrest were fitted was added to produce the same position as that assumed on the boat. Oxygen uptake, the carbon dioxide expired, the pneumotacogram, the pulmonary ventilation, and exercise heart rate were analyzed continuously and calculated by a computer in real time (Fengyes and Gut, Basel).

The test was conducted in an air conditioned room after at least 3 hr after the previous meal and in the course of a whole day.

The subject at work on the kayak ergometer was filmed at high speeds in both lateral and sagittal planes in order to be able to compare these movements with those carried out in the kayak.

During the film recordings shot while the subjects were in the kayak, the camera was fixed on a boat moving at the same speed as that of the kayak.

The ergometer test of the simulated kayak consisted in three 2-min work bouts, increasing respectively, from 800, to 1200, to 1600 kgm/min; for only two subjects the last load was 1800 kgm/min since their muscular strengths were superior to the average team strength. The test was carried out to the point of exhaustion. The same pattern was used in the bicycle ergometer test.

RESULTS

A comparison between the metabolic and cardiocirculatory responses and the work carried out on the kayak ergometer and on the bicycle ergometer is given in Tables 2 and 3. All means were compared by using t-tests.

Table 2, which gives the maximal measures of the parameters examined, shows that there were no substantial differences (except for the ratio VO_2/FC, $0.20<p<0.025$). However, on the bicycle ergometer a greater amount of work was carried out because the load was equal and it took longer before complete fatigue was reached.

A different situation is shown in Table 3 where the variables measured immediately before the conclusion of the simulated ergometer test were compared with those obtained after the same amount of work on

Table 2. Metabolic and cardiocirculatory data

	Kayak Ergometer (max. quantity of work)	Bicycle Ergometer (max. quantity of work)	t-ratios and correlation coefficients
HR (beats/min)	186.90 ± 12.59	185.90 ± 11.03	$t = 0.1878$ $r = -0.113$
$\dot{V}O_2$/kg	42.11 ± 6.84	45.40 ± 7.84	$t = 1.454$ $r = 0.485$
$\dot{V}O_2$/HR	18.76 ± 3.71	20.72 ± 3.87	$t = 2,710$* $r = 0.801$ $0.020 < p < 0.025$
$\dot{V}O_2$ (ml/min)	3,369.45 ± 649.97	3,629.90 ± 733.53	$t = 1.438$ $r = 0.629$
$\dot{V}$ (liter/min)	144.29 ± 30.58	133.61 ± 28.69	$t = 1.425$ $r = 0.653$

*Statistically significant

Table 3. Metabolic and cardio-circulatory data

	Kayak Ergometer (max. work load)	Bicycle Ergometer (the same work load)*	t-ratios and correlation coefficients
HR (beats/min)	186.90 ± 12.59	164.54 ± 11.83	$t = 5.377$† $r = 0.363$ $p < 0.001$
$\dot{V}O_2$/kg	40.93 ± 6.41	40.21 ± 3.62	$t = 0.425$ $r = 0.489$
$\dot{V}O_2$/HR	17.61 ± 3.60	20.08 ± 3.45	$t = 3.819$† $r = 0.815$ $0.001 < p < 0.005$
$\dot{V}O_2$ (ml/min)	3,274.54 ± 606.29	3,272.09 ± 481.85	$t = 0.0200$ $r = 0.743$
$\dot{V}$ (liter/min)	144.02 ± 30.70	93.09 ± 12.92	$t = 5.716$† $r = 0.297$ $p < 0.001$

*After the same quantity of work.
†Statistically significant.

the bicycle ergometer. In this case, the differences found were highly significant (except for oxygen consumption).

Furthermore, the energy cost on both tests as carried out was also evaluated. Table 4 presents the $\dot{V}O_2$, $\dot{V}O_2$/Kg, $\dot{V}O_2$/cm, and $\dot{V}O_2$/P.I. values related to loads of 800 and 1200 kgm/min and to maximal work load, which was the same in both tests carried out with the same individual. Significant differences occurred only at the 800-kgm/min work load, while the results were similar for the higher loads. It is interesting to note that the quantity of oxygen consumption was greater on the kayak ergometer during the test carried out with the 800-kgm/min load.

Figures 1 and 2 show a graphic representation of the course of heart frequency and pulmonary ventilation in relation to the oxygen consumption during the last minute of work in the kayak ergometer test and after the same quantity of work in the bicycle ergometer test.

The cinematographic sequences in Figures 3 and 4 indicate that in the lateral view the movements carried out on the simulator are identical or at least almost identical to those carried out in the kayak, while the sagittal view shows that the muscular masses involved are the same, in spite of the fact that the movement, although clearly similar, is not quite identical.

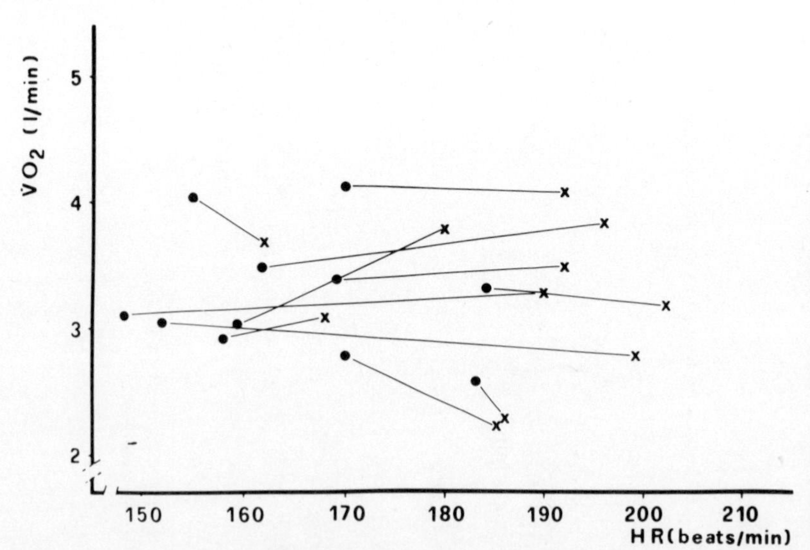

Figure 1. Oxygen uptake and heart rate; comparison between values obtained after the same quantity of work performed on kayak (×) and bicycle (●) ergometers.

Table 4. Kayak and bicycle ergometer energy cost

Kayak ergometer	800 Kgm/min	Bicycle ergometer	Kayak ergometer	1200 Kgm/min	Bicycle ergometer	Kayak ergometer	Max. work-load VO_2 max.	Bicycle ergometer
2,410.27 ± 375.52 3.012 ± 0.469 $t = 2,219$*	ml O_2 ml O_2/kgm $r = -0.247$	2,044.27 ± 315.37 2.554 ± 0.394 $0.05 < p$	2,856.00 ± 551.50 2.379 ± 0.459 $t = 1.698$	ml O_2 ml O_2/kgm $r = 0.751$	2,659.36 ± 292.13 2.215 ± 0.243	3,369.45 ± 649.97 2.110 ± 0.378 $t = 1.438$	ml O_2 ml O_2/kgm $r = 0.629$	3,629.90 ± 733.53 2.185 ± 0.379
30.70 ± 5.50 0.0381 ± 0.0068 $t = 2.362$*	ml O_2/Kg ml O_2/kg/kgm $r = 0.0119$	25.75 ± 4.30 0.0321 ± 0.0053 $0.025 < p < 0.05$	36.27 ± 7.49 0.0301 ± 0.0062 $t = 1.631$	ml O_2/kg ml O_2/kg/kgm $r = 0.656$	33.47 ± 4.25 0.0278 ± 0.0035	42.10 ± 6.84 0.0264 ± 0.0046 $t = 1.457$	ml O_2/kg ml O_2/kg/kgm $r = 0.486$	45.40 ± 7.84 0.0273 ± 0.0044
13.37 ± 2.096 0.0160 ± 0.0027 $t = 2.227$*	ml O_2/cm ml O_2/cm/kgm $r = -0.130$	11.35 ± 1.885 0.0137 ± 0.0025 $0.05 < p$	15.83 ± 3.014 0.0127 ± 0.0024 $t = 1.669$	ml O_2/cm ml O_2/cm/kgm $r = 0.723$	14.76 ± 1.809 0.0118 ± 0.0016	18.65 ± 3.424 0.0110 ± 0.0020 $t = 1.432$	ml O_2/cm ml O_2/cm/kgm $r = 0.580$	20.10 ± 3.835 0.0117 ± 0.0020
100.881 ± 16.157 0.125 ± 0.0201 $t = 2.177$*	ml O_2/P.I. ml O_2/P.I./kgm $r = -0.318$	85.596 ± 12.387 0.106 ± 0.0154 $0.05 < p$	119.66 ± 23.112 0.0991 ± 0.0192 $t = 1.694$	ml O_2/P.I. ml O_2/P.I./kgm $r = 0.764$	111.365 ± 11.196 0.0924 ± 0.0092	141.182 ± 27.002 0.0879 ± 0.0158 $t = 1.451$	ml O_2/P.I. ml O_2/P.I./kgm $r = 0.632$	152.137 ± 30.823 0.0911 ± 0.0161

*Statistically significant.

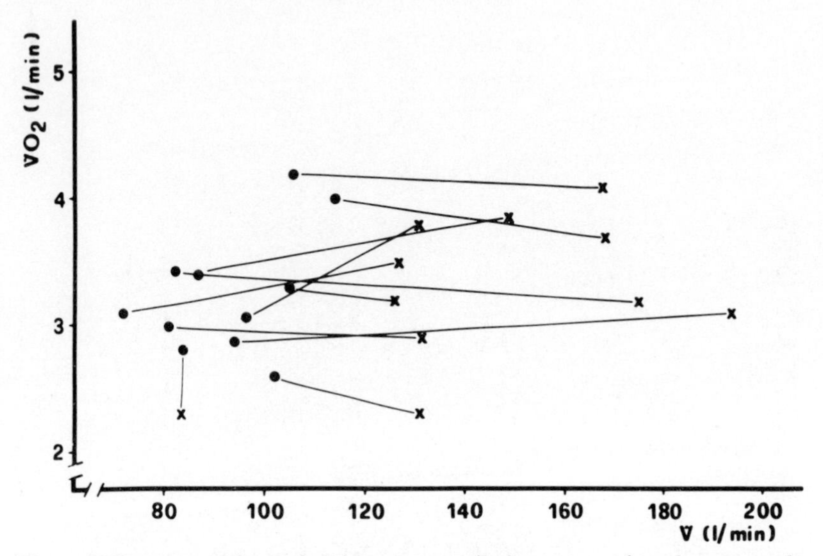

Figure 2. Oxygen uptake and pulmonary ventilation; comparison between values obtained after the same quantity of work performed on kayak (×) and bicycle (●) ergometers.

DISCUSSION AND CONCLUSIONS

The results show that subjects who have engaged in extensive kayak training are able to use, during specific tests, muscular masses comparable to those used in bicycle ergometer tests.

The explanation for the small quantity of work carried out by the upper limbs is to be found in their poor mechanical efficiency.

The remarkable differences of the heart frequency found in the two tests (the quantity of work being equal while the $\dot{V}O_2$ and the $\dot{V}O_2$/kg appear the same) demonstrate that in maximal tests which are very specific the heart frequency is not as valid an indication as in non-specific tests.

The kayak simulating ergometer that we have developed was applied to athletes who had never tried it, but who possessed a high level of training in canoeing.

The values obtained with our ergometer were higher, both compared to the results reported by other authors on different kayak ergometers and in comparison to the bicycle ergometer. These results indicate that this device is capable of assessing (even in the absence of specific training) the functional evaluation of the athletes engaged in kayaking.

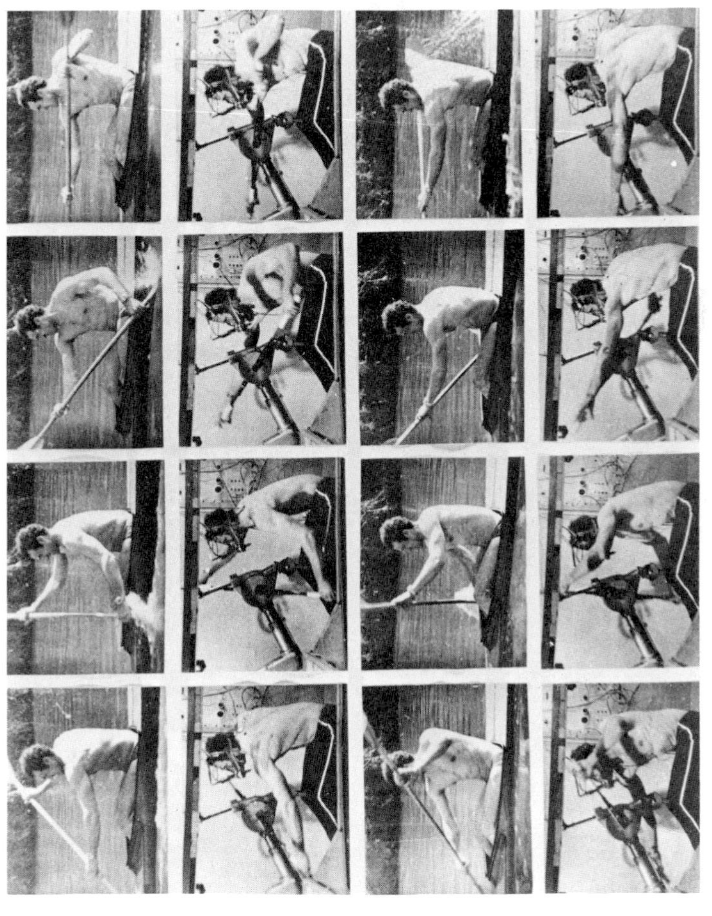

Figure 3. Lateral plane; paddler motion on kayak and simulating kayak ergometer.

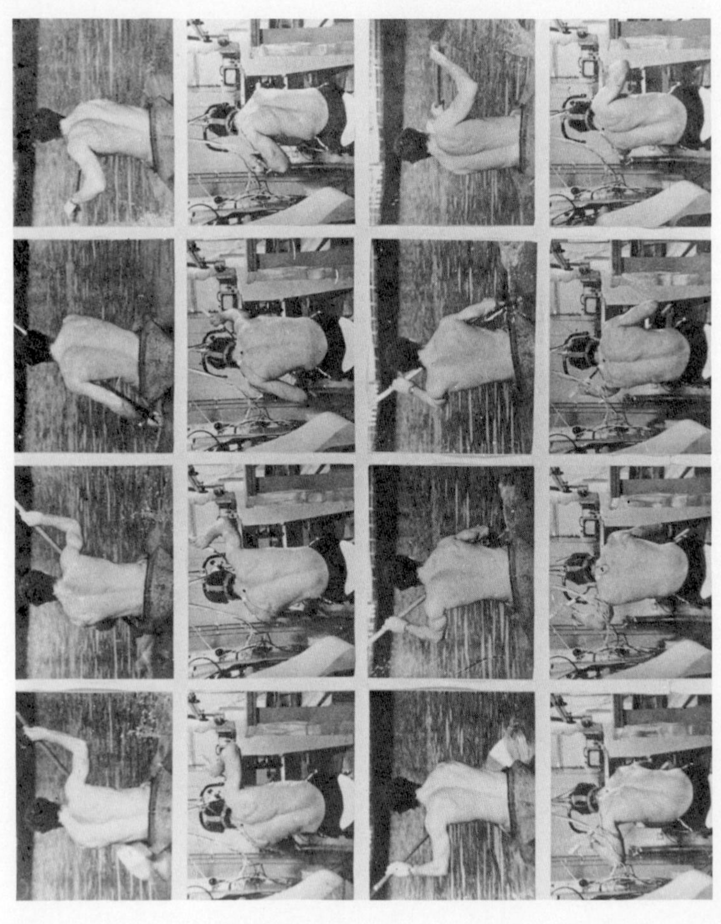

Figure 4. Sagital plane; paddler motion on kayaks and simulating kayak ergometer.

REFERENCES

Asmussen, E., and I. Hemmingsen. 1958. Determination of maximum working capacity at different ages in work with the legs or with the arms. Scand. J. Clin. Lab. Invest. 10: 67–71.

Astrand, P. O., and B. Saltin. 1961. Maximum oxygen uptake and heart rate in various types of muscular activity. J. Appl. Physiol. 16: 977–981.

Bobbett, A. C. 1960. Physiological comparison of three types of ergometry. J. Appl. Physiol. 15: 1007–1014.

Dal Monte, A. 1974. Metodologia della valutazione funzionale specifica degli atleti praticanti attività sportiva di media e lunga durata. (Specific functional evaluation methodology of middle and long-duration sport activities athletes). Paper presented at the Second Stage on Endurance Training, Rome, Italy, May 6–8.

Dal Monte, A., and A. Todaro. 1974. Heart frequency and oxygen input in treadmill exhausting work in 70 endurance athletes. Paper presented at the XXth World Congress in Sports Medicine. Melbourne, Australia. February 4–9.

Pyke, F. S., et al. 1973. Metabolic and circulatory responses to work on a canoeing and bicycle ergometer. Austr. J. Sport Med. 5: 22–31.

Winter sports

Study of the displacement of a skier's center of gravity during a ski turn

H. Sodeyama
Kinjo Gakuin University, Nagoya

M. Miura, K. Kitamura, and H. Matsui
University of Nagoya, Nagoya

There are several factors that affect a ski turn maneuver. They are the movement of the skier's center of gravity (C. G.), edging, characteristics of the skis (flexibility, repulsion, side-cut, etc.), inclination of the slope, and the snow conditions.

Among these, the movement of the skier's C. G. is the most essential factor. For the study of biomechanics as well as for ski instruction, it is important to appreciate the movement of the skier's C. G. during a ski turn (Kinoshita, 1971; Hay, 1973). Regrettably, however, there have been few investigations that experimentally treat the skier's C. G.

The purpose of this study was to investigate the displacements of the skier's C. G. during a ski turn.

PROCEDURE

The experimental slope was made even, with an incline of 13.5 degrees. Two skiers were asked to perform downhill turns on an arc of a circle whose radius was 9.5 m. The arc was divided into five sections of 30 degrees each, and the form of the skiers at each phase was simultaneously photographed by two 35 mm still cameras, positioned at the same distance of 10 m from the skier, one on a line tangent to the arc (directly in front of the skier), and the other perpendicular to the outside of the arc (directly to the left of the skier) (Figure 1).

Of the two skiers who served as subjects in this experiment, Subject A was an excellent skier while Subject B was less skilled but could perform parallel turns. They were asked to perform two kinds of parallel turns, one taking a crouch position and the other with an upward motion at the initiation of the turn. For all five trials and at each phase, photos were taken by both cameras.

In order to obtain the C. G., each picture was enlarged. Matsui's method (1958) was used in determining the skier's C. G. (including the skis, sticks, boots, and clothes). Reference points of the skier on each picture were analyzed by a graph-pen system and the C. G. was calculated by using a computer.

The photos taken from the front view of the skier showed the *Y–Z* displacements of the skier's C. G., while those from the flank gave the *X–Z* displacements (Figure 2). Then the three-dimensional displacement in each phase was obtained by compounding the displacements *Y–Z* and *X–Z*.

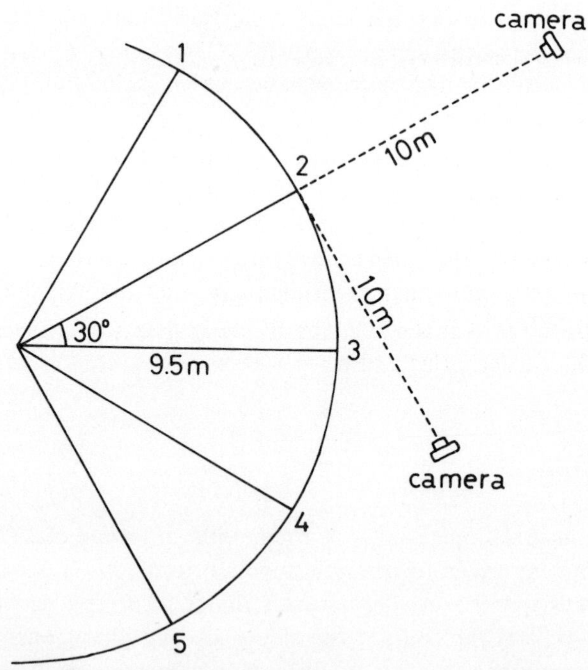

Figure 1. Experimental conditions for photographing. Each number indicates the position of photographing for the skier's form.

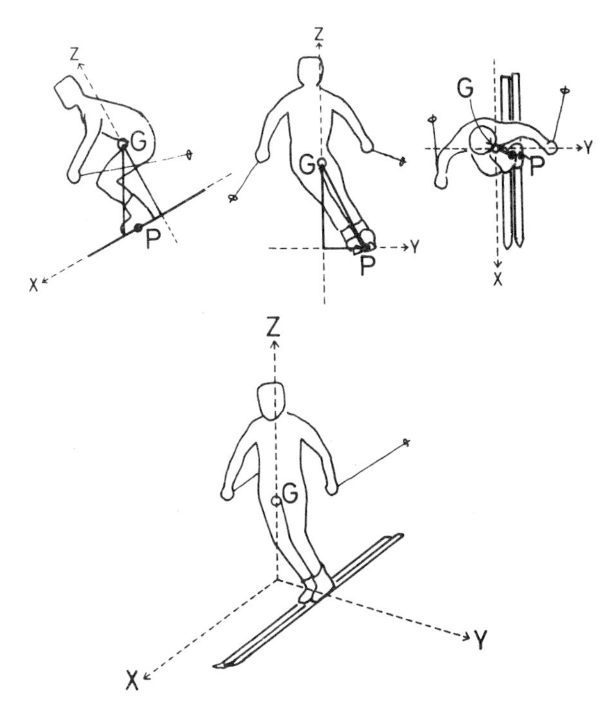

Figure 2. Three-dimensional position of the skier's C.G. G, the skier's center of gravity; P, the center of the rotatory movement of the skis.

In order to analyze the displacement of the skier's C. G. against the arc of the turn, the center of rotatory movement of the skis (point P) was considered. This point was determined on the photo, taken from the downhill direction as the skiers were performing the vertical side-slip, by drawing the perpendicular from the skier's C. G. and intersecting it with the line of the skis. Point P was found to be located at the bottom of the skier's plantar arch in both subjects.

RESULTS

Figure 3 shows the *X–Z* displacements of the skier's C. G., obtained from the flank photos at each phase. This figure indicates the relative positions of the skier's C. G. from the center of the rotatory movement of the skis.

Remarkable differences were observed in the C. G. displacements, between the one taking a crouch position and that with an upward motion at the initiation of the turn. In comparing the performances of

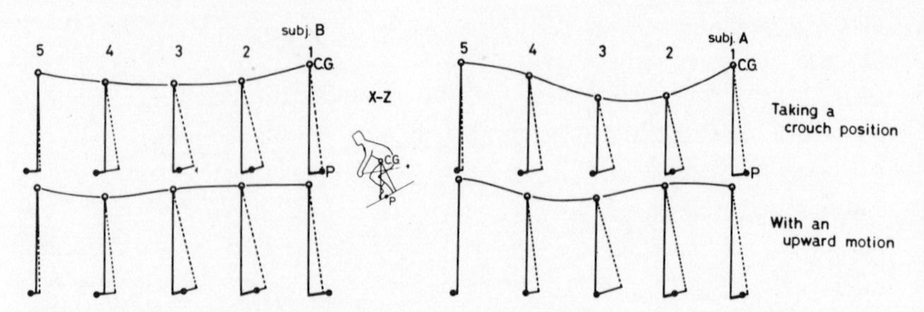

Figure 3. The *X–Z* displacements of the skier's C.G. *A*, excellent skier; *B*, less skilled skier. P, the center of the rotary movement of the skis.

Subjects A and B, the range of upward and downward actions was also quite large. It was apparent that B could not perform active movements because of his inferior skill. The forward and backward movements of the C. G. from point P, however, showed a similar tendency in both subjects. In the early phases of the turn, the skier's C. G. was displaced forward from P, and slightly backward from P in the later phases. In other words, in the early stages of the turn, the skier was putting his weight on the forward portion of his skis, while in the later stages the weight was slightly on the back portion of the skis.

Figure 4 shows the *Y–Z* displacements of the skier's C. G., obtained from the frontal photos at each phase. This figure indicates the relative positions of the skier's C. G. from the arc of the turn.

The horizontal displacements of the C. G. from P, namely, the inward displacements from the arc of the turn, showed opposite tendencies as follows: on the turn taking a crouch position in initiation, the inward displacements were slight in the early phases and increased during the middle to later phases. On the other hand, on the turn with an upward motion in initiation, the inward displacements were greater in

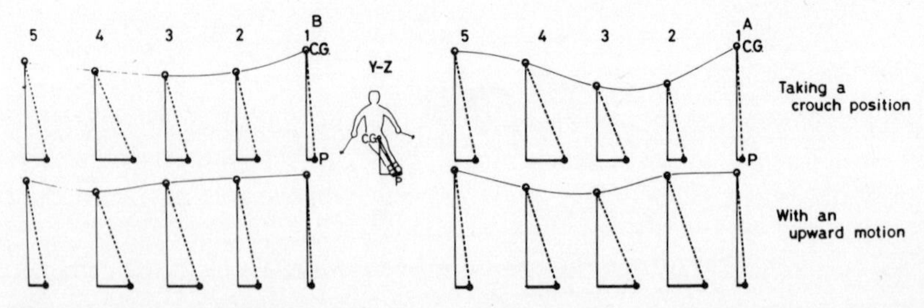

Figure 4. The *Y–Z* displacements of the skier's C.G. *A*, excellent skier; *B*, less skilled skier. P, the center of the rotary movement of the skis.

the early to middle phases compared to those in its later phases. But in the case of Subject B, no clear difference was observed in either of the turns, since B could not clearly distinguish the movements. This tendency also appeared in upward and downward actions.

The *X–Y* displacements of the skier's C. G. were obtained by compounding the *X–Z* displacements. Figure 5 shows the *X–Y* displacements of the skier's C. G. on the arc of the turn. These diagrams indicate the relative positions of the skier's C.G. whose action line falls on the snow surface and the center of the rotary movement of the skis. For the purpose of vivid illustration the displacements of the skier's C. G. are drawn out of proportion with the reduced scale of the arc.

There were some differences in inward displacements between the turn taking a crouch position and that with an upward motion in the initial phase. The relationship between P and the C. G. displacements may, however, be explained as follows: in a ski turn the skier's C. G. was displaced forward against the arc, then inward displacements gradually increased, and then there occurred some backward displacement from P while the inward displacements against the arc lessened.

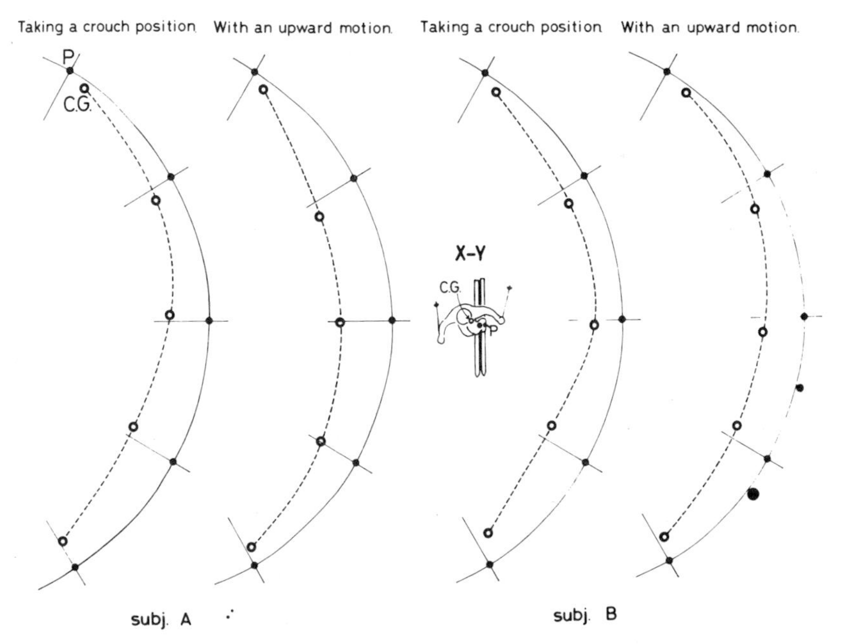

Figure 5. The *X–Y* displacements of the skier's C.G. *A*, excellent skier; *B*, less skilled skier. P, the center of the rotary movement of the skis.

DISCUSSION

It can be reasoned that as the skier performs the vertical side slip, there does not occur any rotary movement since the force of gravity and the resultant force of snow resistance cancel each other out. If the skier's C.G. is displaced either forward or backward, the ski's tip or tail will begin to move downhill because of the displaced force of gravity either forward or backward, respectively. Thus, the rotary movement of the skis occurs.

From these relations, it may be explained that the skier displaces his C. G. forward in the early stages of the turn in order to create the downhill rotary movement, and displaces it backward in the later stages in order to generate the uphill rotary movement.

Furthermore, it suffices to mention that any object needs the centripetal force in mechanics in order to perform a circular movement. The centripetal force is created by the skier's leaning so his C. G. moves inward against the arc of the circle. Thus, in order to gain the centripetal force requisite in completing a turn, the skier displaces his C. G. inward against the arc.

CONCLUSION

From this study the following conclusion can be drawn: in a ski turn the skier's C. G. is initially displaced forward in order to generate the downhill rotary movement of the skis, then the skis are turned with the centripetal force generated by the inward movement of the skier's C. G., followed by the backward movement of the skier's C. G. creating the uphill rotary movement of the skis to complete the turn.

REFERENCES

Hay, J. G. 1973. Analysis of Sports Techniques. Prentice-Hall, Englewood Cliffs, N.J.

Kinoshita, K. 1971. Dynamics of Skiing. *In:* Society of Ski Science (eds.), Scientific Study of Skiing in Japan, pp. 45–56. Hitachi Printing Co., Ltd., Tokyo.

Matsui, H. 1958. The Movement and the Center of Gravity in Man. Science of Physical Education Ltd. (in Japanese).

Matsui, H., and T. Ohnishi. 1959. Biomechanical analysis of the ski turn. Res. J. Phys. Ed. 5 (1) 122–123.

Dynamometric analysis of different hockey shots

R. Doré
Ecole Polytechnique, Montréal

B. Roy
Université Laval, Québec

The purpose of this investigation was to measure the variation in time of the forces applied on a hockey stick by players while shooting at a target. The different shots performed were the sweep, wrist and slap shots, while stationary and in motion. The velocities of the puck for slap shots were also measured and the influence of the shape of the time dependent forces on these velocities was studied.

To our knowledge only one study (Romechevsky, 1974) has dealt with the kinematics and the kinetics of hockey shots. However, he described the methodology without giving any results of the experiment.

During the course of this investigation, nine skillful adult amateur hockey players were used as subjects. Some of their characteristics are listed in Table 1.

METHODOLOGY

The forces produced by both hands and the reaction of the ice and of the puck on the stick were obtained using strain gauges appropriately located along the handle and the blade of the stick (Figure 1). Seven Wheatstone bridges were formed in such a way as to compensate for temperature variations during the tests. The output signals of these

Research supported by the National Research Council of Canada (Grant No. A 7513) and by l'Université Laval.

Table 1. Characteristics of the subjects and puck velocity for slap shots

Subject No.	Age (yr)	Weight (kg)	Height (cm)	Length of stick* (cm)	Side of shooting	Caliber†	Puck velocity‡ (m/sec)	
							Stationary	In motion
1	19	61.2	176.5	145	Left	Junior B	24.6	26.0
2	25	70.3	185.4	160	Left	Senior	28.3	29.8
3	23	81.6	185.4	159	Left	Junior A	25.2	28.7
4	18	72.6	185.4	159	Left	Junior B	28.3	29.4
5	17	65.8	176.5	145	Left	Junior B	28.5	30.8
6	19	79.4	190.5	158	Right	Student §	27.5	28.0
7	21	81.6	194.3	159	Left	Student §	27.6	29.7
8		70.8	180.3	154	Left		26.9	29.6
9		73.5	179.1	159	Right		25.8	29.4

*Length of stick was measured from end of handle to tip of blade.
†Caliber: last level attained while playing in competitive hockey (when applicable).
‡Mean value for five slap shots.
§Undergraduate physical education student specializing in hockey.

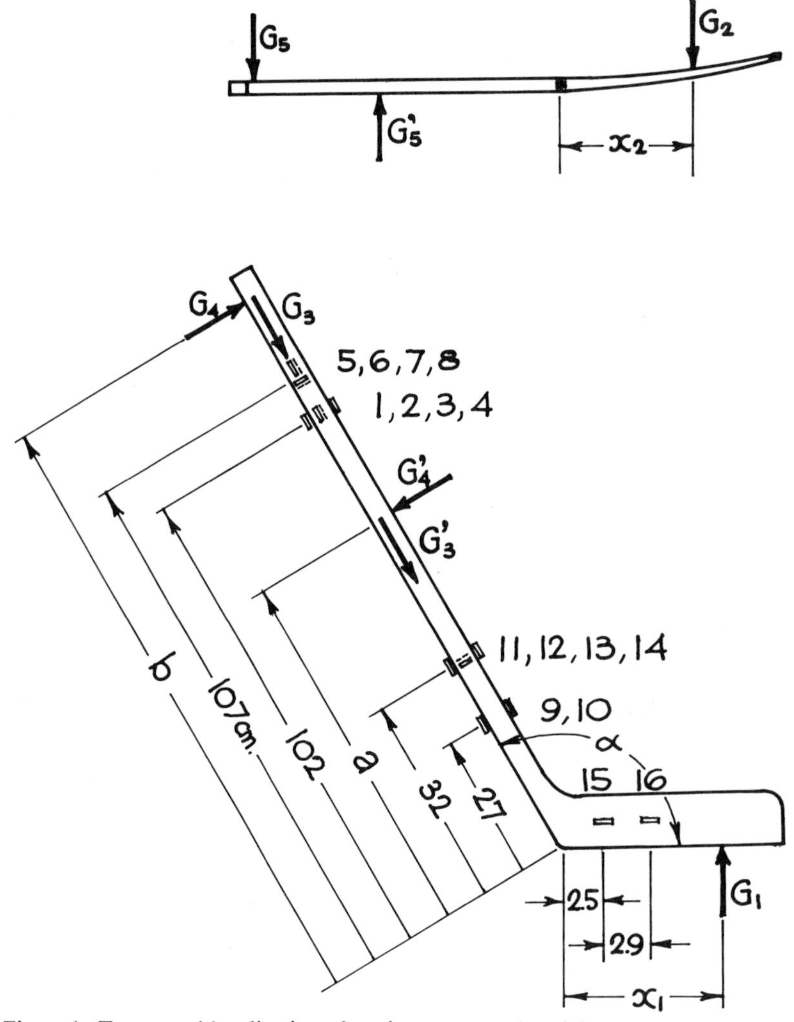

Figure 1. Forces and localization of strain gauges on the stick.

seven bridges and the three equilibrium force equations are, in principle, sufficient to determine the eight forces and the distances X_1, X_2 as shown in Figure 1. G_1 and G_2 represent the reaction of the ice and of the puck on the blade, G_3, G_4, and G_5 are the three components of the action of the upper hand on the handle, and G_3', G_4', and G_5' represent those of the lower hand. The parameters a and b establish the location of the hands and were measured for each subject and for each type of shots. The output signals were amplified and continuously recorded on photosensitive paper (Figure 2). The puck velocity was measured using a digital time counter, triggered and stopped by microphones sensitive to the noise of the impact of the blade on the puck, and the puck on the target.

RESULTS

The puck velocity was measured for slap shots while stationary and in motion. Each subject had five trials of each type of shot. The mean value and the standard deviation for the nine subjects as a group were, respectively, 26.9 and 1.5 m/sec for the stationary slap shot and 29.0 and 1.4 m/sec for the slap shot in motion. The mean values of the velocities for each of the nine subjects are given in Table 1.

This study includes only selected time-dependent force diagrams obtained during this experiment (Figures 3–6). Detailed results are given by Doré and Roy (1975) in the form of a technical report. Forces G_3 and G_3' (Figure 1) could not be obtained accurately. However, the maximum value of these two forces is of the order of 20 kg. It was also difficult to obtain experimentally any meaningful value for the distance X_2. The value of the ice reaction G_1 was computed assuming the distance

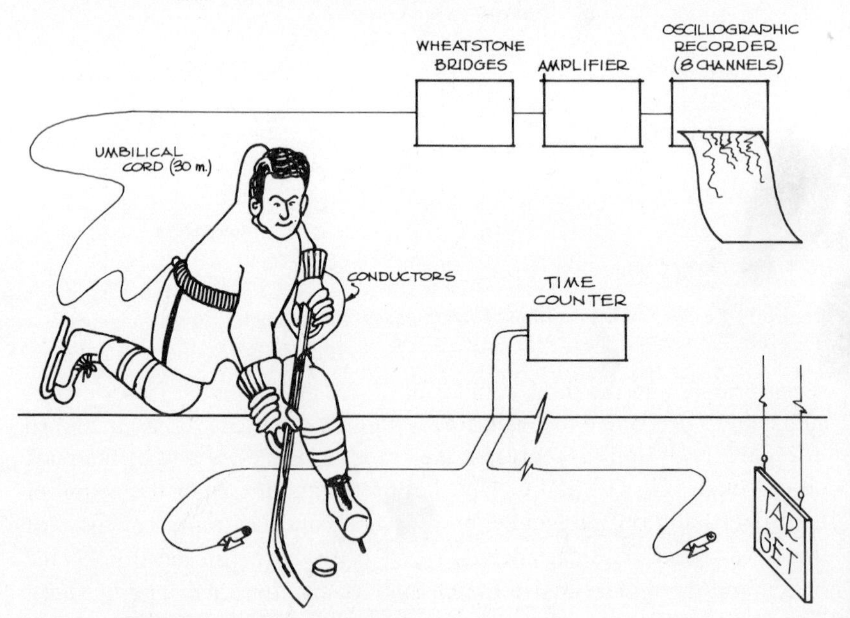

Figure 2. Measuring and recording system.

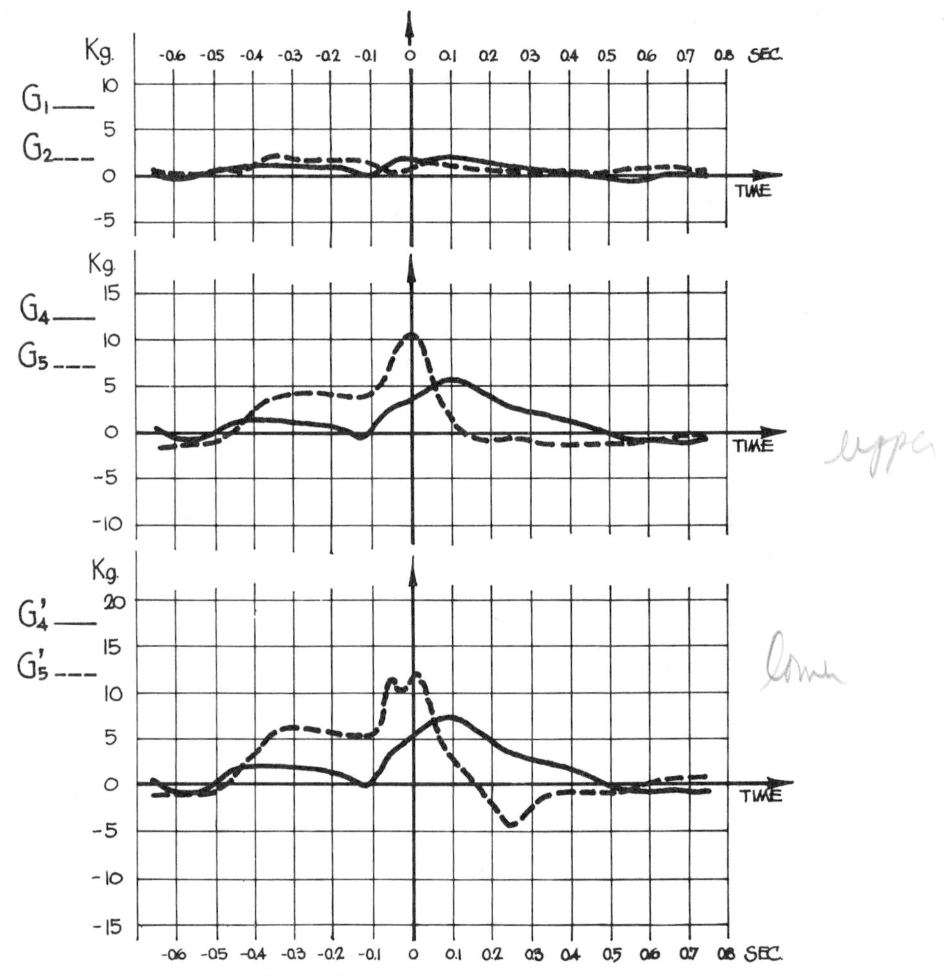

Figure 3. Forces vs time, Subject 1, sweep shot in motion.

X_1 to be 2.5 cm. Of all measured forces, G_5 and G_5' are the most important since they contribute directly to accelerate the puck. For all force diagrams, the time axis has been set such that the peak value of G_5' occurs at 0 sec.

INTERPRETATION OF RESULTS

The analysis of the results revealed that some difference exists in the shape of the force–time diagrams between different type of shots per-

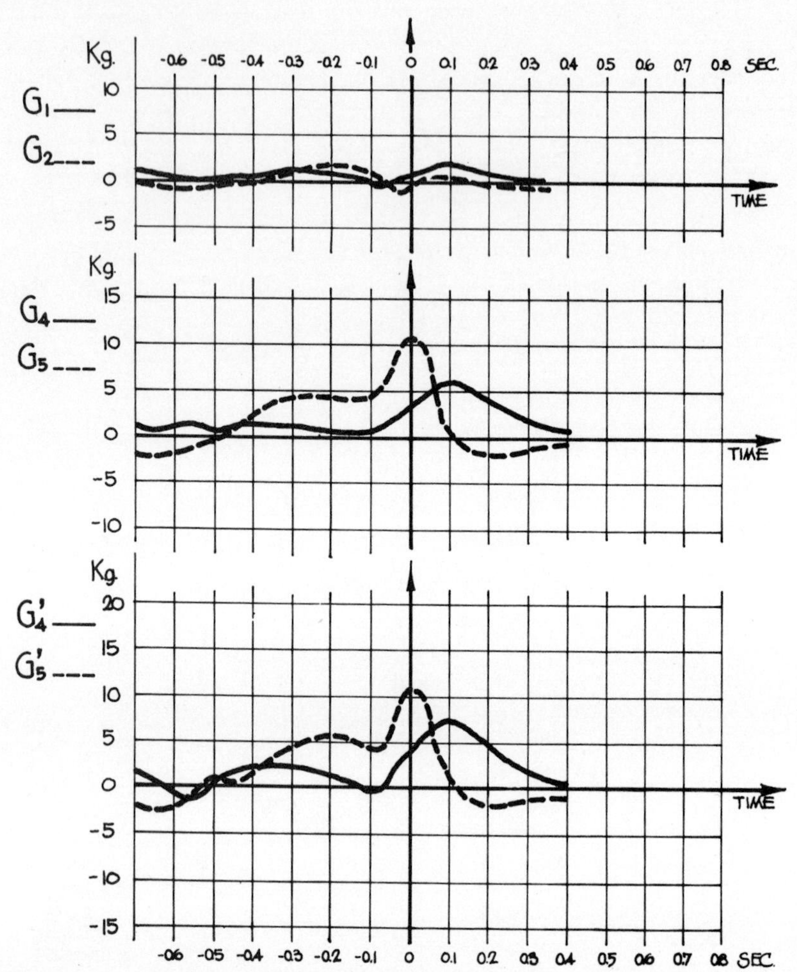

Figure 4. Forces vs time, Subject 1, wrist shot in motion.

formed by the same player. For Subject 1 the maximum value of the most significant force, G_5', was nearly a constant equal to 10 kg except for the stationary slap shot where, oddly enough, this maximum value is somewhat lower. This last result seems to indicate the importance of the impact of the blade on the puck for slap shots, since usually the velocity of the puck for that type of shot is somewhat higher than for the other types. Even though these differences exist between different types of shots performed by the same player, a general pattern can be identified through the various diagrams, making possible the recognition of the "signature" of a player.

However, much dissimilarity exists in the force diagrams of different players performing the sweep shot and the slap shot. Surprisingly, the diagrams were quite repeatable from one player to another for the wrist shot (Doré and Roy, 1975). This similarity could be explained by the fact that the wrist shot is less difficult to perform than the other shots and thus results in a more uniform pattern among all players.

In the case of repeated shots under the same standard conditions by one player, the shape and intensity of the force–time diagrams do not show the similarity one would have expected. Such nonuniformity may

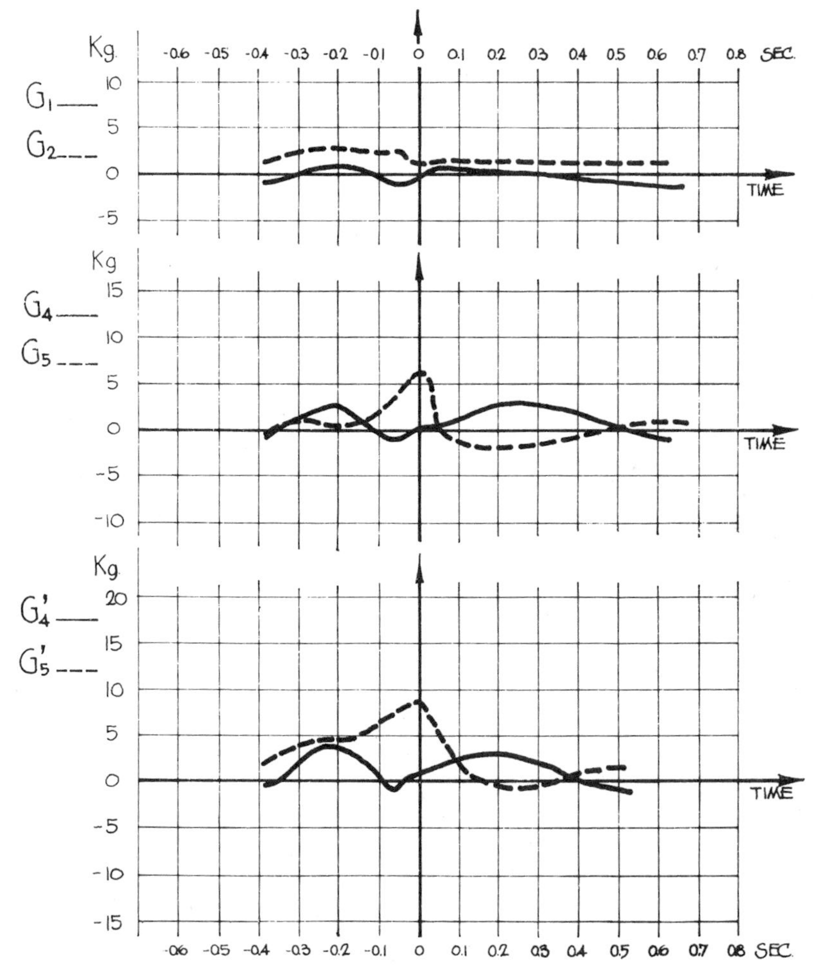

Figure 5. Forces vs time, Subject 1, slap shot while stationary.

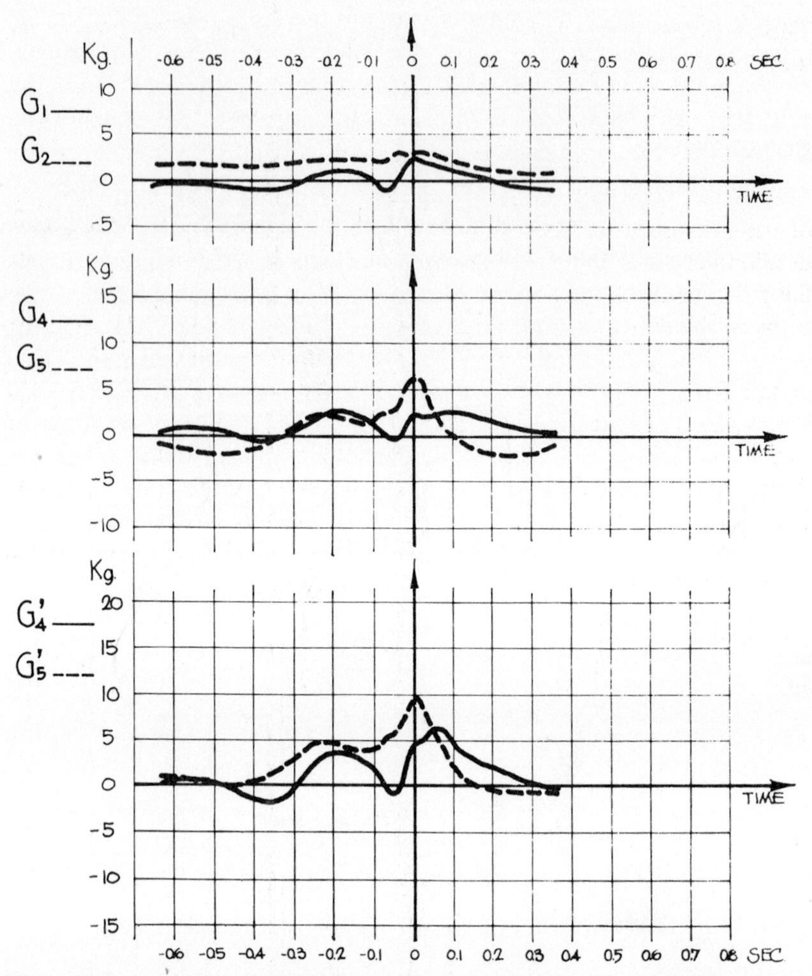

Figure 6. Forces vs time, Subject 1, slap shot in motion.

affect the velocity and the precision of the shots. Again, the pattern was most consistent for the wrist shots.

The puck velocity for slap shots was not directly related to the maximum force a player produced on the stick. Figure 6 shows the maximum value of G_5 and G_5' to be 6 and 10 kg, respectively, while, for slap shots in motion, the mean puck velocity was 26 m/sec for Subject 1 (Table 1). However, these maxima are only 5 and 2 kg for Subject 5, yielding a puck velocity for this subject of 30.8 m/sec (Table 1). Hence, for the slap shot, the puck velocity seems to be highly sensitive to the shape of the diagrams and not so much to the maximum forces produced.

This last observation indicates the importance of the kinematics of the motion performed while shooting.

CONCLUSION

The present investigation has proved the usefulness and the effectiveness of the instrumented hockey sticks for force measurements. The recent development of a system using photoelectric cells to measure accurately the puck velocity for any kind of shot will permit a complete correlation between the dynamometric results and the puck velocities. The coupling of high-speed photography with these methods of measurements will also permit analysis of the influence of the driving motion and of the applied forces on the efficiency of the shots. The use of instrumented sticks of various flexibilities will allow the evaluation of the influence of the characteristics of the stick on the maximum forces applied by the hands and on the puck velocity. Ultimately, these studies will permit a better understanding of the kinematics and kinetics of the different hockey shots in order to develop modern training methods for young hockey players and to design hockey sticks that will be better adapted to the skill and the level of the players and to the type of shots most frequently used.

ACKNOWLEDGMENTS

The authors are grateful to Professor André Bazergui whose advice concerning the arrangement of the Wheatstone bridges and the utilization of the measuring equipment were most valuable and to. Dr. Aouni Lakis for reading and correcting the manuscript.

REFERENCES

Brunelle, R. 1972. Le lancer frappé. (Slap shot). Hockey-Québec. 1: 15–16.

Cotton, C. 1966. Comparison of the ice-hockey wrist, sweep and slap shots for speed. Master's thesis, University of Michigan, Ann Arbor.

Doré, R., and B. Roy. 1975. Results on a kinetic analysis of hockey shots. Technical Report No. EP 75–R–19. Ecole Polytechnique de Montréal, Montréal, Canada.

Hayes, D. 1965. A mechanical analysis of the hockey slap shot. J. Can. Assoc. Health, Phys. Ed. Rec. 31: 17.

Romechevsky, I. 1974. Methodological investigation of the basic techniques of ice hockey (in Russian). Sov. J. Theor. Pract. Phys. Culture. 4.

Roy, B., and R. Doré. 1974. Incidence des caractéristiques des bâtons de hockey sur l'efficacité gestuelle des lancers. (Influence of hockey stick characteristics on the efficiency of shots). Proc. of the 1st Annual Meeting, Can. Soc. for Biomechanics. 1: 1–24.

Kinematics of the slap shot in ice hockey as executed by players of different age classifications

B. Roy
Université Laval, Québec

R. Doré
Ecole Polytechnique, Montréal

In sports in general and in ice hockey in particular, a factor that is frequently disregarded is the optimal adaptation of the equipment to the level of maturity and skill of the individual. In hockey, the ability to shoot the puck with optimal velocity and precision is a decisive factor in the overall performance of a player (Larivière and Lavallée, 1972).

One purpose of this research was to measure the puck velocity in the slap shot, as executed by hockey players of different age classifications. Another objective of this study was to identify some of the morphological, functional, and biomechanical parameters of this skill.

Alexander et al. (1963, 1964), Chao et al. (1973), Cotton (1966), and Furlong (1968) have reported puck velocities for adult hockey players while using the slap shot.

METHODOLOGY

In order to measure the puck velocity, a digital time counter was triggered by a magnetic cell inserted in the ice and stopped by a microphone sensitive to the noise of the puck impact on the target.

The kinematic characteristics of the slap shot (upper torso, shoulder, elbow, and wrist velocity) were obtained from high-speed cinematography.

Standard procedures were used to obtain the following anthropometric measures for each individual: height, weight, and trunk and upper segment lengths. A more complete description of the preceding measures is given in Beaulieu et al. (1974). The subjects of this study were classified into three categories: 11–12-yr-old boys (pee-wee), 15–16-yr-old boys (midget), and 17-yr-old boys and over (adult players).

The cinematographic data were collected from six adult hockey players (undergraduate physical education and junior level players). These players were selected on the basis of their performance in the execution of the slap shot, sweep shot, and wrist shot.

RESULTS AND INTERPRETATION

The puck velocity for the different classifications of players is presented in Table 1. The coefficient of variation was slightly larger for players in the younger category than for the older players.

Figure 1 presents typical curves of the angular velocity of the upper torso and the dominant shoulder, elbow, and wrist joints in the execution of the slap shot. The maximal velocity for the upper torso occurred slightly prior to the impact of the hockey blade with the puck. On the other hand, maximal velocity of the shoulder flexion occurred during the actual impact. This figure shows that these two joints have a relatively greater contribution than the elbow and wrist joints, which were relatively fixed at the time of impact.

A t-test analysis was used to determine if there were statistically significant differences between the three age classifications in terms of puck velocity, morphological, and muscular strength measures. This

Table 1. Mean and standard deviation of puck velocity in the execution of the slap shot for players of different categories

Age classification	Puck velocity (m/sec)	Coefficient of variation (%)
11–12-yr-old boys ($N = 10$)	19.2 ± 2.9	0.15
15–16-yr-old boys ($N = 10$)	26.1 ± 1.6	0.06
17-yr-old boys and over ($N = 19$)	26.7 ± 1.7	0.06

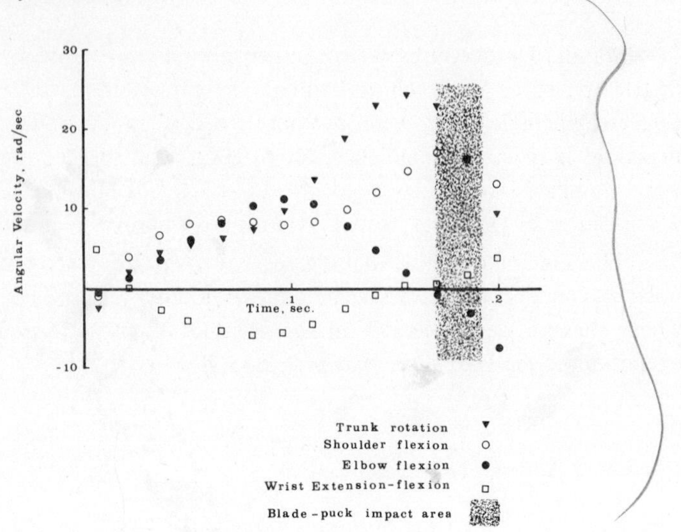

Figure 1. Typical velocity curves of the trunk, shoulder, elbow, and wrist joints. Positive values, trunk rotation, shoulder, elbow, wrist flexion; negative values, elbow and wrist extension.

analysis revealed that the 11–12-yr-old boys were significantly different ($p \leq 0.10$) from the two other classifications in all morphological and strength measures as well as in puck velocity. On the other hand, the 15–16-yr-old boys were not significantly different from the adult group on the majority of these measures.

On the other hand, the coefficients of correlation between puck velocity and morphological and muscular strength measures were higher for the younger players than for the two older groups (Table 2). The dominant arm in Table 2 refers to the segment holding the stick closer to the blade, and the nondominant, to the one located at the upper end of the stick.

In the light of these results, it seems that the younger players must rely more on their morphological and strength attributes than the older players to achieve relatively the same skill. Generally, the younger players use the same type of stick as the older ones. In so doing, they put themselves at a disadvantage especially as far as the weight and flexibility of the stick are concerned. This study would therefore confirm the hypothesis suggesting that more flexible hockey sticks should be used by the less mature individuals (Roy and Doré, 1974). Some preliminary results seem to indicate that among the younger players, the use of a more flexible hockey stick, rather than a more rigid one, is conducive to generally higher puck velocity.

Table 2. Coefficients of correlation between puck velocity versus morphological and isometric strength measures for the different categories

	Dominant arm								Nondominant arm				
Age classification	Height	Weight	Trunk length	Arm length	Forearm length	Grip	Arm adduction	Wrist flexion	Arm length	Forearm length	Grip	Arm adduction	Wrist flexion
11–12 yr old ($N = 10$)	0.92	0.79	0.82	0.79	0.86	0.78	0.60	0.23	0.89	0.86	0.82	0.56	0.40
15–16 yr old ($N = 10$)	0.07	0.28	0.43	0.33	0.16	0.13	0.61	0.36	0.37	0.19	0.23	0.32	0.52
Adult players ($N = 10$)	0.47	0.33	0.50	0.00	−0.22	0.19	−0.14	0.14	−0.01	−0.17	0.10	−0.10	0.10

CONCLUSION

This study has shown that there are specific characteristics in the execution of the slap shot in ice hockey by players of different age classifications.

In the light of the results one could argue that not only the protective equipment (e.g. face mask, and helmet) should be geared to the capabilities of the players but also the implements (i.e. hockey stick) required for the execution of the different skills in the sport.

There are already strong indications from preliminary work in this field that the static and dynamic characteristics of the hockey stick are important factors to take into account in the execution of this skill. Some work in this direction is already in progress.

REFERENCES

Alexander, J. F., J. B. Haddow, and G. A. Schultz. 1963. Comparison of the ice hockey wrist and slap shots for speed and accuracy. Res. Quart. 34: 259–266.

Alexander, J. F., C. J. Drake, P. J. Reichenbach, and J. B. Haddow. 1964. Effect of strength development on speed of shooting of varsity ice hockey players. Res. Quart. 35: 101–106.

Beaulieu, R., S. Cloutier, A. Juneau, and R. Kirallah. 1974. L'influence de la force musculaire et de la longueur segmentaire sur la vitesse du lancer frappé au hockey. (Influence of muscular strength and segmental length on the slap shot). Département d'Education Physique, Université Laval, Québec, Canada (mimeographed).

Chao, E. G., F. H. Sim, R. N. Stauffer, and K. G. Johnson. 1973. Mechanics of ice hockey injuries. Mechanics and Sport, pp. 143–154. American Society of Mechanical Engineers.

Cotton, C. 1966. Comparison of the ice hockey wrist, sweep and slap shots for speed. Master's thesis, University of Michigan, Ann Arbor.

Furlong, W. B. 1968. How science is changing hockey: 80 mph mayhem on ice. Popular Mechanics. February, 110–114.

Larivière, G., and H. Lavallée. 1972. Evaluation du niveau technique de joueurs de hockey de catégorie moustique (Technical evaluation of young hockey players). Mouvement 7: 101–111.

Roy, B., and R. Doré. 1974. Facteurs biomécaniques caractéristiques des différents types de lancers au hockey sur glace (Biomechanical factors of the different types of shots in ice hockey). Mouvement 9: 169–175.

Running economy in long-distance speed skating

A. Kuhlow
Johann Wolfgang Goethe-Universität, Frankfurt am Main

In the present study an attempt was made first to quantify and second to compare running economy of "good" and "poor" skaters performing in the 3000 m event. There are various popular methods employed to evaluate economy of movement, defined as the relationship between energy expenditure and work-output, e.g. direct and indirect measurement of caloric expenditure, heart rate, and pulse rate measurements as well as action potential measurements, etc., but none proved to be practical for competitive speed skating. To evaluate running economy in long-distance speed skating we applied a concept suggested by Michailow (1972) for track and field running events that is based on the following hypothesis: the less the deviation of mean speeds (M) within specified track sections from mean speed ($\overline{M}$) throughout the total distance, the more economic the running movement. In accordance with the above definition, running economy was calculated by means of a mathematical quantity, the coefficient of variation (V), which in this special case may be considered the coefficient of economy.

METHODS

As illustrated in Figure 1 mean speed was recorded during the competitive 3000 m running events by 20 photoelectric cells (PEC) installed at different intervals around the 400-m speed skating track. Altogether, there was a total of 20 track sections of different lengths: four sections of 28.50 m within the straightaways and six sections within either curve which ranged from 15.44 m (outside lane) to 13.34 m (inner lane). When the skater's ankle passed the PEC the time elapsed from the start-

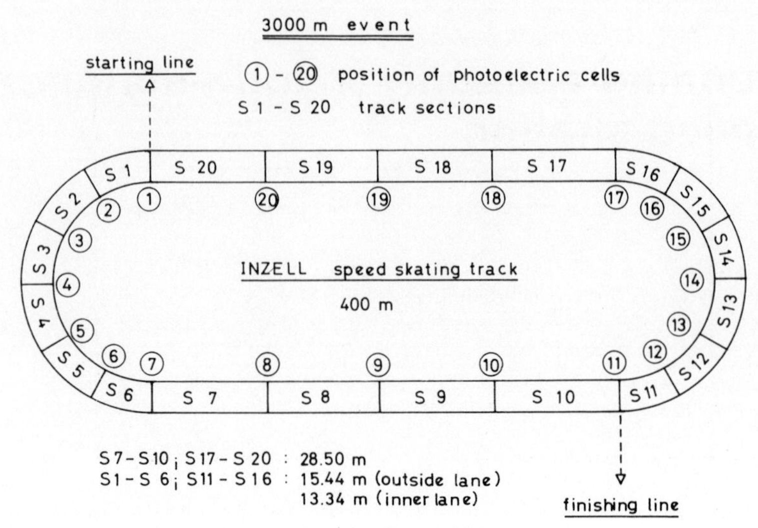

Figure 1. Speed skating track with photoelectric cells.

ing line to the specified position of the PEC was recorded. The electrical impulses from the starting gun and the finish line were transmitted from the official chronograph to the recorder (Kuhlow and Gottschalk, 1974). Measurement error was limited even throughout the shortest track section to an acceptable level of precision (0.05 m/sec).

Twelve male skaters, participants in the Bavarian Championships held in Inzell, West Germany in December, 1972, served as subjects. Their competitive performance times ranged from 4 min 51.71 sec to 5 min 33.56 sec. For comparison of their running economy our sample was classified according to the time for the 3000 m event into two groups of six skaters each: (1) "good" skaters (G_1), 4 min 51.71 sec–5 min 04.22 sec; (2) "poor" skaters (G_2), 5 min 17.20 sec–5 min 33.56 sec. Running economy was presumed to be different between good and poor skaters and therefore was calculated throughout three divisions of track sections (see Figures 2 *a*, *b*, and *c*) as follows:

1. Within each of the seven rounds (R_1—R_7) following the first 200 m after the starting line; limited by PEC 11
2. Within each of the 28 quarter-rounds (QR_1—QR_{28}); first QR limited by PEC 4 and 9, second QR limited by PEC 9 and 14, third QR limited by PEC 14 and 19, fourth QR limited by PEC 19 and 4, etc.
3. Within each of the seven (back-) straightaways (ST_1—ST_7); limited by PEC 17 and 1

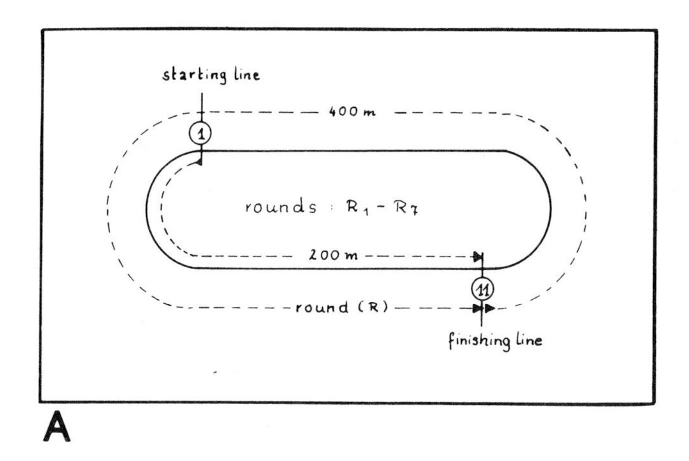

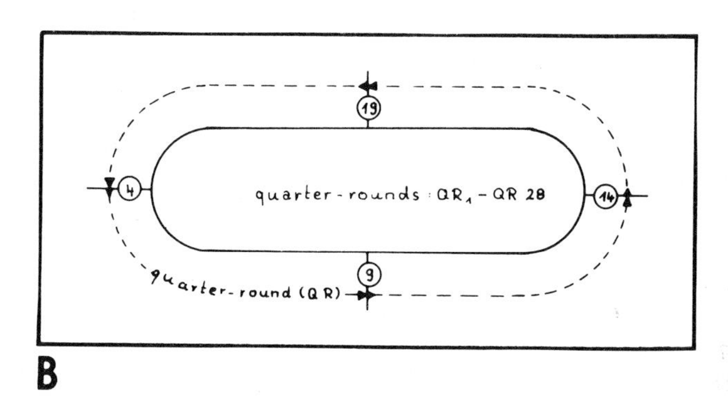

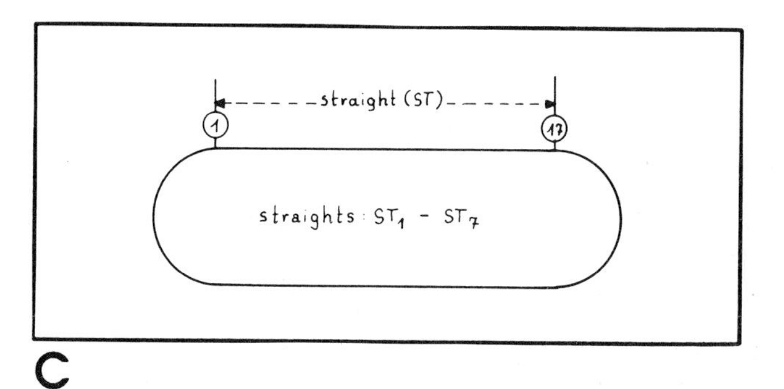

Figure 2. *a*, Track section "rounds"; *b*, track section "quarter-rounds"; *c*, track section "straightaways."

Using these three track sections as criteria of running economy in the 3000 m event the data were subjected to ANOVA to ascertain whether or not there was a significant difference between good and poor skaters.

RESULTS

An examination of the data in Table 1 reveals significant differences in running economy in favor of good skaters: significant differences at the 1 percent level were noted for the criterion running economy within straightaways whereas differences on the 5 percent level were disclosed for the criteria running economy within rounds as well as running economy within quarter-rounds. Figures 3, 4, and 5 clearly demonstrate the observed characteristics.

First Criterion: Running Economy within Rounds (R_1–R_7)

As illustrated in Figure 3 mean speeds (M) of good skaters within the track sections R_1–R_7 show only slight deviation from mean speed ($\overline{M}$) throughout the total distance of seven rounds. This finding indicates that good skaters run with almost constant speed. After a rapid increase of mean speed during the first 200 m following the start the speed reduced only slightly below $\overline{M}$ in the following rounds. Peak mean speed was also reached by the poor group at the end of the first 200 m. Contrary to good skaters this high mean speed during the starting phase was followed by a considerable decrease in mean speed to a value far below $\overline{M}$.

Second Criterion: Running Economy within Quarter-rounds (QR_1–QR_{28})

Figure 4 shows that mean speed ($\overline{M}$) within the track sections quarter-rounds was very similar to the criterion rounds. Moreover, good skaters increased their mean speeds whenever turning from curves to straightaways beginning with the starting section up to the finishing line, whereas poor skaters maintained this change in speed characteristic only until the fifth round.

Third Criterion Running Economy within (Back-) Straightaways (ST_1–ST_7)

As can be seen from Figure 5 mean speeds (M) of good and poor skaters within the track sections (back-) straightaways confirm the findings of the criteria, "rounds" and quarter-rounds. Deviation of mean speeds (M) within straightaways 1–7 from $\overline{M}$ was less for good skaters

Table 1. Mean velocities of groups of long distance speed skaters for rounds, quarter rounds, and straightaways

Criterion track sections	Groups	M_{max} (m/sec)	M_{min} (m/sec)	M_{max}–M_{min} (m/sec)	$\overline{M}$ (m/sec)	Max. deviation from $\overline{M}$ (m/sec)	$\overline{V}$	α
Rounds	G_1	10.4	9.9	0.5	10.0	0.4	0.04	0.05
	G_2	10.2	8.5	1.7	9.3	0.9	0.06	
Quarter-rounds	G_1	10.6	9.8	0.8	10.2	0.4	0.03	0.05
	G_2	10.3	9.0	1.3	9.4	0.9	0.06	
Straight aways	G_1	10.5	9.9	0.6	10.1	0.4	0.02	0.01
	G_2	10.1	8.3	1.8	9.3	1.0	0.06	

G_1, "good" skaters (4 min 51.71 sec–5 min 04.22 sec).
G_2, "poor" skaters (5 min 17.20 sec–5 min 33.56 sec).
α, level of significance (Wilcoxon respectively Aspin Welch test).
$\overline{V}$, arithmetic mean of the coefficient of variation (V) within specified track sections.
M, arithmetic mean of mean speeds within specified track sections.
$\overline{M}$, arithmetic mean of group speed for total distance.

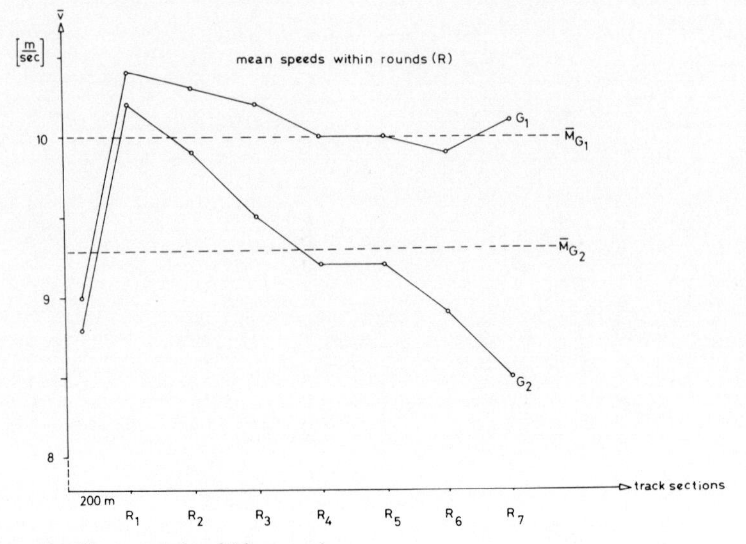

Figure 3. Mean speeds within rounds.

because their mean speeds remained almost constant with even slight increase on the last straightaway. Conversely, poor skaters showed continual decrease of mean speeds (M) from straightaway to straightaway at varying rates with maximum decrease on the last straightaway.

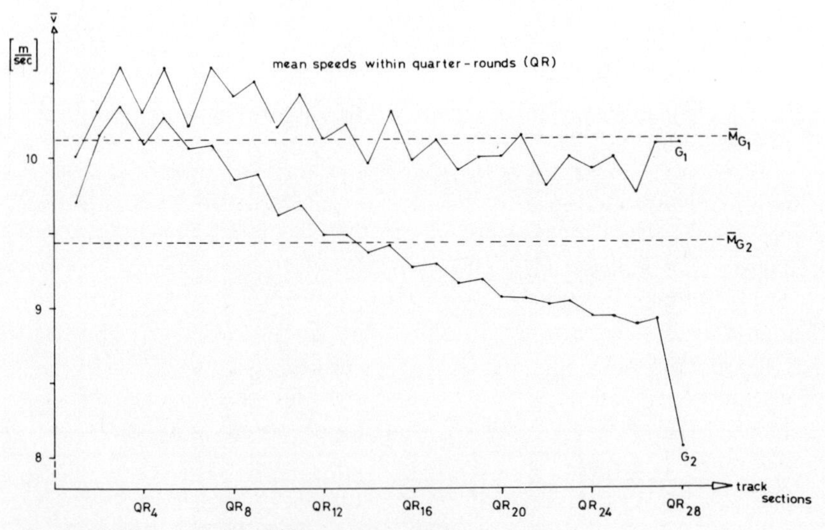

Figure 4. Mean speeds within quarter-rounds.

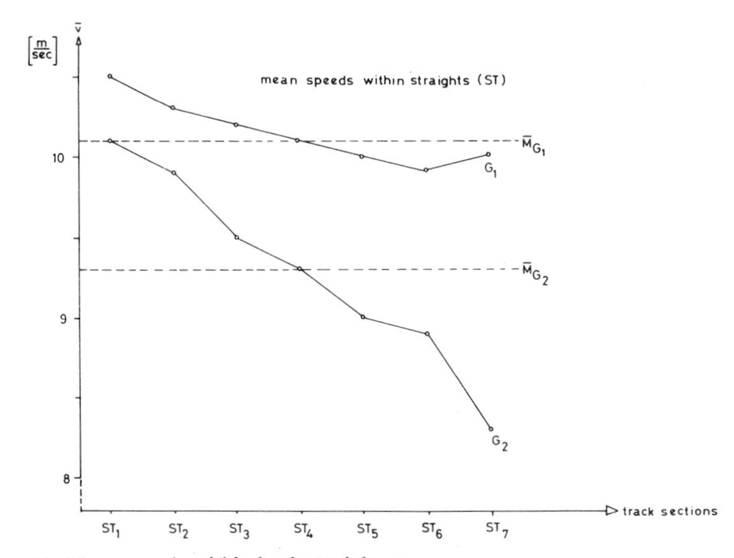

Figure 5. Mean speeds within back straightaways.

DISCUSSION AND CONCLUSIONS

Running with almost constant speed is an essential qualification for successful speed skating. This hypothesis proposed by Michailow (1965) and other experts for track and field running events seem to be verified by the results of our study on the 3000 m running event in speed skating. If we can presume that champion performers have achieved their constant running speed throughout the entire contest by an intensive training program over a couple of years and by competitive experience, they support the theory that constant distribution of mean speed is physiologically most economical.

Taking into account the results of the present study, some recommendations can be given for optimizing running economy in long-distance speed skating by developing economical distribution of energy based upon judgment of pace.

First, provide stationary time tables at different track sections which simultaneously with the running event show the skater's elapsed times at specified points in the race. This information may be advantageous for the athlete setting out to cover the 3000 m in whatever time to run at his known optimum distribution of speed and thus more economically.

Second, keep up with a pacemaker running at the skater's mean speed already achieved by him in competitive or training races.

Third, work on an ergometer; although work output of speed skating movements cannot be measured by means of traditional ergometers, they nevertheless may help the athlete to develop his judgment of constant pace.

REFERENCES

Kuhlow, A., and H. W. Gottschalk. 1974. Zur Bewegungsökonomie bei Ausdauerleistungen im Eisschnellaufen (Economy of movement in long distance speed skating). Praxis der Leibesübungen 3: 51–53.

Michailow, W. M. 1965. Varianten der Kraftverteilung bei Eisschnelläufern (Distribution of energy in speed skating). Theorie und Praxis der Körperkultur 3: 212–220.

Michailow, W. M. 1972. Die Effektivität des Tempowechsels beim leichtathletischen Lauf auf Wettkampfstrecken (The effectiveness of change of speed in competitive track and field running events). Theorie und Praxis der Körperkultur 11: 1013–1017.

Dynamic response criteria for ice hockey helmet design

P. J. Bishop
University of Waterloo, Waterloo

In order to generate design criteria for head protection against cerebral concussion in ice hockey, a mathematical model of an ice hockey helmet under a sideboard collision was developed. Details concerning the assumptions and the behavior of the model (Figure 1) are presented elsewhere (Bishop, 1975). The equation of motion governing this collision was shown by Bishop (1975) to be:

$$\ddot{X} + \frac{cl_2 \sin_2 \theta_0}{I_o} \dot{X} + \frac{kl^2 \sin^2 \theta_o}{I_o} X = 0 \qquad (1)$$

A soluton for 1 was defined for the underdamped case (Bishop, 1975) as

$$X = \frac{\dot{X}_0 e^{-\xi \omega n t}}{\omega n \sqrt{1 - \xi^2}} \sin (\sqrt{1 - \xi^2}\, \omega n t) \qquad (2)$$

Equation (2) now represents the distance through which the helmet liner would deform upon contact with the boards in terms of the initial velocity of contact ($\dot{X}_o$), the angle of head strike (θ_o), the spring constant for the liner material (k), the fraction of critical damping (ξ), the length of the head and neck (l), the moment of inertia of the head (I_0), the duration of the collision (t), and natural frequency of the head (ωn).

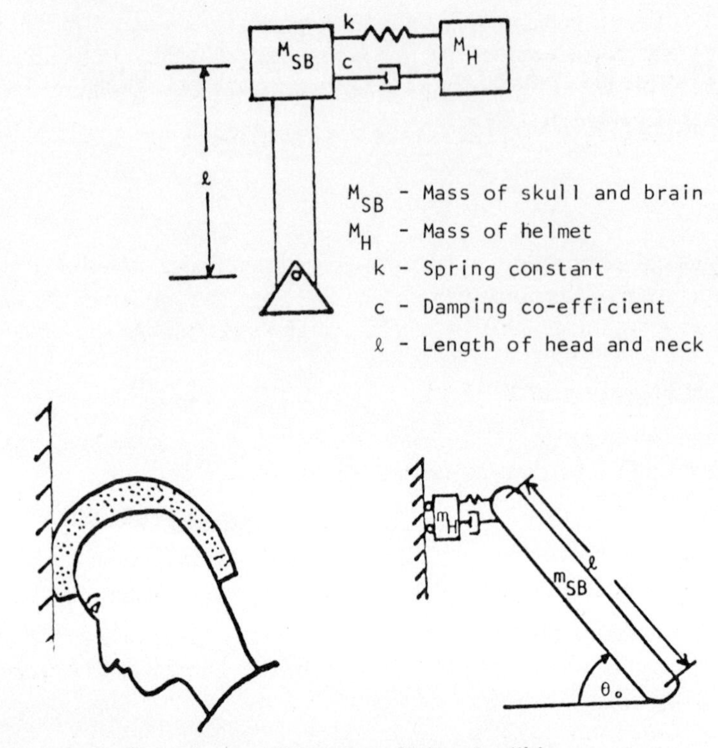

Figure 1. Ice hockey helmet model under a sideboard collision.

PROCEDURES

Suitable design criteria, based upon a solution to equation (2), depend upon reasonable estimates of the input conditions.

A series of static load deflection and dynamic impact tests, conducted on six material specimens used as helmet liners and on two production model ice hockey helmets, revealed that such materials exhibit nonlinear behavior. Values for k_s (static) and k_d (dynamic) were estimated by linearization. k_d was always greater than k_s with a range of 1,050 N/cm $\leq k_d <$ 3,850 N/cm (600 lb/in. $\leq k_d \leq$ 2,200 lb/in.) (Bishop, 1975).

Mindlin (1945) demonstrated that the maximum acceleration in a mass, spring, and damper system occurs after time $t = 0$ for $\xi \leq 0.5$. Since examination of the deceleration-time traces obtained from the material and helmet impact tests indicated that the maximum deceleration occurred after time $t = 0$, values were considered for $\xi \leq 0.5$.

For the purposes of this study an impact velocity of 6.1 m/sec (20 ft/sec) was considered as the maximum against which protection was desired. This is within the range of forward skating and sliding speeds found by Chao et. al. (1973).

RESULTS

For this collision condition, the design characteristics were based on a consideration of the maximum deflection of the liner material, the maximum deceleration of the head, and the total collision time.

Natural Frequency of the Head

The natural frequency of the head, ωn, as it acts as a rigid body against the liner and helmet, was calculated from

$$\omega n = \sqrt{\frac{k_d l^2 \sin^2 \theta_o}{I_o}}$$

I_o was determined by considering the head to be a sphere of radius 8.9 cm (3.5 in.) and weighing 4.5 kg (10 lb). The overall head length (including the neck) was taken as $l = 25.4$ cm (10 in.) and I_o was calculated from the shoulders.

Values for ωn, at $\theta_0 = 65°$, ranged from 32 Hz at $k_d = 1{,}050$ N/cm to 64 Hz at $k_d = 4{,}200$ N/cm. Since Gurdjian, Hodgson, and Thomas (1970) indicated that the head undergoes rigid body acceleration in the frequency range below 150 Hz, the head was considered to undergo rigid body acceleration under this collision condition.

Maximum Head Displacement

To determine the thickness of the padding material required in a helmet, equation (2) was evaluated for the maximum displacement the head would undergo in coming to rest after the collision.

The relationship between the maximum head displacement at impact and the spring constant (k_d) for each value of ξ is presented in Table 1. The results demonstrate that the softer the cushioning material (low k_d) the greater the maximum displacement required to bring the head to rest. At $\xi = 0.25$ ($\dot{X}_0 = 6.1$ m/sec), the distance required to stop the head is 1.52 cm for $k_d = 2{,}100$ N/cm and 2.13 cm for $k_d = 1{,}050$ N/cm. If these materials could deflect to only 70 percent of their original thickness without bottoming, the thickness of the liner material for the above situations would be 2.17 cm ($k_d = 2{,}100$ N/cm) and 3.04

Table 1. Maximum head displacement (cm) and maximum head deceleration (g) for several values of ξ, ωn and k_d at impact velocity of 6.1 m/sec and impact angle of 65°

k_d	ωn	ξ = 0.05		0.15		0.25		0.35		0.45	
(n/cm)	(Hz)	X	$\ddot{X}$	X	$\ddot{X}$	X	$\ddot{X}$	X	$\ddot{X}$	X	$\ddot{X}$
1050	32	2.79	116	2.44	100	2.13	88	1.93	80	1.73	72
1400	37	2.44	134	2.13	116	1.88	102	1.63	92	1.52	82
1750	41	2.25	150	1.98	130	1.63	114	1.52	102	1.32	92
2100	45	1.98	164	1.73	142	1.52	126	1.32	112	1.22	100
2450	49	1.83	178	1.57	154	1.42	136	1.27	122	1.12	108
2800	52	1.73	190	1.52	164	1.32	144	1.12	130	1.02	116
3150	55	1.63	202	1.42	174	1.22	154	1.12	138	0.97	124
3500	58	1.52	212	1.32	184	1.17	162	1.02	146	0.91	130
3850	61	1.47	222	1.27	192	1.12	170	1.02	152	0.91	136
4200	64	1.42	232	1.22	200	1.07	176	0.97	158	0.86	142

X measured in cm. $\ddot{X}$ measured in g.

cm ($k_d = 1{,}050$ N/cm), respectively. As ξ is increased to 0.45, and under the same initial velocity and bottoming conditions as stated, the thickness of padding is reduced to 1.74 cm ($k_d = 2{,}100$ N/cm) and to 2.50 cm ($k_d = 1{,}050$ N/cm).

Maximum Head Acceleration

Two of the factors associated with cerebral concussion are the linear acceleration experienced by the head during impact and the time over which the acceleration acts (Goldsmith, 1966; Gurdjian, Roberts, and Thomas, 1966).

From equation (2)

$$\ddot{X} = -\dot{X}_0 e^{-\xi \omega n t} \omega n$$

$$\left[2\xi \cos(\sqrt{1-\xi^2}\,\omega n t) + \sin(\sqrt{1-\xi^2}\,\omega n t) \left(\frac{1-2\xi^2}{\sqrt{1-\xi^2}} \right) \right] \quad (3)$$

Equation (3) was then evaluated for the relationship between the deceleration of the head during impact and the dynamic spring constant (k_d) for each value of ξ considered. Examination of Table 1 indicates that the deceleration of the head increases with increasing stiffness of the liner material and decreases with increasing values of ξ. Again, at $\xi = 0.25$ ($\dot{X}_0 = 6.1$ m/sec) the maximum head deceleration expected is 88 g for $k_d = 1{,}050$ N/cm and is 126 g for $k_d = 2{,}100$ N/cm. As ξ is increased to 0.45, and under the same impact conditions as above, the level of deceleration is reduced to 72 g ($k_d = 1{,}050$ N/cm) and 100 g ($k_d = 2{,}100$ N/cm).

Since the time over which the deceleration acts is a critical factor associated with cerebral concussion, a determination of the total time of the collision was made by considering the time required for the force to become zero, i.e. by finding the time for which

$$F = c\dot{X} + kX = 0$$

The time durations ranged from 4.2 msec ($\xi = 0.05$ and $\omega n = 64$ Hz) to 14 msec ($\xi = 0.45$ and $\omega n = 32$ Hz).

DISCUSSION

If the tolerable level of head acceleration were known, the choice of the proper padding material could be made by selecting one that would maintain head acceleration below some critical level at impact velocities which are likely to be encountered in ice hockey.

Gurdjian, Lissner, and Patrick (1962), Gurdjian et al. (1964), Gurdjian, Roberts, and Thomas (1966), and Hirsch (1966) have all estimated that high peak head accelerations could be tolerated for short time periods (≤ 5 msec) while peak accelerations of 50 *g* were tolerated for up to 40 msec. It was further estimated that the protected head could tolerate 80 *g* for 20 msec (Gurdjian et al., 1964) and data from football studies indicated peak accelerations of 40–230 *g* lasting 20–420 msec without injury (Reid et al., 1971). Hirsch (1966) estimated that for an impact to the unprotected head lasting 5–20 msec, the tolerable level of head acceleration ranges from 50 *g* to 100 *g*. Since the time durations calculated for the collision model fall within this range, the maximum tolerable level of head deceleration was considered to be 100 *g*.

From Table 1, in order to maintain the maximum level of head deceleration at or below 100 *g* for an initial impact velocity of 6.1 m/sec, soft liner materials (1,050 N/cm $\leq k_d \leq$ 2,100 N/cm) that afford a high fraction of critical damping are required. The effectiveness of these soft liner materials must be weighed, however, against the thickness criterion of the materials to prevent helmet bottoming. If 70 percent is taken as the allowable deflection rate before bottoming, and if 2.5 cm is used as the maximum allowable thickness of the liner material, then the maximum displacement permitted to prevent bottoming is 1.75 cm. At $\xi = 0.25$ ($\dot{X}_0 = 6.1$ m/sec), the softest liner material yielding a deflection of 1.75 cm has a $k_d = 1{,}750$ N/cm. This material will yield a head deceleration of about 126 *g*, slightly above the considered tolerable level. At $\xi = 0.45$, however, and under the same conditions as above, the softest material has a $k_d = 1{,}050$ N/cm, which yields a head deceleration of about 70 *g*, well below the considered tolerable level. If the critical impact velocity against which protection is desired is increased to, say, 9 m/sec, the requirements discussed above will also change.

CONCLUSION

In order to effectively provide protection for the head (in the form of prevention of cerebral concussion) in ice hockey at maximal initial impact conditions of 6.1 m/sec, the helmet liner should provide a reasonably high value of critical damping ($0.15 \leq \xi \leq 0.45$), should be approximately 2.5 cm thick, and should have a dynamic spring constant of 1,050 N/cm $\leq k_d \leq$ 2,100 N/cm.

REFERENCES

Bishop, P. J. 1975. Head protection in sport with particular application to ice hockey. Ergonomics. (In press).

Chao, E. Y., F. H. Sim, R. N. Stauffer, and K. J. Johnson. 1973. Mechanics of ice hockey injuries. *In:* J. L. Bleustein (ed.), Mechanics and Sport. Proceedings of the ASME Winter Annual Meeting, Detroit, Michigan.

Goldsmith, W. 1966. The physical process producing head injuries. *In:* W. F. Caveness and A. E. Walker (eds.), Head Injury: Conference Proceedings, pp. 350–382. J. B. Lippincott, Philadelphia.

Gurdjian, E. S., H. R. Lissner, and L. M. Patrick. 1962. Protection of the head and neck in sports. J.A.M.A. 182: 509–512.

Gurdjian, E. S., V. R. Hodgson, W. G. Hardy, L. M. Patrick, and H. R. Lissner. 1964. Evaluation of the protective characteristics of helmets in sports. J. Trauma 4: 309–324.

Gurdjian, E.S., V. L. Roberts, and L. M. Thomas. 1966. Tolerance curves of acceleration and intracranial pressure and protective index in experimental head injury. J. Trauma 6: 600–604.

Gurdjian, E. S., V. R. Hodgson, and L. M. Thomas. 1970. Studies on the mechanical impedance of the human skull: Preliminary report. J. Biomech. 3: 239–247.

Hirsch, A. H. 1966. Current problems in head protection. *In:* W. F. Caveness and A. E. Walker (eds.), Head Injury: Conference Proceedings, pp. 37–40. J. B. Lippincott, Philadelphia.

Mindlin, R. D. 1945. Dynamics of package cushioning. Bell System Tech. J. 24: 353–409.

Reid, E. E., J. S. Tarkington, H. M. Epstein, and T. J. O'Dea. 1971. Brain tolerance to impact in football. Surg. Gyn. Obs. 133: 929–936.

Other sports

Biomechanical study of forward and backward swings

J. Borms, R. Moers, and M. Hebbelinck
Vrije Universiteit Brussel, Brussels

The teaching process of a forward or backward giant swing on the high bar requires from young gymnasts the acquisition of a certain basic motoric and physical potential. Besides such qualities as strength and flexibility, the primary sensorimotor schemes have to be well developed. At each moment during a forward or backward rotation, the performer should be able to locate his body in space. Beginners, especially, observe the surrounding space as moving. Through training, however, they learn that this space remains immobile.

Another prerequisite of a correctly performed giant swing is that, from the beginning, the teaching process should eliminate errors occurring in each learning process; often these are of a different nature for different performers. The purpose of this study was to investigate certain biomechanical characteristics and specific aspects of forward and backward giant swings in order to reduce learning time, to establish techniques, and generally to contribute to our knowledge of these movements.

METHODS

A top-class gymnast was used as the subject in this study. He was asked to perform forward and backward giant swings in a most esthetic way and with a maximum of body extension.

The following anthropometric reference points were marked on the gymnast: ankle point, trochanteric point, and shoulder point.

Cinematographic data were taken with a 16mm, Arriflex camera, positioned perpendicular to the plane of motion, i.e., in the axis of the high bar.

The film images were projected onto a table screen, where they were analyzed through the use of a computerized electronics graphics system that fed the required coordinate points via an interface into a minicomputer.

The coordinate points of each reference mark were taken with respect to an origin, coinciding with the axis of the high bar.

For each of the reference points, the following parameters were calculated: time (t), the life-size coordinates (X, Y), segmental angular velocities and accelerations, and the body's center of gravity.

RESULTS AND DISCUSSION

All movements started from the handstand position. This position was defined as 90 degrees where time (t) equalled zero. For uniformity and to avoid confusion, all movements were analyzed in a counterclockwise direction.

Angular changes of arm–trunk and legs–trunk during the backward giant swing are shown in Figure 1.

The trunk-arm angle remained positive throughout the movement, i.e., the shoulders were never hyperextended. The maximal hyperextension of the hip angle occurred at the point 0.83 sec. Between 0.83 sec and 1.09 sec, the legs were preparing for the hip swing. The legs–trunk angle was altered from hyperextension into pronounced flexion. The angle between arm and trunk decreased, resulting in maximal flexion in the shoulder at 1.29 sec.

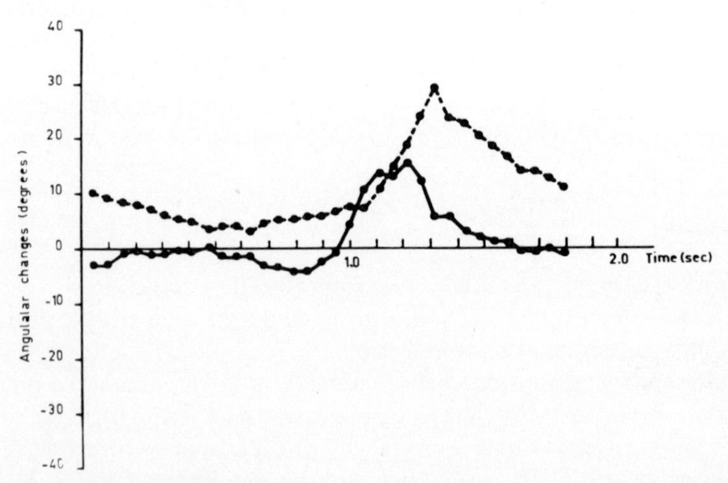

Figure 1. Angular changes of arm-trunk (*dotted line*) and legs-trunk (*solid line*) during the backward giant swing.

In Figure 2, angular changes of arm–trunk and legs–trunk during the forward giant swing are shown. In the hip a flexion-extension was observed first. By transfer of impulse the same flexion-extension action occurred for the shoulder–trunk action.

Figures 3 and 4 show the angular velocity during a backward giant swing of three body reference points: ankle point (upper), body center of gravity (middle), and shoulder point (lower).

The velocity curves during the backward swing in the first quadrant had a very regular course. In the second quadrant of the movement, both shoulders and the body center of gravity reached their maximal velocity after 0.88 sec. These velocities were 6 and 3.5 m/sec, respectively. The velocity of the feet was increasing steadily and reached its maximum (12.5 m/sec) in the third quadrant after 1.03 sec. This coincided with the flexion in the hip joints. Correspondingly, the velocities of the body center of gravity and shoulder decreased to a minimal value at 1.24 sec. These velocities were 4.7 and 1.9 m/sec, respectively.

In the fourth quadrant, the velocity of all reference points decreased continuously, but there remained sufficient velocity to reach the top of the swing and to start a new giant swing.

The fact that there was no overlapping of velocities of the body center of gravity and ankles confirmed that the movement was performed with maximal extension.

In the forward giant swing, the velocity curves of the three reference marks showed a more or less similar pattern in the first and second quad-

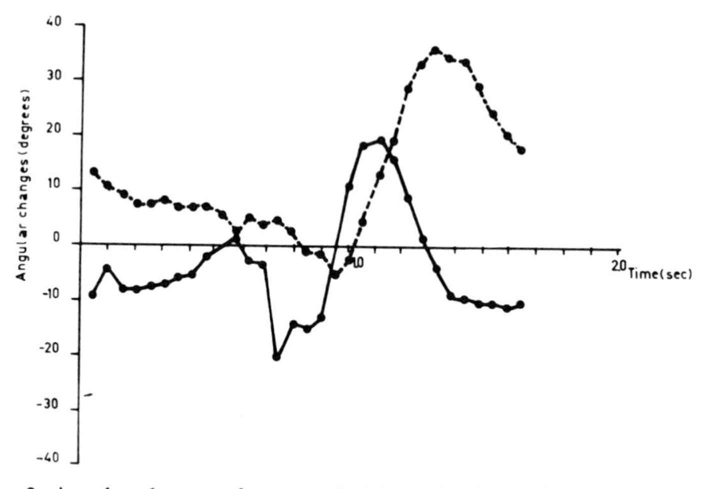

Figure 2. Angular changes of arm-trunk (*dotted line*) and legs-trunk (*solid line*) during the forward giant swing.

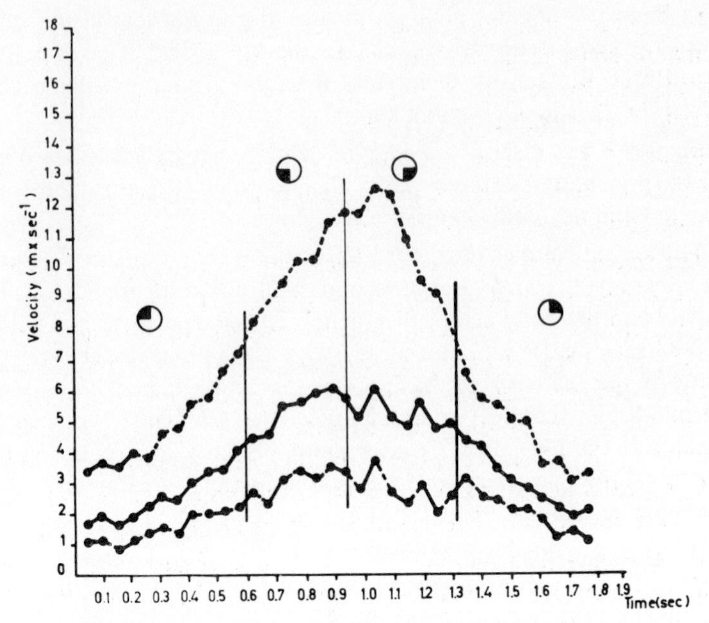

Figure 3. Angular velocities of three body reference marks during the backward giant swing. *Top line,* malleolus lateralis; *middle line,* body center of gravity; and *lowest line,* acromion process.

rants. However, the maximal velocity of the ankles occurred at the end of the second quadrant (0.78 sec; velocity of 13 m/sec). Between 0.81 and 1.14 sec, a sharp decrease in the velocity was noticed (0.98 sec; velocity 5.3 m/sec). Through a transfer of impulse, the velocities of the body center of gravity and of the shoulders increased and reached a maximal value, thus indicating a discharge of the upper part of the body. Immediately after the lowest velocity of the ankles, this velocity increased again, while the velocity of the two other reference points decreased until the end of the fourth quadrant. It was noted that the velocity curves of the body center of gravity and the shoulders had about the same magnitude, thus indicating a greater pulling force of the arms in order to accelerate the shoulders upward.

CONCLUSIONS

From the biomechanical analysis the importance of the partial movements has been revealed. The relative changes of position of the body segments determine the course of the movement. They have to meet the biomechanical prerequisites in the best way possible. Small differences

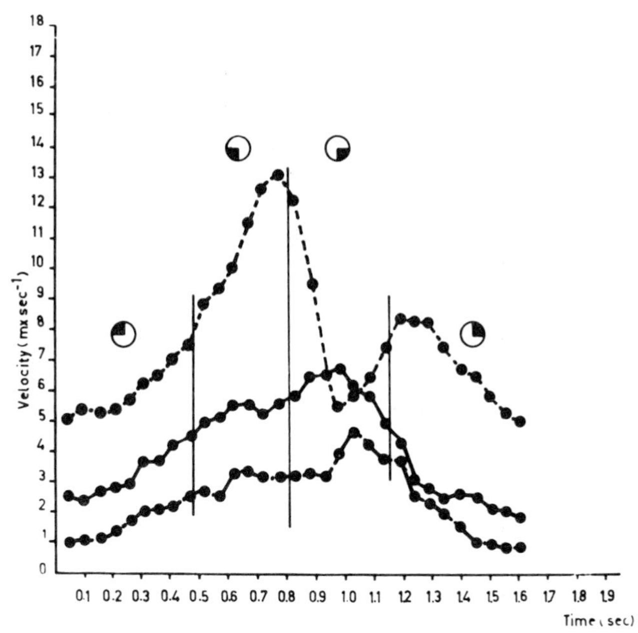

Figure 4. Angular velocities of three body reference marks during the forward giant swing. *Top line,* malleolus lateralis; *middle line,* body center of gravity; *lowest line,* acromion process.

lead to slightly different techniques of execution, which differ only in details. However, some relevant characteristics seem to justify to some extent the following generalizations.

For the forward giant swing: In the first quadrant of the movement the body is fully extended while it is hyperextended in the second quadrant. The kinetic energy of the body's center of gravity is maximal. Maximal velocity is reached by the feet in the third quadrant. Hips are then flexed maximally. The body is extended again in the fourth quadrant.

For the backward giant swing: At the end of the first quadrant of the movement, the body is slightly hyperextended in order to support the subsequent leg swing. The flexion of the hips increases in the second quadrant, just before the start of the third quadrant. The velocity of the legs increases in the first half of the third quadrant while the velocity decreases again in the second half. After passing the horizontal line in the fourth quadrant, the arms exert a forceful pull on the bar, which provokes first an upward, then a forward movement of the shoulders.

Comparison of the take-off forces in the flic flac and the back somersault in gymnastics

A. H. Payne
University of Birmingham, Birmingham

P. Barker
Walsall Technical College, Walsall

The flic flac and the back somersault are two basic skill movements required of any gymnast of reasonable standard. Usually these activities are performed as parts of sequences of movements, but for this comparison they were started from a standing position. In teaching the flic flac and the back somersault, the coach will issue instructions to the gymnast which seem to be quite different for the two movements: "Fall backwards, leaving the feet behind while reaching and looking for the floor" and "Strive for height jumping forwards and upwards before throwing the head back and tucking," respectively.

This study was an attempt to see if the coach's subjective judgments and instructions could be supported by objective measurements of the forces evoked during the take-off phases of the movements, since the performance of any airborne activity is very largely determined by the take-off—the only control during flight being that of change in moment of inertia, which is not always permitted by the rules or the technique requirements of the movement.

This work was made possible by a grant from the Science Research Council.

METHOD

Four good British club level gymnasts each performed several flic flacs and back somersaults from a force platform while being filmed at 32 frames per second by a 16 mm motion picture camera. It was possible later to relate the movements on the processed film to the force records by means of a continuous motion clock that was wired to give pulses on the force records (see Payne, Slater, and Telford, 1968). The force platform, which has been described by Payne (1974), was used to measure the following forces exerted by the gymnasts during the take-off phases of the movements, vertical (Z), horizontal forward and backward (Y), and moment of force about a transverse horizontal axis through the center of the force platform top surface (M_x) (see Figures 1 and 2). The films and records of the best flic flac and best somersault of each gymnast were selected for detailed analysis.

Determination of Center of Gravity and Moment of Force

In order to obtain reasonably clear enlargements from the motion picture film, color film was used and sequences of 35 mm black and white negatives were made of selected frames, which were then enlarged to a size convenient for making measurements. The locations of the centers of gravity of the gymnasts in the instantaneous positions shown in these selected frames were determined by placing the gymnasts in these respective positions, on a center of gravity board supported by four weighing scales. [This method was described by Hay (1973).] The locations of the centers of gravity were drawn in on the appropriate prints.

The resultant force acting at the instant in time represented by each frame of each sequence was calculated from the Z and Y values. The point of application of the resultant force was calculated from the M_x and Z values after first correcting for the Y contribution to the M_x value (Payne, 1974). Then the corresponding vector line was drawn to scale on each print (see Figures 3 and 4).

The moment of force about the center of gravity (M_{cg}) was calculated from the known magnitude of this force and the measured perpendicular distance from the center of gravity to the line of the force vector.

Finally the values of M_{cg} were plotted on the same time scale as Z, Y, and M (see Figures 1 and 2). Straight dotted lines were used to join the calculated coordinates to give an approximate shape to the curve.

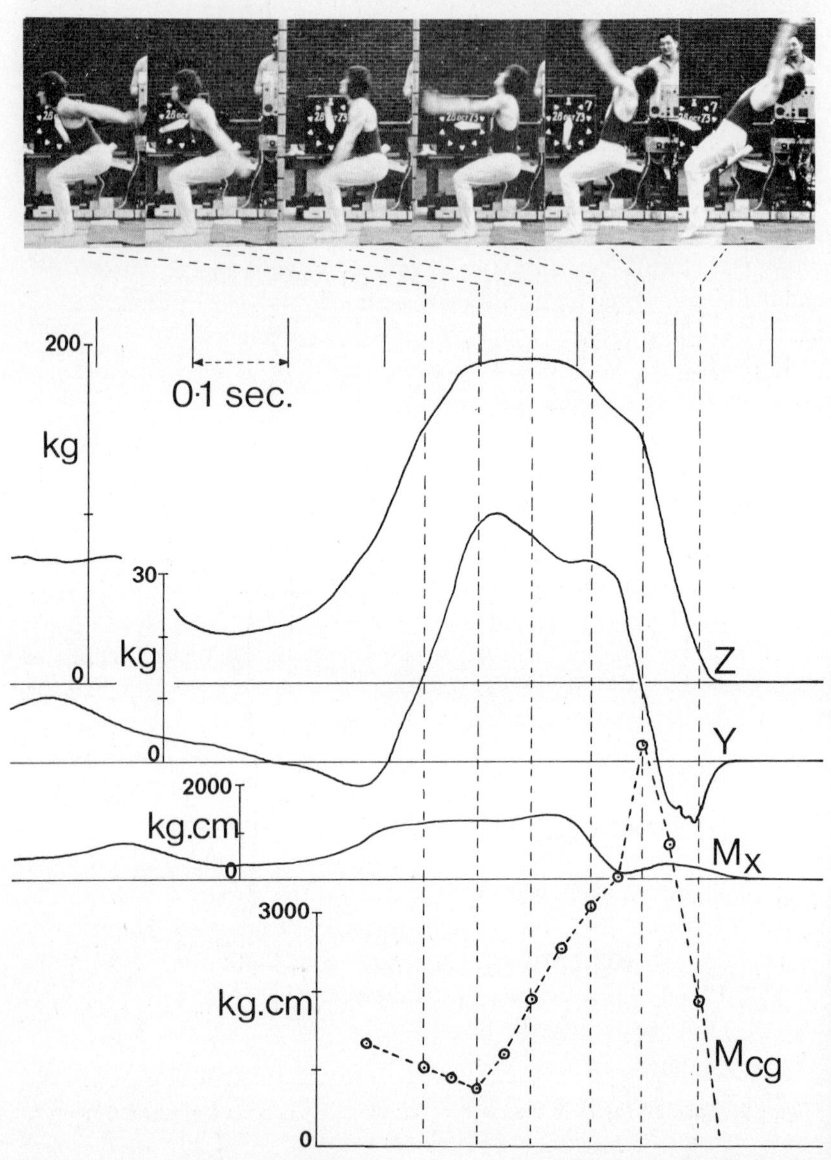

Figure 1. Take-off forces in the flic flac. Z is the vertical force, Y is the horizontal force in the line of the movement, M_x is the moment of force about the platform's transverse horizontal axis, and M_{cg} is the moment of force about the gymnast's center of gravity (which is marked ⊕ in Figure 3). All forces are in kilograms. The exact times marked by the dotted lines refer to the pictures above.

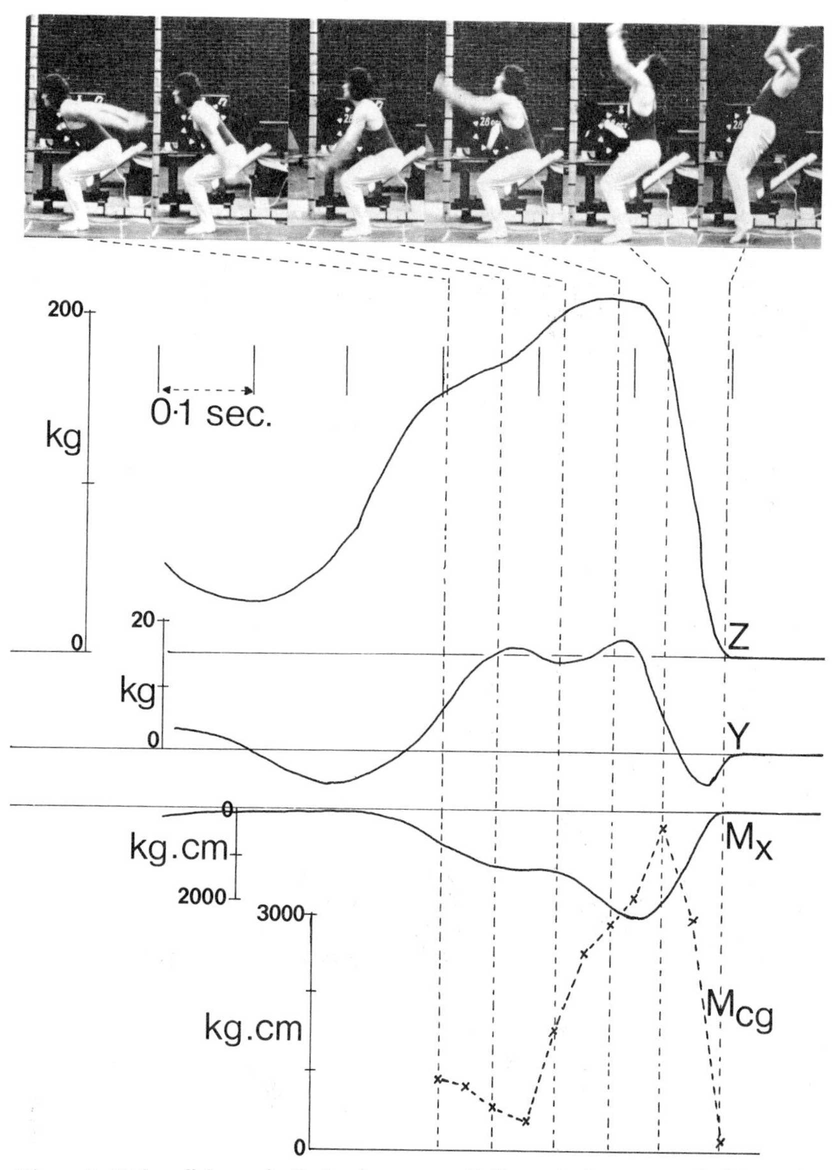

Figure 2. Take-off forces in the back somersault. Legends the same as in Figure 1.

Accuracy of Results

The continuous motion clock made 5 revolutions per second. The face of the clock was divided into ten divisions, each large enough to allow estimation to the nearest tenth; therefore times could be judged to nearly

Figure 3. The flic flac. The resultant force vector is represented by the line in each frame. The horizontal scale in the background is equivalent to 18 cm actual distance per division in the plane of the gymnast's movement. The magnitude of the force vector can be represented by this scale as well; 1 division is equivalent to 35 kg.

Figure 4. The back somersault. Compare the body positions in this sequence with those of the flic flac.

0.002 sec. A similar accuracy was possible in reading the time scale on the force record. However, this accuracy was reduced somewhat by the exposure time of each frame, approximately 0.01 sec, to something nearer

0.005 sec. It is doubtful that the center of gravity reaction board method can locate the human body's center of gravity to closer than ± 1 or 2 cm and this limitation accounts for the highest possible error in the experiment, averaging approximately 11 percent for values of M_{cg} above 1,000 kg · cm, although this reduces to 2 percent for the peak values. For a discussion of the limitations of the force platform measurements see Payne (1974).

RESULTS AND DISCUSSION

Figures 1 and 2 show the take-off forces and moments of force for a flic flac and a back somersault performed by the same gymnast.

Recordings

At first glance the patterns of the force traces have a remarkable similarity, and this was generally true of the records produced by the other gymnasts as well. The moment of force about the transverse horizontal axis at the platform surface was different in Figures 1 and 2 because the foot position relative to the center line of the platform was different.

On deeper scrutiny it can be seen that the more upright body position and harder drive in the later take-off period of the back somersault add a second peak to the vertical component (Z) which ensures that the gymnast's center of gravity rises high enough to allow him time to perform the complete rotation. Flight time in the somersault is almost twice that of the flic flac. Horizontal travel is not required in the back somersault, and this is revealed in the horizontal force trace (Y), where the horizontal impulse (area under the curve) is only about one-half of that obtained in the flic flac.

Moment of Force about the Center of Gravity

Again, the patterns of M_{cg} against time for the two movements are remarkably similar, with most of the angular momentum being developed during the second half of the take-off phase in each case.

In nearly all the gymnasts studied the integral of the M_{cg} curve for the flic flac was slightly greater than that for the back somersault. This indicates that the larger moment of inertia of the in-flight body position in the flic flac requires a higher angular momentum even though the angle of rotation to landing on the hands (about 130°) is less than half of the angle of rotation to the feet in the back somersault (about 310°).

CONCLUSION

If the force–time and M_{cg}–time curves are so similar for the flic flac and back somersault take-offs, one may well ask where, besides the difference in the in-flight body position, do the resulting differences arise in the two movements? The answer is to be found by an inspection of the angle of the gymnast's body to the ground in the film sequences (see Figures 3 and 4). Although the changes in body posture are almost identical, the general body backward lean is much more pronounced in the flic flac (about 48° to the ground as the feet lose contact) than in the back somersault (70°). It would seem, therefore, that the usual instructions from the coach in teaching these movements have a sound mechanical basis.

ACKNOWLEDGMENTS

The authors would like to thank Mr. S. K. Joshi, Miss M. G. Antonucci, and the gymnasts who acted as subjects.

REFERENCES

Hay, J. G. 1973. The Biomechanics of Sports Techniques. Prentice-Hall, Englewood Cliffs, N.J.

Payne, A. H., W. J. Slater, and T. Telford, 1968. The use of a force platform in the study of athletic activities. A preliminary investigation. Ergonomics 11, 2: 123–143.

Payne, A. H. 1974. A force platform system for biomechanics research in sport. *In:* R. C. Nelson and C. A. Morehouse (eds.), Biomechanics IV, pp. 502–509. University Park Press, Baltimore.

Biomechanical analysis of the formation of gymnastic skill

A. V. Zinkovsky
Polytechnical Institute of Leningrad, Leningrad

A. A. Vain
University of Tartu, Tartu

R. J. Torm
University of Tartu, Tartu

One of the essential purposes of the performance of gymnastic exercises is to achieve the body's motion in accordance with the description of the exercise. The main aspects of the uprise technique are the coordination of the muscle contractions in order to use the kinetic energy for the displacement of the body to its final position. The required energy for the performance of the uprise is achieved by the swing. The value of the kinetic energy for the rise depends on the displacement of the body in the vertical plane. In the case of an excessive value of kinetic energy, it is not possible to perform the uprise by means of muscle contractions only. Too great a value of kinetic energy disturbs the coordination of the movement of the body links.

If the abilities of the gymnast are sufficient, success in performing exercises on a high technical level is acquired. In such a case the data obtained by the recording of the body movement in the development of the skill reflect the regularities in the acquisition of the exercise technique.

The purpose of the present study was to analyze different stages during the development of the skill in performing the long underswing upstart on the horizontal bar.

METHODS

Films of all the experiments were made by means of a high-speed camera. The experiments were carried out in two series. In the first series, the performance technique was explained to ten sportsmen. The second series was carried out after skill was acquired in performing the upstart.

The vertical and horizontal components of the push exerted by the body on the bar were measured by means of the changes in the tension of strain-gauge elements on the bar. The tracings of the accelerometer output (fixed on the back of the gymnast), EMG, and tension of strain gauges recorded on the paper of an oscillograph were synchronized with the motion picture camera shutter. This was done by disconnecting the current's chain in the cable between the oscillograph and the camera. By means of the frames of the film (coordinates of the center of the body and link axis), biomechanical characteristics were calculated (Vain, 1973). A statistical analysis of all the characteristics was performed by a generally known method.

RESULTS AND DISCUSSION

We divided the exercise into three phases: Phase 1, the acquiring of mechanical energy during the swing; Phase 2, taking the initial position for the application of muscle energy to perform the uprise; and Phase 3, performing the uprise. In Phase 1, the athlete's body acquires mechanical energy (E) during the swing revolving with a certain amplitude around the axis of the horizontal bar. In Phase 1, we considered the athlete as a pendulum revolving around Point O (Figure 1). When the athlete

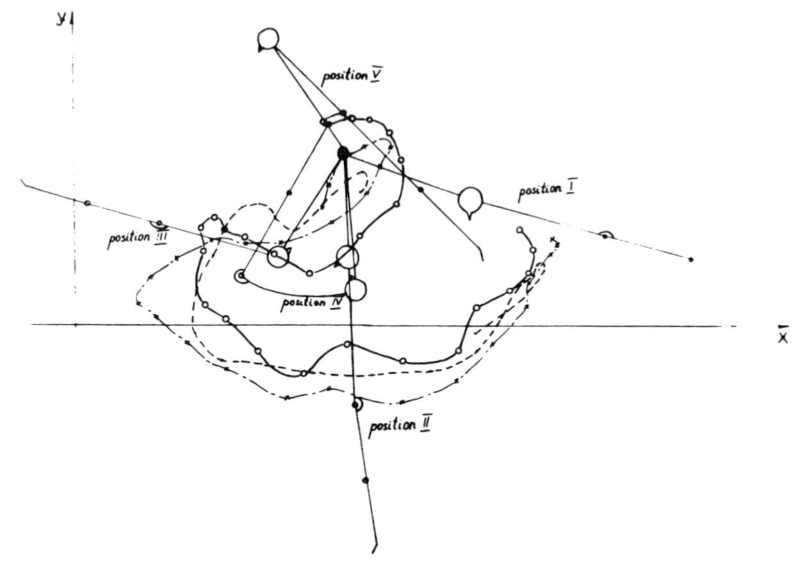

Figure 1. Successive positions in executions of the uprise on the horizontal bar. Paths of body center of gravity: initial attempt (— — —); later trial (x—•—x); final success (o —— o).

reached Position 1 (Frame 1) his kinetic energy was at a minimum and potential energy at a maximum. In Position 2 (Frame 8) $E_{kin} = E_{max}$ and $E_{pot} = E_{min}$ while in Position 3 (Frame 14) $E_{kin} = E_{min}$ and $E_{pot} = E_{max}$ again. Now Phase two begins. In this position one must start to take the initial position for the application of muscle energy in order to perform the uprise. The velocities of parts of body in the nonmoving plane are minimum and thus the condition for flexion of the hip joints is most favorable.

The flexion in the hip joints must be completed at the instant when E_{kin} is maximum and E_{pot} is 0. It is essential to have assumed the starting position at this moment; otherwise it is not possible to use the kinetic energy for trunk rotation around the bar. In Position 3 the formation of the second pendulum begins; this pendulum is short and accordingly the period of swing is shorter as well. As shown in Figure 2 the kinetic energy in Position 4 (Frame 19) is equal to that in Position 2. In Phase 4, the trunk is moving counterclockwise around the bar, this radius being R, while the legs are extended at the hip joints. Consequently, the momentum from the motion of the legs is opposite to the momentum of the motion of the trunk. The effective force in the hip joint creates the moment of rotation ($F_{in} \cdot R$). Therefore, the gymnast performs the rise. In Figure 1 we can see that in the first period of teaching the mechanism of the second

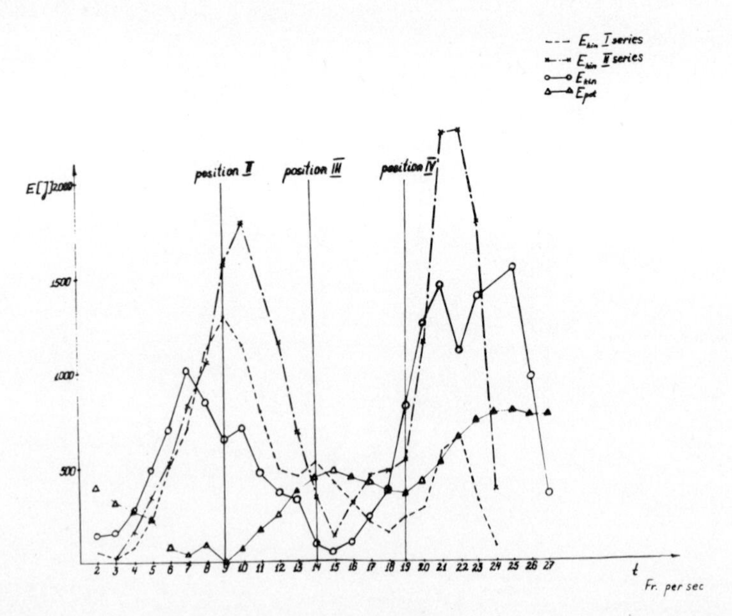

Figure 2. Curves of kinetic (E_{kin}) and potential (E_{pot}) energy during initial and final stages of learning the uprise on the horizontal bar.

pendulum has degenerated, since the athlete makes an attempt to bring the hips to the bar. Therefore, the center of gravity does not move as a pendulum and it is impossible to perform the movement of extension. Thus, the mechanism of the technique of uprise consists of the formation of the second pendulum and the storing of the kinetic energy obtained in Phase 1.

The analysis of the results of the EMG revealed that body movement is related to muscular activity. The activity of antagonistic muscles coordinates the movement of the links for better use of mechanical regularities described. The obtained results are in accordance with earlier studies (Torm, 1974). When teaching gymnastic techniques, it is essential to consider both mechanical aspects of movements and also muscle coordinations.

REFERENCES

Torm, R. 1974. Biomechanical analysis of the regularities of the formation of gymnastic skills technique. Dissertation, University of Tartu, Tartu, Estonian SSR (in Estonian).

Vain, A. 1973. Treatment of biomechanical information by electronic computer. *In:* S. Cerquiglini, A. Venerando, and J. Wartenweiler (eds.), Biomechanics III, pp. 104–107, S. Karger, Basel.

Electromyographic and cinematographic study of the volleyball spike

H. Oka
Osaka Kyoiku University, Osaka

T. Okamoto
Kansai Medical School, Osaka

M. Kumamoto
Kyoto University, Kyoto

The straight spike, which is one of the most fundamental spikes of volleyball, has not been studied adequately in terms of the functional mechanism of the muscles. So, we attempted to record the electromyograms and movement patterns simultaneously, and to analyze both of them in detail electromyographically and cinematographically.

METHODS

The subjects employed in the experiments were 20 skilled male adults including a 1972 Olympic gold medalist. They were required to approach from the left sideline, and to spike with full force the ball which was tossed from center front by a skilled setter. Some of the subjects were also requested to spike a hanging ball from various directions.

Electromyograms were recorded from 14 muscles of the upper extremity, shoulder girdle, and trunk, with an 18-channel multipurpose electroencephalograph using surface electrodes 10 mm in diameter. Movements were filmed with 16 mm high-speed cameras from the side (64 f./sec and 200 f./sec), the front (64 f./sec), and the top (64 f./sec) of the subjects. A signal of each frame of the film was simultaneously recorded with the electromyograms. In some cases, electrogoniograms of the wrist and elbow joints were simultaneously recorded with the electromyogram.

RESULTS AND DISCUSSION

It is apparent from the motion analysis that the straight spikes have two styles. One is an elevation style, in which the spiker's humerus is elevated in a diagonally lateral direction above the horizontal during the backswing, as in Figure 1A, and the other is the backswing style, in which the humerus is swung below the horizontal, as in Figure 1B. Most of the subjects used the elevation style, while some utilized the backswing style. The Olympic gold medalist was chosen as a typical example of the elevation style, and a player who belongs to the national volleyball league represented the backswing style.

In Figure 2, the left and right columns show the representative EMGs of the elevation style and backswing style spikes, respectively.

In the backswing phase of the elevation style spike, a set discharge pattern was observed. The marked discharges obtained in the biceps brachii long head, the anterior deltoid, and the clavicular portion of the pectoralis major indicate that the humerus was elevated in a plane more than 30 degrees lateral from the sagittal plane, and the discharge of the right serratus anterior indicates that the humerus was elevated in a plane medially from the frontal plane, which was suggested by Okamoto, Takagi, and Kumamoto (1967) and Okamoto (1968).

At the period of transition from backswing to forward swing, the discharge pattern of the elbow and shoulder joint muscles tended to disappear, which indicates that these muscles were completely relaxed.

While the humerus was elevated, at the early part of forward swing, the discharges of the flexor carpi radialis, the pronator teres, and the triceps brachii lateral head indicate that the flexion of the wrist joint and the extension of the elbow joint were performed with the forearm pronated. From the results of this electromyographic analysis of the fundamental movements of the arm, which was suggested by Okamoto, Takagi, and Kumamoto (1967) and Okamoto (1968), the discharge pattern of the anterior deltoid and the clavicular portion of the pectoralis major indicates that the humerus was elevated with the force exerted medially. As the right humerus was elevated and the left humerus was flexed in this phase, the discharge of the right serratus anterior shows that the scapula was rotated upward, and the discharge of the left serratus anterior shows that the scapula was shifted upward. This flexion of the left humerus in this phase would serve as a prevention against left-hand rotation of the trunk.

Figure 1. Comparison of forms from backswing to forward swing in elevation style spiker (*upper, A*) and backswing style spiker (*lower, B*) in the straight spike.

While the marked discharge was observed in the teres major, which is a prime mover of the shoulder in extension, at the latter part of forward swing in which the humerus was extended downward, the discharges of the clavicular portion of the pectoralis major and the anterior deltoid were seen throughout the forward swing. These discharge patterns seem to show that the humerus was extended with downward and medial force, consistent with the results of the electromyographic analysis of the fundamental movements of the arms, which was reported by Okamoto, Kumamoto, and Yamashita (1972). In order to analyze the movement in detail, other subjects were required to perform the straight spike while exerting a medial force using a hanging ball. In these situations, there was no perceptible change in the spiking movements. The EMGs recorded showed similar discharge patterns to that of the gold medalist. This fact would show that the humerus was extended while exerting a downward and medial force. The marked discharge observed in the rectus abdominis throughout the forward swing, and the distinct trunk flexion in this phase, indicate that additional force was probably transmitted to the ball.

Just before contact, the discharge of the rectus abdominis disappeared and the sacrospinalis showed a burst of activity. The discharge patterns and the analyzed motion indicate that the trunk was actively flexed laterally to the left.

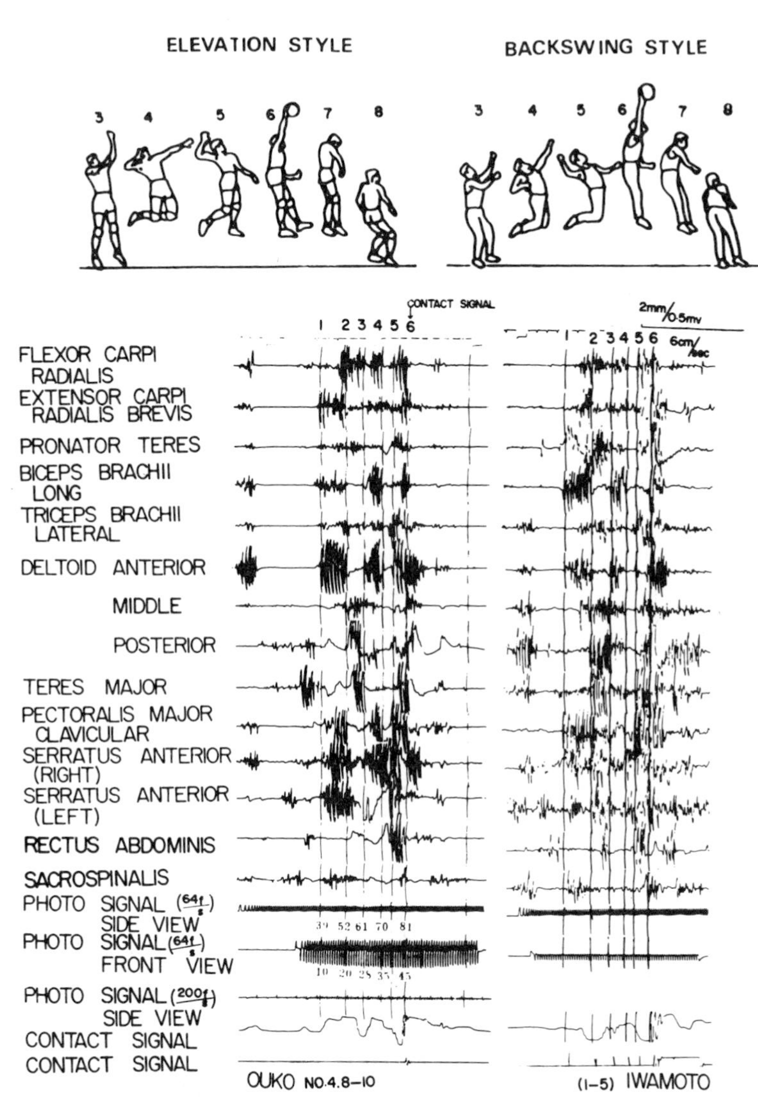

Figure 2. Comparison of EMGs of the upper extremity and the trunk in elevation style spiker (*left*) and backswing style spiker (*right*); the backswing phase (pose 3 ∼ 4), the transition phase from backswing to forward swing (pose 4 ∼ 5), the forward swing phase (pose 5 ∼ 6), and the contact (pose 6).

As to the discharge patterns of wrist joint muscles, the marked discharges appeared just before contact in the flexor carpi radialis and then suddenly shifted to the extensor carpi radialis brevis, immediately after contact. These distinct discharges that appeared right after contact would prevent the additional flexion of the wrist, as was suggested by Basmajian (1974).

In the follow-through phase, the marked discharges observed only in the anterior and middle deltoid and the serratus anterior indicate that the scapula was rotated upward while the humerus was pushed forward during this period.

During the backswing, the discharge patterns in the shoulder girdle muscles of the backswing style spiker were different from the one of the elevation style spiker. That is, the discharge of the clavicular portion of the pectoralis major was not observed, but the discharges of the teres major and the posterior deltoid were seen. These discharges and the motion analyzed indicate that the humerus was horizontally abducted below the horizontal. During the period of transition from backswing to forward swing, discharges of the elbow and shoulder joint muscles were very slight, as in the elevation style spiker.

Throughout the period from forward swing to contact, the discharge patterns of the three parts of the deltoid varied from time to time to some extent. In most cases, the marked discharges were observed in the middle and the posterior deltoids. However, in some cases, the marked discharge appeared in the anterior portion and a slight discharge was noted in the middle and the posterior portions, as in Figure 2 (right).

The variations of the discharges observed in the three portions of the deltoid during the spiking movements could be caused by the difference in the direction of load according to Okamoto, Kumamoto, and Yamashita (1972). When the straight spike was performed with the slightly medial or lateral force exerted, remarkable changes were observed in the discharge pattern. The discharge pattern that was often observed in the backswing style spiker, as mentioned previously, indicated that the extension of the arm was performed with lateral force being exerted. However, the spike was not always performed with the force exerted in the same direction. Immediately after contact, the discharge patterns of the wrist joint muscles were the same as seen in the elevation style spiker.

In the follow-through phase, pronounced discharge patterns occurred in most shoulder joint muscles, which was not the case with the elevation style spike. This would indicate that unnecessary contraction was occurring.

COMPARISON OF EMGS IN THE TWO STYLES OF SPIKE

During the backswing, the discharge pattern of the shoulder muscles of the elevation style spiker indicated that the humerus was elevated between a plane 30 degrees lateral from the sagittal plane and the frontal plane, and in the backswing style spiker, the humerus was horizontally abducted beneath the horizontal.

At the period of transition from backswing to forward swing, the discharges of the elbow and shoulder joint muscles were very slight in both styles of spikers.

At the latter part of forward swing, the set discharge pattern of the three portions of the deltoid was apparent in the elevation style spiker, but no uniformity existed in the backswing style spiker.

In the follow-through phase, marked discharges were observed only in the muscles participating in upward rotation of the scapula in the elevation style spiker, and in all shoulder joint muscles tested in the backswing style spiker.

REFERENCES

Basmajian, J. V. 1974. Muscles Alive. Williams and Wilkins, Baltimore.

Okamoto, T., K. Takagi, and M. Kumamoto. 1967. Electromyographic study of elevation of the arm. Res. J. Phys. Educ. 2(3): 127–136.

Okamoto, T. 1968. A study of the variation of discharge pattern during flexion of the upper extremity. J. Dept. Lib. Arts Kansai Med. School 2: 111–222.

Okamoto, T., M. Kumamoto, and N. Yamashita. 1972. Electromyographic study of fundamental movements of the upper extremity. Sixth International Congress of Physical Medicine, Barcelona, Spain. 2: 288–291.

Biomechanical analysis of the volleyball spike

J. Samson and B. Roy
Université Laval, Québec

The spike is one of the most important and yet one of the most difficult techniques a volleyball player has to master. Understanding the mechanisms by which a player attains maximum height is especially important to teachers and coaches. Technically, the movement is made up of five phases: the approach, the take-off, the suspension, the hitting motion of the arm, and the recovery. The purpose of the present study was to analyze the kinematic characteristics of the approach and take-off phases of the spike.

PROCEDURES

The data were obtained from films taken at 250 frames per second from a distance of 30 meters. The subjects were 11 players of national caliber invited to the selection camp of the Canadian Men's National Team for the year 1973. Eight of these men were eventually chosen on the 1973 National Team and six of them were on the national team that participated in the 1974 World Championships in Mexico.

RESULTS

Table 1 shows some of the morphological characteristics of the subjects. The vertical jump represents the difference between the maximum height attained by the center of gravity and the height of the center of gravity in the normal standing position. The angles of approach and of flight were obtained from the values of the horizontal and vertical velocity of the

Table 1. Morphological and biomechanical characteristics of subjects ($N = 11$)

	Mean	S.D.
Age (yr)	24.4	4.3
Weight (kg)	80.2	8.3
Height (cm)	184.9	4.9
Duration absorption (sec)	0.142	0.063
Duration impulsion (sec)	0.229	0.073
Angle of approach (degrees)	7.5	4.1
Angle of flight (degrees)	78.7	6.7
Vertical jump (cm)	71.5	4.8

center of gravity at the moment the feet touched the ground in the take-off phase and immediately after the feet left the ground, respectively.

In terms of vertical displacement of the center of gravity, the take-off phase was divided into two sub-phases; during the first part, the subject must reduce his forward motion considerably and take a semi-crouch position in order to generate the maximum impulse for the upward motion. The time elapsed from the first contact with the ground to the position of maximum flexion is the duration of the absorption sub-phase. In the second sub-phase, the subject makes a forceful extension at the hips, knees, and ankles in order to raise his center of gravity as high as possible. The time from the beginning of this extension to the moment when his feet leave the ground represents the duration of the impulse.

RESULTS

Figure 1 presents the horizontal and vertical displacement of the center of gravity (top graph) during the approach and the take-off phase. Technically, the approach is usually made up of two to three steps. In this figure, the approach represents only the last step. The lower graph illustrates the horizontal and vertical velocities of the center of gravity during the same two phases. From this figure, we note that the lowering of the center of gravity begins during the last step of approach and continues over approximately 40 percent of the take-off phase. The mean

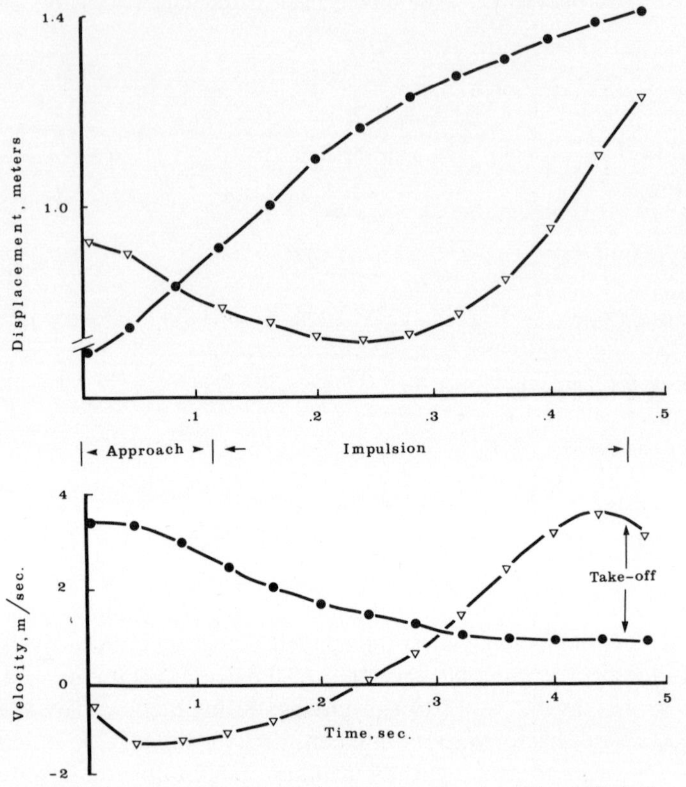

Figure 1. Typical curves of the horizontal (—●—) and vertical (—∇—) displacement and velocity of the center of mass in the approach and impulsion phases of the spike in volleyball.

lowering of the center of gravity during the flexion sub-phase of the take-off was 11 cm and the mean elevation 42 cm. The horizontal velocity decreased sharply at the moment of foot contact with the ground and reached its lowest value shortly after the beginning of the impulsion. On the other hand, it is somewhat surprising to note that the maximum vertical velocity was reached nearly 0.02 sec before the take-off. The decrease in vertical velocity during that part of the take-off phase was in the order of 0.248 m/sec.

This phenomenon was observed in the majority of the subjects and indicates that the jumpers did not profit as much as they should from a complete and powerful extension of the ankle joint.

When comparing these results with those of Figure 2, it can be seen that the maximum velocity of the hip and knee joint was attained nearly

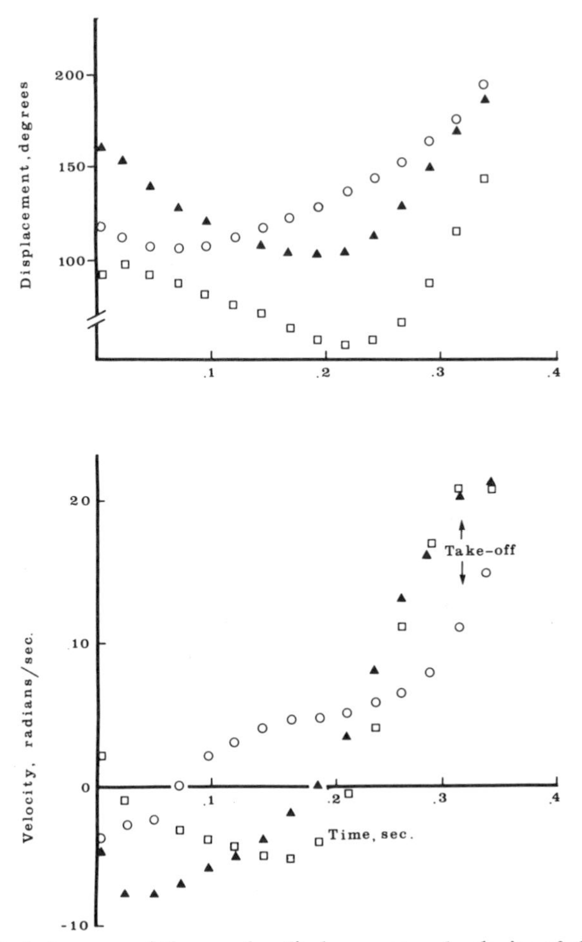

Figure 2. Typical curves of the angular displacement and velocity of the hip (○), knee (▲), and ankle (□) joints during the approach and the impulsion phases of the spike in volleyball.

at the moment of take-off whereas that of the ankle was not reached until after take-off.

The displacement patterns (Figure 2, top graph) for the knee and ankle joint are very similar, contrary to that of the hip joint, which shows that this particular subject extended the hip over most of the take-off phase. This was also observed in most of the other subjects.

The velocity curves (lower graph) also show the same asynchrony with the knee and ankle curves representing the flexion and forceful extension sub-phases. However, the hip velocity curve shows that extension of the hip was more forceful at the beginning and at the end of the take-off phase. However, the curve shows a constant velocity during that part of the curve where the most forceful extension should have occurred just prior to take-off.

This would seem to indicate that the subject did not make proper use of the extension force of the hip.

INTERPRETATION

Although these subjects were of national caliber, the mean scores for the vertical jump were below the average for such groups as reported by Wielki (1972) and Toyoda (1974).

The angles of the knee joint during the take-off phase corresponded to those suggested by Nicholls (1973) but those of the ankle and hip were somewhat different. As can be seen from Figure 2, the subject contacted the ground in an almost fully flexed position at the hip and initiated the extension of the hip almost immediately, even though the knee and ankle were still flexing. This caused the trunk to remain straight and the knees to move farther over the feet thus reducing the angle at the ankle. The main disadvantage of this position is that only the forceful extension of the knees can be effective in the take-off.

When comparing the displacement and velocity curves for the hip, knee, and ankle joints with similar curves for the standing broad jump (Roy, 1972), it was noted that the pattern of the hip joint in the long jump was similar to that of the knee and ankle joints. Although the spike requires a vertical jump, it would seem advisable to make use of the power of the extension of the hip as well as that of the knee and ankle joints.

REFERENCES

Nicholls, K. 1973. Modern Volleyball. Henry Kimpton Publishers, London.

Roy, B. 1972. Jump in 7, 10, 13 and 16 year old boys. *In:* Taylor, A. W. and M. L. Howell (eds.), Training: Scientific Basis and Application, pp. 290–304. C. C. Thomas, Springfield.

Toyoda, H. 1974. The lists of physical performances of volleyball players. University of Tokyo (mimeographed).

Wielki, C. 1972. La Hauteur du Filet et la Circonférence du Ballon de Volley Ball en Fonction des Mesures Prises sur des Participants aux Championnats du Monde à Varna et Sofia, 1970 (The height of the net and circumference of the ball in relation to measures taken during the 1970 World Championship in Varna and Sofia). Munich. Personal communication.

Simple mathematical model of weightlifting

M. A. Ranta
University of Technology, Helsinki

Weightlifting competition has been the subject of many recent studies, both in the medical field and in the area of physical education. Yet, comparison between results in different weight classes and between different lifting techniques is rather cumbersome. Furthermore, it is no easy chore to examine previous weightlifting records and determine where marks can be improved significantly. The simple mathematical model of weightlifting provided in this presentation, along with the resulting performance index, suggests a new approach to analyzing and comparing quantitative results of weightlifting contests.

DERIVATION OF THE MATHEMATICAL MODEL

The following mathematical model describes the first part of the so-called "clean" process in weightlifting. This process involves the two stages: (a) raising the barbell from the platform with a powerful lifting motion, and (b) shifting the body weight underneath the bar. The second part of the "clean" movement, that of straightening to an erect position, is ignored in the development of the model. In addition, the factor of balance has not been considered.

The barbell is represented by one or more plates at either end of the bar (Figure 1.) The maximum radius R of any plate is 22.5 cm. The barbell's center of gravity is described as being located at a particular height above the platform. The weightlifter pulls the moving barbell of mass M with varying force F to a distance s from the platform. The coordinate y is measured vertically upwards from the platform. Consequently, the barbell rises an additional distance Δs during a short time

Figure 1. Diagram of clean process in the squat snatch.

interval Δt. Within the same time interval, the lifter must shift himself underneath the bar. The lift is most efficiently accomplished when the lifter reaches the lowest point of his dive while the barbell is at its uppermost height. Thus, the lifter must alter his center of gravity by a vertical distance l at an average speed V. Now, by letting h represent the maximum height of the barbell, mathematically equivalent to $(s + \Delta s)$, the principle of the conservation of energy gives equation (1):

$$(h - R)\, Mg = \int_R^{s = h - \frac{gl^2}{2V^2}} F(y)\, dy \tag{1}$$

where h = maximum height of barbell
R = radius of largest plate
M = mass of barbell
g = acceleration due to gravity
F = lifter's real force
l = shift of lifter's center of gravity
V = average speed of shift l
y = bar's momentary vertical distance

Equation (1) is then nondimensionalized according to the expressions listed in Table 1 and converted into the universal pair of relationships denoted by equations (2) and (3). All distances enumerated in the table are referenced with respect to the lifter's height L, while the force F is described in terms of both the lifter's mass m and lifter's height L. The dimensionless performance index (PI) indicates the relative work performed during the lift:

$$\text{PI} = (\eta p - 1)\, M/m \tag{2}$$

Table 1. Nondimensional symbols utilized in the performance index formulas, equations (2) and (3)

Symbol	Description	Mathematical equivalent	Practical range
z	Relative distance	$z = y/L$	0.1–0.7
p	Lifter's relative height	$p = L/R$	6–9
$f(z)$	Relative force	$f(z) = Fp/mg$	15–30
η	Technique parameter	$\eta = h/L$	0.5–0.6
λ	Technique parameter	$\lambda = l/L$	0.3–0.5
u	Speed parameter	$u = V/\sqrt{gR/2}$	1–3
q	Lumped parameter	$q = (\lambda/u)^2$	0.01–0.1

$$\mathrm{PI} = \int_{1/p}^{\eta - qp} f(z)\, dz \tag{3}$$

ANALYSIS OF PERFORMANCE INDEX

The validity of the performance index in equations (2) and (3) can be ascertained by observing results of a selected group of weightlifters. Those lifters in the world record category comprise a relatively homogeneous set, and there is sufficient evidence to suggest a close correlation between their physical characteristics. Subsequent paragraphs compare equations (2) and (3) with world performances listed on October 15, 1974. These equations are further analyzed to identify conditions for achieving the best lifting results.

Performance indices for world record lifts are tabulated in Table 2 according to the various weight classes. The first column gives the mass m of the lifter, rounded off to the upper limit of the weight class. The second and third columns list the body height L or p, estimated on the basis of the body mass being proportional to the cube of body height. Mathematically, this becomes: mass (in kg) is approximately $0.19\ p^3$. The fourth and sixth columns provide world records in the *snatch* and *clean and jerk*, respectively. The fifth and seventh columns show the corresponding performance indices. Finally, technique parameters η and λ are assumed to be constant. A reasonable estimate for η is 0.59 in the *snatch* and 0.50 in the *clean and jerk*.

Table 2. Performance indices $PI = (\eta p\text{-}1)M/m$ for world record lifts in different weight classes for Snatch and for Clean and Jerk; world records as of October 15, 1974

Weight class	Height		Snatch $\eta = 0.59$		Clean and jerk $\eta = 0.50$	
m (kg)	L (m)	p	M (kg)	PI	M (kg)	PI
52	1.48	6.6	106.5	5.93	140	6.19
56	1.50	6.7	117.5	6.20	151	6.34
60	1.55	6.9	126	6.45	158.5	6.47
67.5	1.60	7.1	137.5	6.50	177.5	6.70
75	1.65	7.3	152.5	6.72	190	6.71
82.5	1.70	7.6	163.5	6.90	202.5	6.87
90	1.75	7.8	175	7.00	215	6.93
110	1.85	8.2	177.5	6.19	227.5	6.41
135	1.95	8.7	187.5	5.74	241.5	6.00

Optimum Condition for η

In order to determine the optimum condition for η, both equations (2) and (3) are differentiated with respect to η. Elimination of $\partial PI/\partial\eta$ yields the expression given in equation (4):

$$f(\eta - qp)[1 - \partial(qp)/\partial\eta] = pM/m \tag{4}$$

This equation simplifies into the dimensional form of equation (5):

$$F(h - \Delta s)\ (1 - \partial\Delta s/\partial h) = Mg \tag{5}$$

Here, the partial derivative $\partial(pq)/\partial\eta$ *or* $\partial\Delta s/\partial h$ is normally much smaller than one and can be ignored.

Optimum Height of Lifter

The performance index (PI) has a maximum value at the point where $\partial PI/\partial p = 0$ and $\partial^2 PI/\partial p^2 < 0$. Consequently, the PI curve at this location is convex, and the slope of its tangent vanishes. The degree of convexity can be discerned from the values of PI in Table 2. The vanishing aspect of the tangent's slope leads to equation (6) for determining the optimum value of p:

$$\frac{1}{p} f(1/p) = qp\, f(\eta - qp) \tag{6}$$

In principle, the solution of equation (6) is readily found graphically if the $f(z)$ curve is known. The shaded areas under the curve of Figure 2 must be equal in the optimum case.

However, the solution of equation (6) is much simpler to obtain if the lifter initially performs at the optimum value of η as described in equation (4). The force $f(1/p)$ that begins the pulling action must exceed the weight of the barbell, pM/m, by a multiplier n. Otherwise, the lift is impossible. Thus, in conjunction with these assumptions and the notation $q = (\lambda/u)^2$, the approximate solution of equation (6) becomes

$$p = \frac{u}{\lambda} \sqrt{n} \tag{7}$$

where n is the multiplier of lifting force at start of pull.

The validity of this simple formula can be verified by approaching the optimum height in a different manner. The relative length of the pull is the difference between the limits of integration of equation (3). Since a longer relative length of pull is advantageous in accelerating the barbell, the optimum combination of p and q is achieved when this length reaches its maximum value. This is attained by the following:

$$p = u/\lambda \tag{8}$$

Equations (7) and (8) for the optimum height differ from each other by only the factor $\sqrt{n}$. Since the barbell cannot be jerked at the beginning of the pull for physiological reasons, the multiplier exceeds unity by only

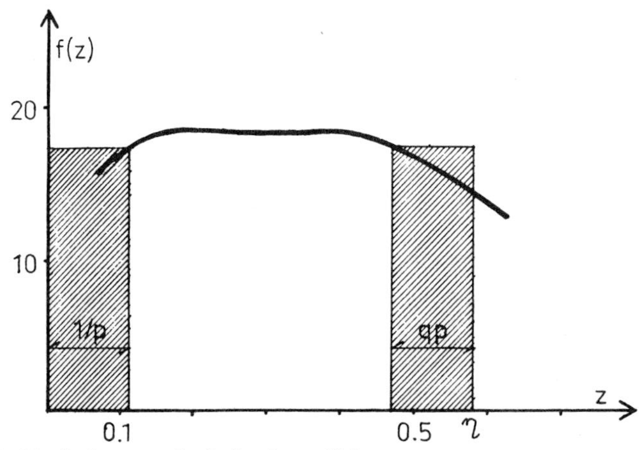

Figure 2. Typical curve of relative force $f(z)$.

a slight amount. Therefore, both equations (7) and (8) yield practically the same result.

As an example, consider the following plausible parameters for a world champion lifter: $u = 2.7$, $\lambda = 0.36$, $n = 1.1$. The respective values of p from equations (7) and (8) are $p = 7.9$ and $p = 7.5$. This is in close correlation with values shown in Table 2, although it is intended as only an example, not a rigorous proof of the exact location of the optimum.

RESULTS

Examination of Table 2 reveals some interesting points. The average of the performance indices in the Snatch category is 6.40; the average of the performance indices in the Clean and Jerk category is 6.51. Excluding the outermost weight classes, those of the flyweight and superheavyweight levels, the Snatch average is 6.57 ± 0.24 and the Clean and Jerk average is 6.63 ± 0.21, yielding a probability of 0.95. Thus, the performance index can be an invaluable means for comparing results.

DISCUSSION

The mathematical model of weightlifting provided in this paper has not been tested sufficiently as yet, mainly since little information is known about lifters' heights and lifting parameters. In the future, it would be desirable to record additional data concerning the lifter's physical characteristics. With reliable and comprehensive statistics, it will be possible to estimate the values of lifting parameters and to solve the stochastic integral equation, equation (3), with respect to the force function $f(z)$. In order to improve or to check the lifting technique, important estimators such as $f(z)$ can serve as a point of comparison along with the normal measured values.

In spite of insufficient data from past performances, numerous direct conclusions can be drawn from the model. The maximum lift can be achieved by satisfying the following conditions: (a) the start of the pull must be sufficiently powerful, (b) the force must equal the weight at the end of the pull, (c) the barbell's uppermost height must be as small as possible, (d) the body shift underneath the bar must be fast and short, (e) taller, faster lifters can more readily attain a supreme performance index, and (f) the relative increase in lift results is directly proportional to the relative increase in body weight for a constant performance index.

Most of these facts are already known from qualitative experience. Their quantitative values can only be found by using an appropriate mathematical model.

Within the period of ten years, the best performance index has been moved up from the lightweight class. This phenomenon can best be explained on the basis of equations (7) and (8). The increase in speed parameter u has been the main reason for the transition of the optimum point towards the heavier classes, where lifters are usually taller as well as heavier. Introduction of the squat lifting technique with resulting smaller λ has had a similar effect.

Performance determinants

Optimization of construction of systems of movements in sports

D. D. Donskoi
Central Institute for Physical Culture, Moscow

H. H. Gross
Pedagogical Institute, Tallin

Man's motor apparatus is characterized by a large number of degrees of freedom of movements and of groups of muscles functioning under variable conditions and entering into complicated interactions. This gives rise to many variations when we are faced with the solution of a motor problem. We should find the *optimum solution* each time corresponding to specific conditions, which is of special significance in sports. To construct the optimum "model of the future" (Bernstein, 1967) means to set the purpose without which control is impossible. This requires not only a formalized (mathematical) model, but also a substance model. Such a model shows not only the mechanical aspects, but also the biological and psychological aspects, and permits one to understand the tasks and ways of training in sport and to single out the leading elements in each training stage. The aim of the present study was to evolve and check experimentally the methodology of *optimizing the construction* of systems of movements in locomotor sports.

BASIC DEFINITIONS

The construction of the system of motions has two aspects: (a) the method of uniting subsystems and their elements into a system of movements with the given technique of performing the exercise and (b) the way of shaping a system of movements in the course of training. The *optimization of the technique* implies the choice of such a structure of the system of movements which leads best of all to the attainment of the purpose (the subject of training). The *optimization of the method of training* envisages the most rational succession of the chosen auxiliary exercises and the technique of their use in training (the method of training). The optimum model of the technique can be constructed only in the optimum way. Because sportsmen's individual features, the degree of their physical and technical preparedness, and particularly the conditions of their activities vary greatly, the *optimum model* is not the precise and single standard, but such a standard which sets the limits of permissible deviations (the range of individual and adaptation versions).

THE METHODOLOGY

To control the construction of a system of movements it is necessary to set (a) the purpose of the whole system and the subpurposes of its components (subsystems), (b) the indices required in the model and checked out in real movements, and (c) the modes of correction during training. The biokinematic and biodynamic characteristics of subsystems serve (a) the system analysis (Donskoi, 1973) (the determination of subsystems —temporal phases and spatial elementary actions) and (b) the *system synthesis* (determining of the interconnection of subsystems in step-by-step training aimed at mastering the system of movements and during its practical use). In the course of the study the characteristics were simultaneously recorded by methods of cinematography, goniography, tensodynamography, and electromyography. During mathematical processing of the characteristics, use was made of correlation models (Donskoi, Nazarov, and Gross, 1974) to determine the degree of the correlation between characteristics of subsystems.

THE OPTIMIZATION OF MODELS

Proceeding from the specific features of sport techniques and the rules of contests, the purpose of the system of movements was determined (in cyclic locomotion its optimum speed corresponding to the length of distance). Using characteristics, a system analysis was conducted—the division of the system of movements into its elements in time (phase). Each phase was singled out by strict rules by means of qualitative biodynamic analysis and the *subpurpose* was set that would lead to the attainment of the common purpose. With the change of the phase, subpurposes also changed and the chain of successively changing leading subpurposes was determined. Then through the comparison of data pertaining to outstanding sportsmen and the analysis of the mechanism of movements, the *aims* of movements in this phase were determined (what should be achieved) and the requirements for the performance of movements (how to do this). Thus, the subpurposes, aims, and requirements for movements in each phase were determined.

On this basis in the same system of movements either subsystems (in space) were determined in the form of *elementary activities*. These subsystems were developed in several phases simultaneously and consecutively. They also have subpurposes, tasks, and requirements for movements. It is easier to describe and explain them than to verify their correctness and to rectify deviations in them. Using mathematical models, their interconnections, interaction, and coordination were established in phases and elementary activities. Thus, a *substance model* of the system of movements on the whole and in detail (Gross and Donskoi, 1974) was constructed. Finally, proceeding from the correct understanding of the mechanisms of movements at the mechanical and biological levels, the *optimum way* of constructing the system of movements during training was determined.

RESULTS OF INVESTIGATIONS

The techniques of skiing uphill, skating, swimming, and other types of locomotions of the best sportsmen were studied. The efficiency of the results obtained has been confirmed by high achievements of sportsmen and the practical use of the optimum technique in training.

CONCLUSION

The investigations into the optimization of constructing systems of movements in sport represent a new stage in the development of the theory of the structure of movements. This stage is characterized by: (a) the optimization of systems of movements, (b) the optimization of building their structure, and (c) the solidification of abstract models in practice (the substantiation of the elements of instruction, the determination of their importance and succession). The sportsman is regarded not merely as a mechanical system, but as an organism (the biological aspect) and as a personality (the social aspect). The substance (logical and sensual) model of the sport technique permits the sportsman when he is performing mechanical movements to adequately prepare for contests, and to act consciously and vigorously since he knows, understands, and feels what brings him top results.

REFERENCES

Bernstein, N. 1967. Coordination and Regulation of Movements. Pergamon Press, London.

Donskoi, D. 1973. Control exercised over the reconstruction of the movement's systems. *In:* S. Cerquiglini, A. Venerando, and J. Wartenweiler (eds.), Biomechanics III, pp. 124–128. S. Karger, Basel.

Donskoi, D., V. Nazarov, and H. Gross. 1974. Control of biodynamic structures in sport. *In:* R. C. Nelson and C. A. Morehouse (eds.), Biomechanics IV, pp. 276–278. University Park Press, Baltimore.

Gross, H., and D. Donskoi. 1974. The rationalization of the sport technique on the basis of the simulation of systems of movement. Teor. Prakt. Fiz. Kult. No. 11.

Optimization of sports effort during training

K. Fidelus, L. Skorupski, and A. Wit
Academy of Physical Education, Warsaw

Various exercises are usually performed during training. The aims of these exercises are to improve motor skills or the trained condition. This trained condition allows the competitor to obtain better sports results. One can distinguish the following interdependences in the process of training: training → trained condition → sports result.

It is necessary to apply various measurements of phenomena of a biochemical, physiological, biomechanical, and psychological nature and tests of motor skills, in order to analyze the entire training process. One can acquire the most basic and detailed information when applying biochemical and physiological measurements while tests of motor skills provide more general information.

Biomechanical phenomena, as they can be precisely defined and measured, served as the starting point in the elaboration of our model of training (Figure 1). It is necessary to distinguish three separate factors, technique, physical features, and tactics, since the improvement of each factor is related to a different biological process. The fourth factor, psychic features, is difficult to define and impossible to measure. This factor influences the sports result either positively or negatively through the other three factors.

Technique training consists of repetitive performances of the main exercise (high jump, putting the shot, striking the ball, etc.) and in performing exercises similar to the main one. Both our research and the research of others (Zatziorsky, 1969) allow us to state the following hypothesis: the result of technique training is a function of the number and kind of exercises that are performed (n_e) and of the number and

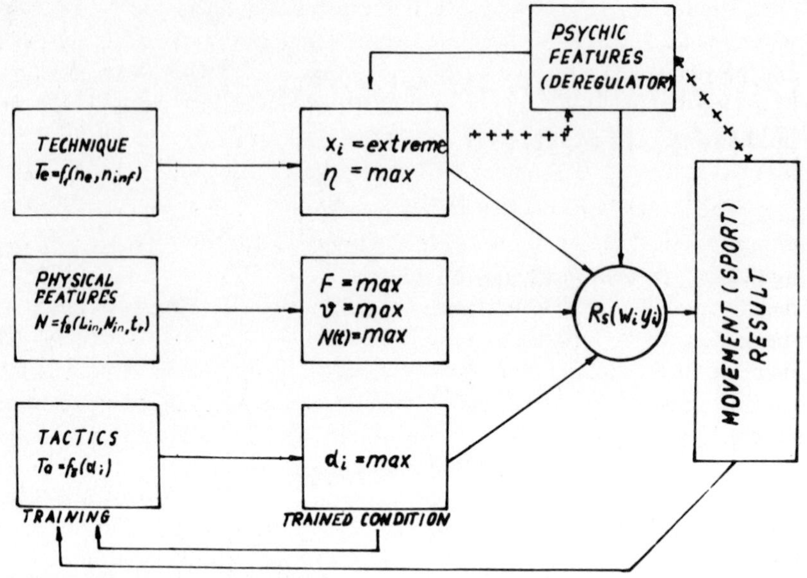

Figure 1. Model of sports training.

kind of information (n_{inf}) that the competitor is provided with during his training.

Using the criterion of technique effectiveness one can assess the trained condition with respect to technique (Fidelus, Kania, Stanoszek, 1973; Fidelus, 1974a). At the same time we must stress that there are different criteria for each exercise (x_i, Figure 1). The value of mechanical efficiency (η), which indicates to what extent internal work (L_{in}) is turned into external work (L_{ex}), may also serve as a criterion of technique. The methods of measurement of internal work and power were published by Fidelus (1974b).

Such physical features as force (F), velocity (v), and endurance (N/t) characterize man's potential abilities and they form a basis for determination of man's power ($N = F \cdot v$). Endurance is characterized by power decrease in the function of time. The training of man's power is the function of internal work (L_{in}), internal power (N_{in}), and optimum time for the period of rest (t_r) that depend on when the maximum of the supercompensation phase (t_s) occurs (Fidelus, Wit, and Pawlikaniec, 1974). The values of t_s for sprint running with maximal intensity can be calculated on the basis of the following equation:

$$t_s = 0.056\, t_e^2 - 0.73\, t_r + 6.75$$

where; t_e is the time of exercise and t_r is the period of rest.

Tactics training consists in defining and making use of decision variables (d_i), while the degree to which tactical activities are mastered can be determined by means of the criterion of activity optimization, based on the mathematical theory of games. We do not deal with the problems of tactics in this article, assuming that in the case of weight-lifters and throwers, tactics are of limited importance.

The level of psychic features is always measured together with technique, physical features, and tactics. One assumes that psychic features influence the trained condition and sports results either positively or negatively. They are linked together by a positive feedback loop and at the same time they play the part of a deregulator in a given system. An increase in the trained condition or a better sports results improves our psychic state, while the opposite situation exerts a negative influence on psychic features, which in turn diminishes the trained condition and the sports result. Each parameter of the trained condition (y_1) influences the sports result (R_s) in definite weights (w_i) that are different for strength sports, endurance sports, and technique sports.

MATERIALS AND METHODS

The investigations were carried out in three groups: (a) six throwers of the Polish National Team, aged 20–25, with training practice of 5–7 yr; (b) nine junior weightlifters, 14–17 yr of age, with 3 yr sports practice, and (c) the control group consisting of ten persons who did not practice sport, aged between 14 and 18. The subjects were chosen so as to determine the increase in power, velocity, and force (muscle torques), which were treated as physical features in the function of training duration and changes connected with biological, independent development of boys. In the case of first two groups we could also observe how sports results and the technique of movement improved under the influence of training. The investigations were carried out according to the methods described in the Symposium in the Theory of Sports Technique (1970). During a whole year (May 1973–May 1974) we measured the value of the sum of muscle torques ΣM_m of the extensors and flexors in the trunk, hip, knee, shoulder, and elbow joints—ten muscle groups altogether.

In the same period of time, velocity of extension of the leg in the knee joint with a variable external force was determined, the value of which was from 45 N to 405 N (radius of force, 0.38 m). Also measured was external work (L_{ex}) performed during the training by means of a film using the following equation:

$$L_{ex} = (Q_w + Q_b) \cdot h$$

where Q_w is the weight of the barbell, Q_b is the weight of the lifter's body, and h is the vertical distance of the center of both masses.

Internal work (L_{in}) was calculated on the basis of the following equation:

$$L_{in} = \frac{L_{ex}}{\eta} .$$

Internal power (N_{in}) was calculated on the basis of the following equation:

$$N_{in} = \frac{L_{in}}{t} ,$$

where t is the time of the exercise.

RESULTS OF TRAINING IN WEIGHTLIFTERS AND THROWERS

During the 1-yr training period the value of ΣM_m in the throwers increased from 3,042 to 3,265 N · m or about 7.3 percent; in junior weightlifters from 2,110 to 2,563 or about 21.4 percent; in members of the control group from 1,755 to 2,000 or about 13.9 percent. Comparing the last two groups one obtains the increase of 7.5 percent, which is comparable with the increase in throwers.

In the case of throwers, velocity of extension of the leg in the knee joint (the resistance force $F = 225$ N) increased from 3.75 m/sec (May 1973) to 4.00 m/sec (May 1974) or about 6.6 percent (see Table 1). In weightlifters with the same value of F, velocity of extension increased from 3.08 to 3.75 m/sec (21.7 percent) and in the control group from 1.70 to 2.02 m/sec (18.8 percent). The increase as a result of the effect of training in the case of the weightlifters was 21.7 — 18.8 = 2.9 percent.

We calculated the maximum power developed by the extensors of the knee joint multiplying resistance force ($F = 405$ N) by velocity of shank movement (v) (Table 1). In the case of throwers this power was 1,117.8 W in 1973 and 1174.5 W in 1974, which was an increase of 5.1 percent. Junior weightlifters' power was, respectively, 830.2 and 1044.9 W, so the increase was about 25.8 percent. The control group values were 374.8 and 472.5 W, which gives an increase of 26 percent. However, the greatest power of the control group was developed at resistance force of $F = 315$ N and it constituted 50 percent of power developed by junior weightlifters.

Table 1. Mean value of physical features of throwers (A), lifters (B), and untrained (C)

	1973	1974	Δ	Δ%
	N_{max} of knee joint extensors (W)			
A	1117.8	1174.5	56.7	5.1
B	830.25	1044.9	214.65	25.8
C	374.85	472.5	97.65	26.0
	$V_{F = 225\ N}$ of knee joint extensors (m/s)			
A	3.75	4.00	0.25	6.6
B	3.08	3.75	0.67	21.7
C	1.70	2.02	0.32	18.8
	M_m, 10 flexors and extensors (N • m)			
A	3042	3265	223	7.3
B	2110	2563	453	21.4
C	1755	2000	245	13.9

Figure 2 presents proportional increase of changes in parameters that were studied during the period of 27 months, from February 1972 until May 1974. During that time junior weightlifters performed internal work of about 30 MJ at mean internal power ranging from 55 to 75 W.

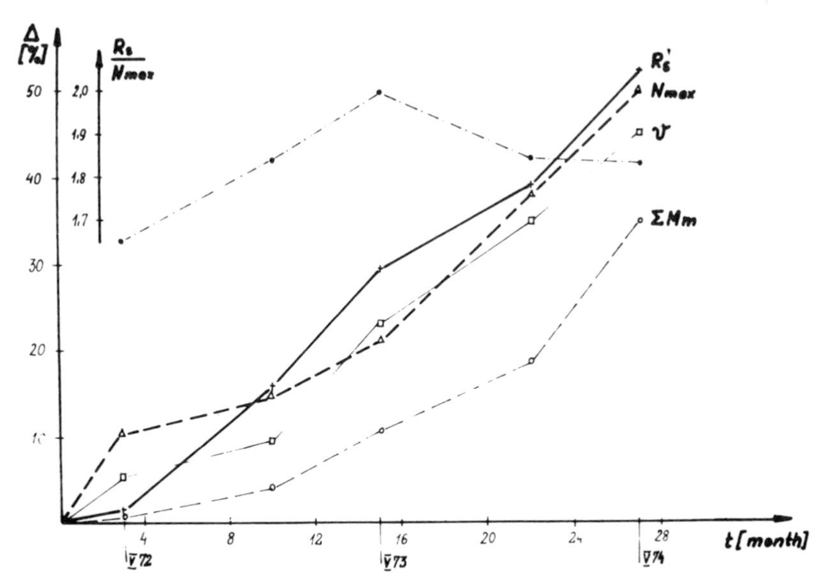

Figure 2. Percentage increase during the 27-month training period for the physical features (M_m, V, N_{max}) sports result (R_s) and the changes in sports technique (R_s/N_{max}).

We observe a faster increase of power (N_{max}) and velocity (v) compared with static muscle strength (ΣM_m) and the sports result (R_s) in the first period of training. In the second year sports result increased most rapidly, and in the third year physical features increased proportionally to sports results. The least increase was observed in static muscle strength.

Figure 2 presents also the change of value of R_s/N_{max}, which was chosen as an indicator of the effectiveness of the movement technique. The acceptance of this indicator is based on the hypothetical assumption that the result obtained in weightlifting is proportional (when taking into consideration appropriate weights w_i) to the product of physical features and the technique of movement. Framing of such an assumption allows one to determine the contribution of technique to sports result and permits one to explain an irregular development of the parameters under investigation here.

SUMMARY

This article deals with a model of the training process and with results of some investigations on throwers and weightlifters carried out in accordance with the assumptions of this model. The proposed model allows for optimization of physical effort when using definite measurement data. Variables measured were: internal work and power developed during the training and force, velocity and power as physical features, and criteria of evaluation of technique effectiveness that characterize the trained condition of the competitor.

REFERENCES

Fidelus, K., H. Kania, and W. Stanoszek. 1973. Effectiveness of the high jump in Fosbury-Flop and straddle technique. Wych. Fiz. Sport 3: 33–41.

Fidelus, K. 1974a. Technika ruchów ludzkich (Technique of human movements). Podst. Prob. Wspólczesnej Tech. PAN 18: 219–237.

Fidelus, K. 1974b. Propozycje jednolitego pomiaru obciażenia treningowego (Suggestions on uniform measuring of the training load). Sport Wyczy. 9: 3–10.

Fidelus, K., A. Wit, and K. Pawlikaniec. 1974. Próba określenia czasu wystepowania maksymalnej sity mieśni po wysitkach szybkościowych i sitowych (An attempt at determining the time of occurrence of the maximum muscle strength after speed and strength efforts). Sport Wyczyn. 11: 3–9.

Sympozjum Teorii Techniki Sportowej. 1970. (Symposium on the Theory of Sports Technique). Sport i Turystyka, pp. 128–175.

Zatziorsky, V. M. 1969. Kibernetika, matiematika, sport (Cybernetics, Mathematics, and Sport). Fizkultura i Sport, Moscow.

Regularities of interdependence of levels of activity in the muscular system and their reflection in athletic motion

I. P. Ratov
All-Union Scientific Research Institute of Physical Culture, Moscow

The purpose of the investigation was to study the functional mechanisms of the muscular system and to determine the causes and conditions of deterioration in intramuscular coordination during athletic exercises.

METHODS

Using multichannel electromyography (up to 12 channels) we studied the amplitude and time characteristics of bioelectric activity of the muscles during different sports and modeled exercises. Simultaneously with the recordings of myograms by an oscillograph a number of other biodynamic characteristics were recorded depending upon the aim of the experiment (force, acceleration, velocity, changes in joint angles, duration of phases). For the automatic treatment of experimental data we used a rapid analysis method based on the use of analog computers and devices for graphic vectoral indication of relationships between different characteristics of movements.

Our experiments can be divided into two main groups. The first group comprised experiments in which changes in phases of movements were modeled artificially. During these experiments the subjects who sat in an armchair had to do exercises with maximum effort involving as many muscles as possible. During this task they had to produce, in succession, additional accentuated tensions in those muscles from which electromyograms were being recorded. The second group consisted of experiments in which athletic exercises were studied. In experiments of this group electromyograms and some biodynamic characteristics were recorded during athletic exercises (sprint running, hurdling, middle distance and long distance running, cycling, skiing, skating, swimming, rowing, high jumping, long jumping, pole vaulting, discus throwing, javelin throwing, shot putting, weightlifting). Characteristics were recorded when the athletes were trying to show the best results.

In analysis of the electromyograms recorded during modeled and athletic exercises relative changes in the levels of activity of muscles provoked by changes in accentuated muscle tensions in each phase of a movement were studied.

Experiments were carried out from 1963 to 1974 (altogether 1,050 experiments). Subjects were highly trained male athletes from 20 to 31 yr old (370 men).

RESULTS

It was shown that during exercising with maximum tension of as many muscles as possible, additional tension of one of the muscles while provoking an increase in the level of its electrical activity, at the same time caused a decrease in electrical activity of other muscles.

Changes in accentuated muscle tension caused relative changes in the levels of activity of the system of muscles in accordance with the increase of levels of activity in the accentuated muscles. Amplitude characteristics of electrical activity in other muscles were simultaneously decreased.

Thus, an original pattern of changes in movement phases was realized, which permitted observation and study of the basic mechanism of coordination changes in levels of activity of the system of muscles.

It was also learned that the degree of possible influence of either muscle on tension levels of other muscles was predetermined by such characteristics as the ability for a quick development of activity or in other words to become "easily excited." Classification of muscles according to their capability for quick development of activity showed that "small" muscles become activated very quickly while "big" muscles become excited more slowly. At the same time the accentuated tension of quickly activated muscles leads to a considerable decrease in the levels of activity of other muscles which are less important in comparison with accentuated tension of muscles that develop activity slowly.

A very important consideration for the understanding of the functional mechanism of the locomotor apparatus lies in the fact that excessive activation of muscles that can be called quick muscles takes place, at the expense of large muscles. This situation causes a decrease in their tension level and consequently, a decrease in the effective external force of a movement.

Regularities obtained in model experiments were confirmed by the results of the second group of experiments in which myograms and biodynamic characteristics were recorded during different athletic exercises. It was discovered that excessive or untimely activation of some small muscles which, according to their ability for a quick development of activity, can be classified as "quick" muscles, provokes considerable changes in characteristics of electrical activity of other muscles and such negative changes in biodynamic characteristics as a decrease in the resultant force developed during athletic movements.

In order to show more clearly and under stricter conditions the reflection of changes in correlation of muscular tensions levels in characteristics of athletic exercises, we used a method of "artificial mistakes." The essence of the method was that the conditions for the fulfillment of an exercise being studied were programmed in such a way that one of the quickly activated muscles would be activated untimely or to a greater degree than was necessary at a certain moment planned beforehand.

Under such conditions we followed the influence of excessive tensions of quickly activated muscles (upper extremities, neck, and facial muscles) on changes in the characteristics of movements in sprint running, middle distance running, and pedaling on the bicycle ergometer, and also the influence of untimely and excessive tensions of the trapezius and muscles of the upper extremities during weightlifting and shot putting.

It should be noted that we used especially constructed devices for visual and sound indication of the muscle's electrical activity. With the help of these devices it was possible to control the dosage of "artificial mistakes."

After we had proved experimentally that the reason for the majority of mistakes in athletic exercise techniques was a consequence of excessively sharp deviation in the relationship of levels of tension in the system of muscles, these devices for instrumental control of electrical activity of muscles were used to create such conditions for fulfilling exercises under which the possibility of excessive or untimely activation of muscles was lowered.

CONCLUSIONS

Regularities in the functioning of the system of muscles make it possible to explain the reasons for most of the mistakes in techniques of athletic exercises. These errors were the result of deterioration in intramuscular coordination. Since any additional increase in the level of activity and tension of any muscle during exercise performed with maximal intensity is inevitably accompanied by a corresponding decrease in the activity level of other muscles, it becomes clear why an excessive activation of some muscles leads to the decrease of the working effect of other muscles.

Excessive tension of relatively small but quickly activated muscles is more likely to cause a decrease in activity level of "big" muscles than activation of "big" muscles would influence "small" muscles. This explains the often observed decrease in the efficiency of "big" muscles which is caused by excessive activity in "small" muscles.

From these results it follows that to ensure appropriate intramuscular coordination in order to achieve high results, we should find variations of exercises in which the probability of excessive or untimely activity of quickly activated muscles will be very low.

Biomechanics of athletic shoe design

G. B. Ariel
University of Massachusetts, Amherst

In all athletic performances, whether team games such as football, soccer, basketball, volleyball, or in individual sports such as running, jumping, throwing, or cycling, where gravitational forces play a major role, the shoes on which the weight bearing athlete plays are probably one of the most important contributing factors in the execution of efficient human performance. The interaction between the uniqueness of each specific activity and the athlete is influenced by factors such as the force of impact, the weight of the athlete, the surface upon which the activity is performed, the individual's stride and gait, etc. Athletic footwear cannot be evaluated separately from the "athlete in the shoe." Therefore, biomechanical factors of the athletic performance must be considered when designing shoes.

For a number of years, repeated attempts have been made by the writer to determine the scientific methods used by various shoe companies in the United States, Germany, Finland, and other countries in the development of their athletic shoes. However, it is evident that shoe design and safety have been the province of both the shoe designer and safety engineer. The engineer has been concerned with the physical components of the shoe such as the coefficient of friction and durability factors influencing the shoe. Often esthetic features receive the greatest consideration. However, it is contended that athletic performance cannot be thoroughly researched without taking into account "the athlete in the shoe." Unfortunately, the shoe designers have overlooked the fact that shoe efficiency, safety, and performance are inextricably tied to the biomechanics of the particular activity and the style of whoever is under scrutiny.

At present, however, there are few reliable approaches for measuring these factors which are essential in athletic shoe design. In the present study, a new approach to athletic shoe design will be discussed.

METHOD

Biomechanical Measurements

A computerized biomechanical analysis system was used in assessing the data. This sophisticated method is described in detail elsewhere (Ariel, 1973). In general, slow motion cinematography is used in conjunction with a force plate to record any desired human motion. This technique permits an undetected recording of an individual's performance under actual conditions. The Model GP-3 Graf-Pen digitizer enables precise determination of the coordinates of the joint centers of the body. A digital processing oscilloscope permits the storage of all forces and moments of force on the contact points between the shoe and the surface. Location of the joint centers provides the measurement of the segment lengths and angles, while determination of the segment masses, centers of gravity, and radii of gyration allows the calculation of forces and moments of force around each body joint center. Appropriate programming (CBA, 1975) results in a segmental breakdown of information of the whole motion, including the total body center of gravity, segment velocities and accelerations, and the timing or coordination of motion among the body segments. The combination of cinematographical data with force-plate data yields the instantaneous forces on the shoe as a function of time.

Shoe Measurements

Vibration testing of the various shoes was performed utilizing the force data obtained from the cinematographical and force-plate data. The force applied to the shoe material at various locations was plotted against the displacement of the material, yielding an elastic hysteresis where the increasing and decreasing forces resulted in curves that did not coincide. The area bounded by this plot determined the energy loss and the slope of the curves determined the energy stored. The sole and inner-sole of the shoe produce a force that is constantly in contact with the test piece and the hysteresis loop takes place during every cycle of vibration. The loop is velocity dependent and yields the damping effect of the materials. The loop also yields the stiffness or restoring force, which is displacement dependent. If the elastic properties are poor, the bottom curve does not return to zero in time for the beginning cycle of the first curve of the next cycle.

RESULTS

The present experiment resulted in two types of data. The first results were the biomechanical behavior of the athlete. The second results were the data obtained from the various athletic shoes based on the dynamic testing of the shoe itself. Figure 1 illustrates the interaction between the surface, shoe, and body joints. A unique approach in this study was that the dynamic characteristics of the shoes were obtained by utilizing the biomechanical data from the athlete as he performed rather than the common trial and error method of observation alone.

Anatomical Consideration

Foot movements are important considerations in the design of the athletic shoe. However, because of the numerous articulations, the foot is quite complex. For example, the foot consists of the following joints: (a) the ankle joint, between the tibia and talus; (b) the talocalcaneal joint, between the talus and calcaneus; (c) the mid-tarsal joints, calcaneus to cuboid, and talus to navicular; (d) the tarsometatarsal and

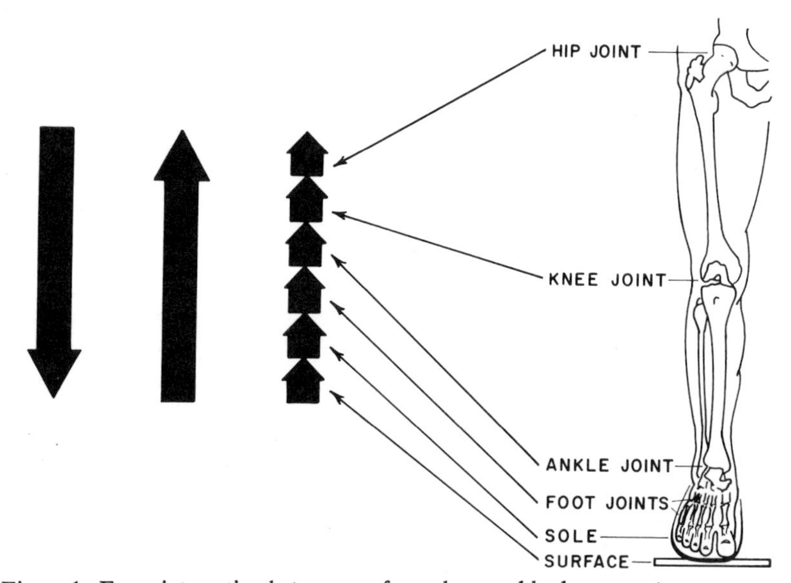

Figure 1. Force interaction between surface, shoe, and body segments.

intermetatarsal joints; and (e) the metatarsophalangeal joints. Cinematographical analyses of 10 athletes running bare-footed yielded the following movements of the foot: (a) the ankle joint allows flexion and extension in the sagittal plane; (b) the talocalcaneal joint permits inversion and adduction combined with foot extension, and eversion and abduction combined with foot flexion; (c) the mid-tarsal joints allow motion in all planes with distinctive supination and pronation of the foot; (d) the tarsometatarsal and intermetatarsal joints permit gliding motion for weight bearing; and (e) the metatarsophalangeal joints permit motions in all planes.

The complexity of foot movements and the specificity of each activity demonstrate the need for biomechanical considerations in designing athletic shoes. A total body analysis is necessary in order to design an athletic shoe that accurately conforms to the needs for the particular event.

Anatomical Factors Relating to Shoe Design

The cinematographical data revealed that the position of the foot is influenced by the mechanical axis of the femur relative to the mechanical axis of the tibia. Normal posture causes the position of the foot to angle outward in a range of 21–30 degrees depending upon individual differences as well as sex differences. The position of the foot was further influenced by motion at the hip joint. Flexion at the hip joint caused medial rotation of the foot, while extension at the hip joint caused lateral rotation. When the foot is in contact with the surface this rotation causes torsional forces on the sole of the shoe as indicated from the force–plate data. Flexion of both the hip and knee joints caused the foot to turn inward, and extension of both joints resulted in outward rotation of the foot.

Shoe Considerations

Figure 2 illustrates the shoe positions on the surface at various positions for two Olympic runners, Wottle and Ryan. One may observe that the placement of the shoe is different for the two runners.

Forces obtained from different activities with various magnitudes were tested for the present study. Figure 3 illustrates the computer output of the dynamic testing for these shoes. The slope of the hysteresis curves shows the firmness of the material, while the area within the loop shows the damping characteristics. The minimum dynamic indentation shows the amount of indentation of the material after a steady-state loop has

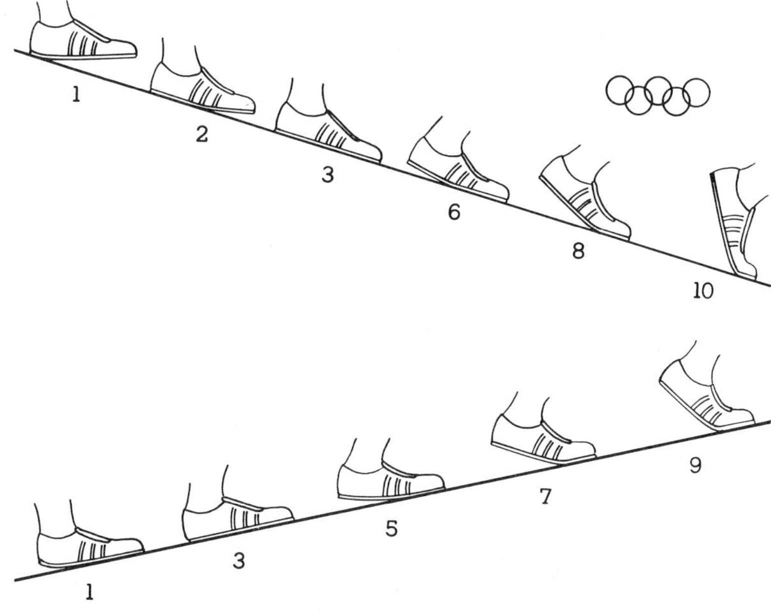

Figure 2. Athletic shoe positions at various contact phases.

been attained from a 10-cycle/sec loading. The slope of the curve and the point at which any abrupt changes in the slope take place must be compared to the impact loading of a runner contacting the ground. The force per unit area on the shoe was calculated for different speeds of running. This information, used in conjunction with the hysteresis loops, showed that the shock absorption quality of the foot and ankle was fully used during the time period that the softest part of the shoe (inner sole) was being compressed. The slope of the hysteresis loop changed sharply upward after a loading of approximately 22–33 kg (10 to 15 lb), and neither the shoe, nor the foot, nor a material with a lower slope can absorb any more vertical forces. From the hysteresis loops in Figure 3, it was revealed that a leading cross country shoe (A) showed little shock absorption that might result in the transmission of high forces to the ankle, knee, and hip joints. Shoe (F), which is a leading jogging shoe, had maximum shock absorption when the load was minimal and minimum shock absorption when the load was at a maximum. This phenomenon illustrates energy loss during running. Shoe (C) is a leader among basketball shoes. Its characteristics are opposite to the desired ones. Energy loss is great, which may contribute significantly to the athlete's fatigue, and the shock absorption characteristics are not consistent with those desired in this game.

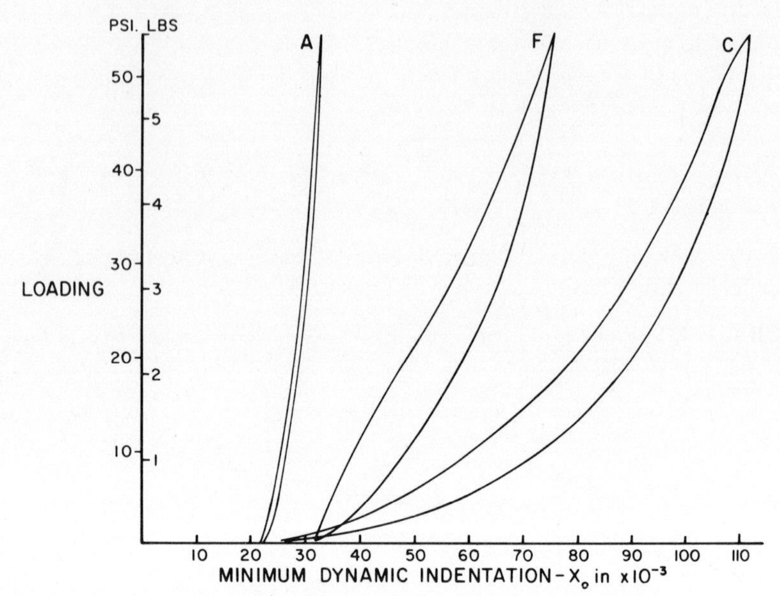

Figure 3. Hysteresis loops obtained from three different athletic shoes.

DISCUSSION

Analyses of more than 35 different running shoes revealed that human factors were not taken into consideration when designing these shoes. The dynamic characteristics of these running shoes yielded results which are opposite to the desired ones. The hysteresis loops obtained demonstrated insufficient shock absorption characteristics or too much absorption at the wrong time. In other words, when the loading was at its minimum the greatest shock absorption occurred while at the point of greatest loading, when the shock absorption is needed, the material does not respond properly. None of the shoes in this study demonstrated results that are considered desirable for the given activity.

From the evaluation of the shoe material and its interaction with the shock absorbing qualities of the foot, it was concluded that the heel and the outside edge of a shoe should have a hysteresis slope of about 45 degrees with no abrupt changes in the slope. The section of the shoe at the ball of the foot should have a hysteresis slope of about 75–80 degrees with no abrupt changes so that the runner or thrower can push off from a solid surface.

At the present time, there are no athletic shoes available which consider the "athlete in the shoe." In fact, some of the shoes may contribute to the injury-risk factor. Some of the best running apparatus used in the past, and currently available, belong to the marathon runner from Ethiopia who ran bare-footed in the 1960 and 1964 Olympic games—and won the gold medal.

REFERENCES

Ariel, G. 1973. Computerized biomechanical analysis of human performance. Mechanics and Sport, pp. 267–275. The American Society of Mechanical Engineers, New York.

CBA. 1975. Biomechanic Package–Body Series. Computerized Biomechanical Analysis Incorporated, 316 College St., Amherst, Massachusetts, U.S.A. 01002.

Computer applications and modeling

Dynamic interactive computer graphics package for human movement studies

D. R. Riley and G. E. Garrett
Purdue University, Lafayette

Since the late 1960s there has been an effort by an interdisciplinary team from Purdue University's School of Mechanical Engineering and Department of Physical Education for Women to pool knowledge in order to make better use of the computer as a tool in analyzing human motion (Garrett, Boardman, and Garrett, 1973). The first attempts at utilization of the computer for animation purposes included filming static CALCOMP plots and graphical output from a Control Data Corporation Graphic Display Console (Garrett et al., 1971). These computer techniques made possible the generation of satisfactory films, but were a long way from real-time animation.

The research performed to date, however, has been aimed at applying dynamic, interactive computer graphics techniques to the process of obtaining and analyzing cinematographical data. The emphasis has been on facilitating the collection and handling of human motion data and, more importantly, providing the movement researcher and educator a new tool with which to analyze complex movement skills. The researcher can now conveniently work at one station for the entire process of digitizing and analyzing the data and visualizing the results. The interactive, real-time, dynamic animation capabilities of the system provide the instantaneous feedback in visual form that is necessary to ensure the accuracy of this process.

THE SYSTEM

The data collection, storage, and retrieval, and dynamic display system is centered around an IMLAC PDS-1D (8,000-word memory) minicomputer (Figure 1) with a refresh-type cathode ray tube (CRT), located in the Computer Aided Design Laboratory, School of Mechanical Engineering. Interfaced to the IMLAC is a magnetic disc storage device and a Graf-Pen sonic digitizer, consisting of a pen, 35-cm (14-inch) tablet, and controller, with a resolution of 0.17 mm (0.007 inches). Digitizing consists of merely touching the pen to the desired point, and the Cartesian coordinates are automatically stored in the memory of the IMLAC. At the same time, the digitizing routine uses the digitized point to display a stick-figure outline on the CRT screen for a quick visual check of the data.

The input data to the system consists of 16 mm film of the motion to be analyzed, which is then projected by a film reader onto the tablet surface from above. The tablet contains a frosted glass surface and plans are to implement rear projection onto the tablet surface so that the user's hand will not interfere with the projection.

When beginning to digitize, the user is first prompted to type in the height and weight of the subject, time between film frames, and a calibra-

Figure 1. IMLAC and Graf/Pen configuration.

tion length to be used in analyzing the data. Each frame to be digitized then must contain two fixed reference points, which are used by the analysis routines to perform origin translation and rotation to compensate for frame-to-frame misalignment during digitizing. Following these two points, 17 joint locations are digitized.

When the user has finished digitizing, the data, including the additional typed-in information, are stored on the magnetic disc storage device for later use. The user may then digitize another subject or go on to manipulate or analyze the data.

The data manipulation routines perform the functions of retrieving the data from the disc and allowing the user to view the data and then transmit the data for analysis either to a PDP 11/40 (24,000-word memory) floating point processor located in the same laboratory, or to a large CDC 6500 computer at the Purdue University Computing Center via the remote terminal system.

In the viewing mode the data are used to cause a stick-figure person to move across the screen of the IMLAC. The user is able to view the data continuously in real time or at any one of infinitely variable slower speeds or in a frame-by-frame stop action mode, both forward and reverse. Figure 2 is a time exposure photograph of the data for a 22-month-old infant running, viewed in the continuous mode. (Note: This and all succeeding photographs were taken directly from the CRT device.) In addition, the user may specify only certain limbs of the body that he would like to isolate for viewing. These computer animations can be filmed directly off the screen for presentations and educational purposes.

One additional viewing option of a slightly different nature has also been provided. If the user desires a "hard-copy" of the figure on the CRT screen, he may cause the current frame or a series of frames to be plotted on a CALCOMP 502 flat-bed incremental plotter in the laboratory.

The analysis routines for the CDC 6500 and PDP 11/40 are identical in the computations performed. The analyses include center of gravity, linear and angular displacements, velocities, and accelerations and kinetic energies of the individual segments and of the body center of gravity (Garrett, 1970; Garrett and Widule, 1971). On the CDC 6500, output includes various selectable CALCOMP plots of any of the above parameters.

For more immediate, interactive viewing of the results, the analysis package on the PDP 11/40 allows user selection from a "menu" of various plots, with viewing either on the IMLAC or on a Tektronix 4014

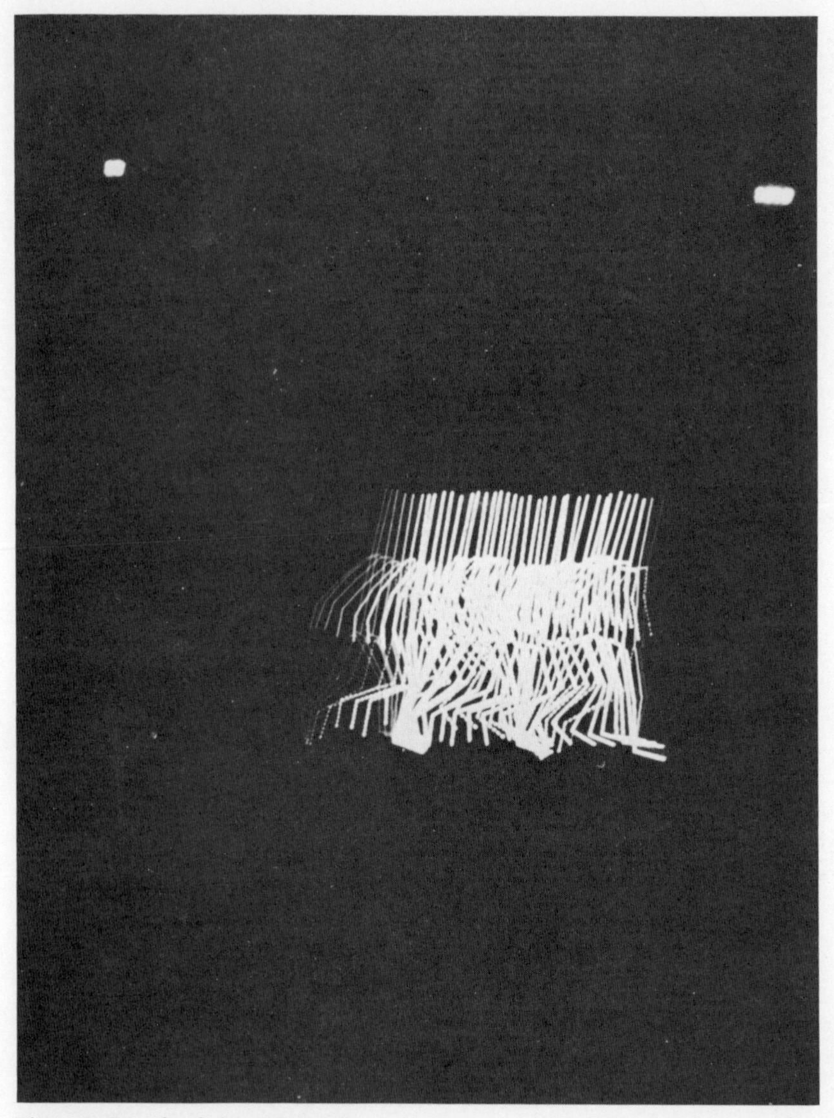

Figure 2. Running infant (IMLAC).

storage-type CRT (35×27-cm screen). Figures 3 and 4 show the stick figure and center of gravity plot for the running infant on the Tektronix. On the Tektronix, the user selects the desired option from the menu by positioning a cross-hair on the screen over the appropriate box and hitting a key to enter his selection. The appropriate plot is then displayed

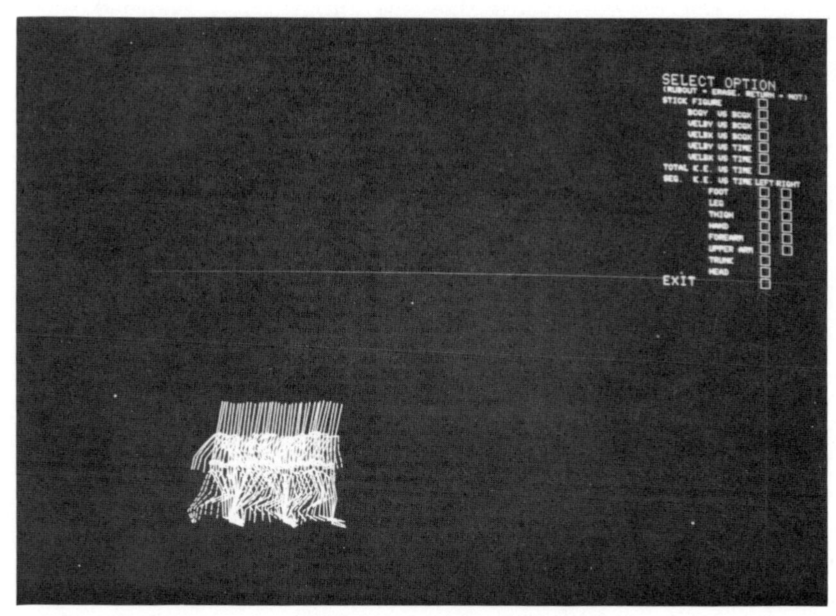

Figure 3. Running infant stick-figure and menu (Tektronix).

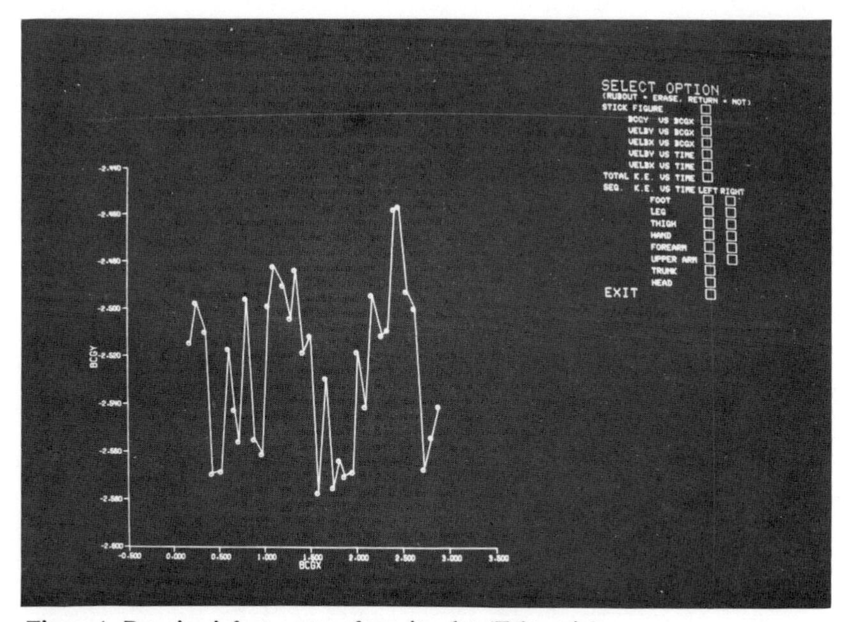

Figure 4. Running infant center of gravity plot (Tektronix).

on the screen. Figures 5 and 6 show the corresponding Tektronix stick figure plots for an adult sprinter and an adult glide kip, respectively. In addition, the user has the option of either erasing the screen first or of superimposing one plot on another for comparison. Figure 7 illustrates this feature for the kinetic energies of the lower legs of the running infant.

What has just been described is the current state of development of the system. Many other options are currently being implemented so that the system is in a continual state of development.

At this point it should be emphasized that the kind of system just described is not so much the "impossible dream" for most researchers that it was a few years ago. Recent technological advances in minicomputer design and manufacturing methods have produced a significant reduction in the costs of acquiring such equipment. Where five years ago equipment capable of performing such complex tasks would have cost at least $250,000, today the costs range from $4,000 for a small Tektronix to $40,000 for an IMLAC with magnetic disc.

APPLICATIONS

This dynamic graphics package has been utilized in two master of science theses in the Department of Physical Education for Women at Purdue

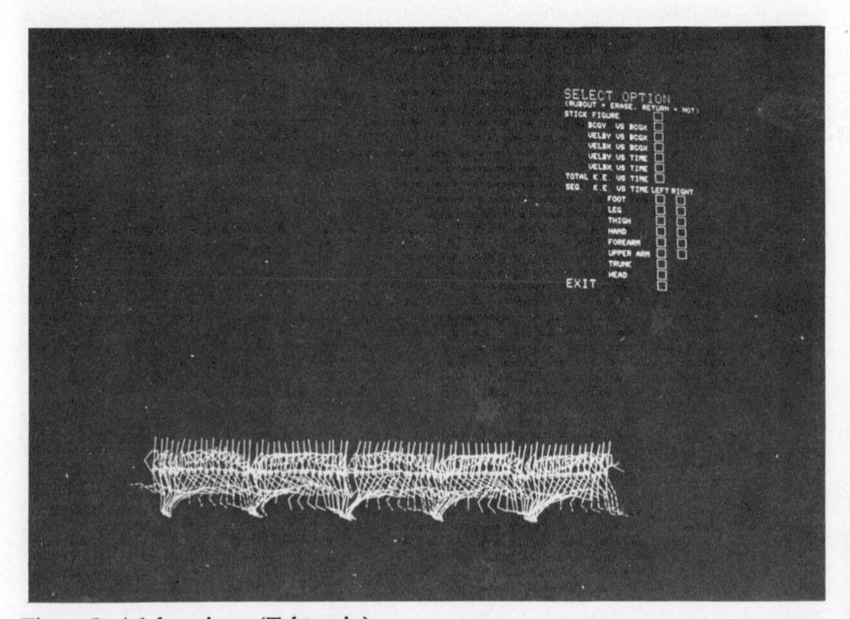

Figure 5. Adult sprinter (Tektronix).

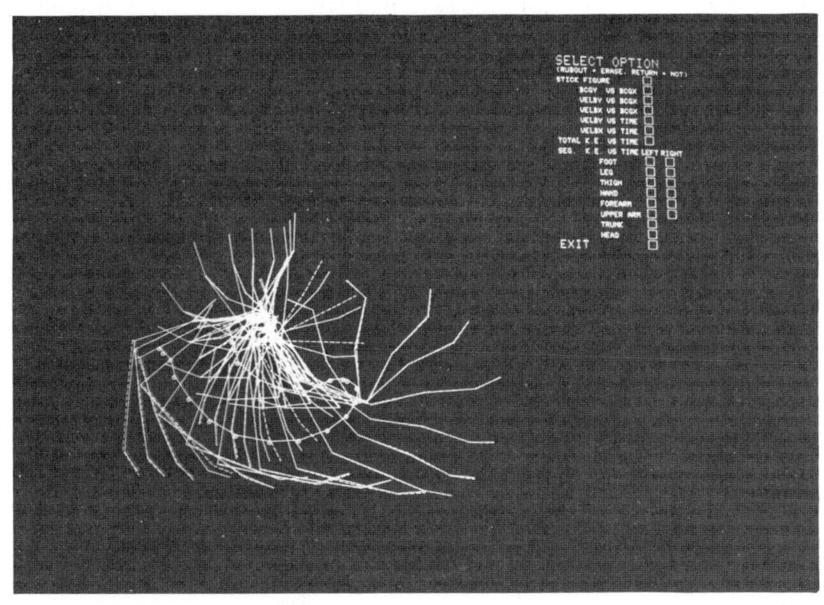

Figure 6. Glide kip (Tektronix).

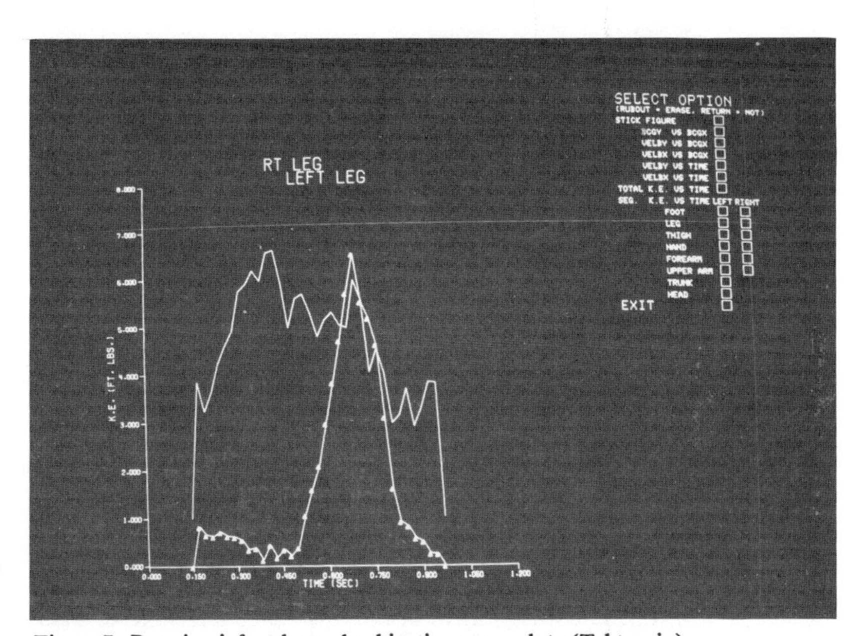

Figure 7. Running infant lower leg kinetic energy plots (Tektronix).

(Mersereau, 1974; Ulibarri, 1974). In both of these studies dynamic graphical output provided the means for comparing female subjects' running skills at various ages and skill levels, bringing to "life" what would otherwise be an overwhelming number of inanimate data points.

Superposition of one plot over another provides the researcher with the capability for making developmental and skill level comparisons, as may be seen in Figure 7. Simultaneous viewing of more than one subject, illustrated by Figures 8 and 9 (Mersereau, 1974), provides the researcher with an accurate visual tool for developmental comparisons. These figures show four female children running at 22 and 25 months of age. The researcher now has instantaneous feedback as he attempts to unravel developmental changes in skill patterns of individual subjects and also within groups.

This dynamic, interactive, computer graphics capability shows promise of becoming the vehicle for providing the research team with the ability to develop and maintain a permanent library or "atlas" of movement history that can be constantly updated. The possibility of such a library opens new vistas for both the researcher and the educator in movement analysis.

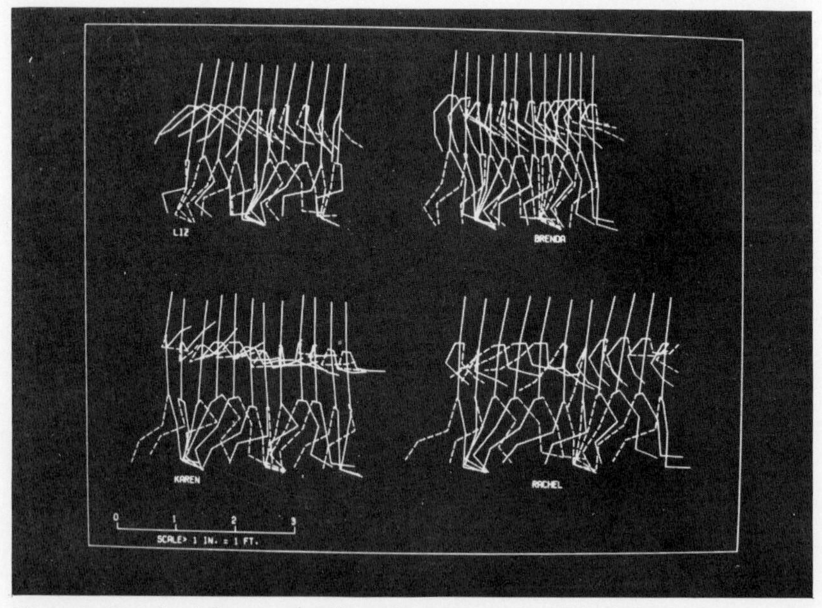

Figure 8. Running infants comparison plots, 22 months (Tektronix).

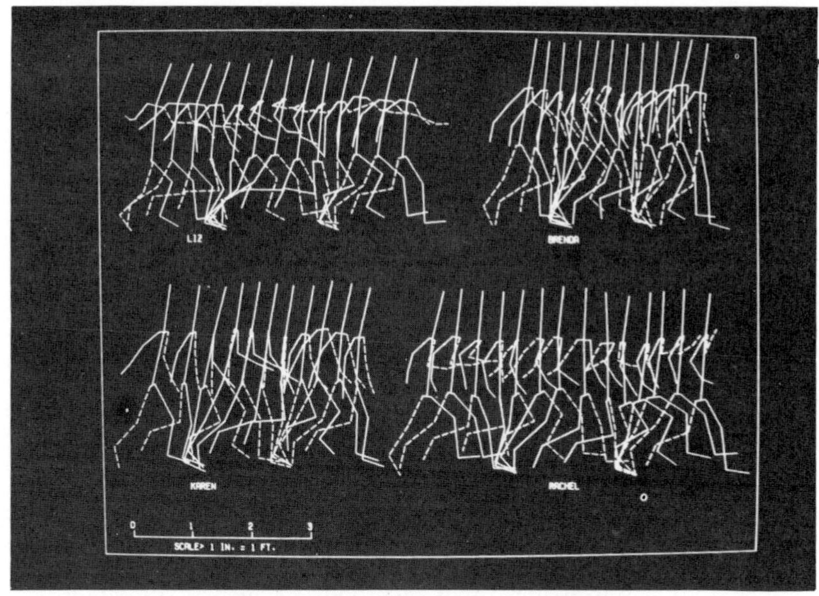

Figure 9. Running infants comparison plots, 25 months (Tektronix).

REFERENCES

Garrett, G. E. 1970. Methodology for assessing movement individuality. Ph.D. Dissertation, Purdue University, West Lafayette, Indiana.

Garrett, G. E., and C. J. Widule. 1971. Kinetic energy: A measure of movement individuality. Kinesiology Rev. 2: 49–54.

Garrett, G. E., W. S. Reed, C. J. Widule, and R. E. Garrett. 1971. Human motion: Simulation and visualization. *In:* J. Vredenbregt and J. Wartenweiler (eds.), Biomechanics II, pp. 299–303. University Park Press, Baltimore.

Garrett, R. E., T. Boardman, and G. E. Garrett. 1973. Poor Man's Graphics. *In:* S. Cerquiglini, A. Venerando, J. Wartenweiler (eds.), Biomechanics III, pp. 74–83. University Park Press, Baltimore.

Mersereau, M. R. 1974. A cinematographic analysis of the development of the running pattern of female infants at 22 and 25 months of age. Master's thesis, Purdue University, West Lafayette, Indiana.

Ulibarri, V. D. 1974. A cinematographical analysis of mechanical and anthropometric characteristics of highly skilled and average skilled female sprinters. Master's thesis, Purdue University, West Lafayette, Indiana.

Model for body segment parameters

R. K. Jensen
Laurentian University, Sudbury

The size and mass distribution of the segments of the human body are fundamental to biomechanical analysis. Simplifying assumptions have to be made in order to determine the properties of the segment. In the present study it was assumed that each body segment is composed of a number of elliptical zones of known uniform density. The purpose was to: (a) develop a model to give the segmental and the whole body volume, mass, buoyancy, mass center, and moment of inertia; (b) develop apparatus to determine dimensions and masses; and (c) compare the calculated parameters with parameters obtained from the model developed by Hanavan (1964).

The assumption of elliptical zones has previously been made by Weinbach (1938), who calculated the body center of gravity and moment of inertia using graphical techniques. Dempster (1955) evaluated the accuracy for volume and found it to be very good with the exception of the shoulders. The commonly used geometric model, by Hanavan (1964), requires few input parameters and calculations. A major limitation of the model is that segment shapes ". . . cannot be duplicated by a geometric solid of revolution . . ." (Hanavan, 1964, p. 39), a problem that was minimized in the present study.

MODEL

Each segment can be sectioned into zones in an inertial reference frame. The radii of the elliptical zone, r_x and r_y, are given by (from Figure 1)

$$r_{xk}' = \frac{1}{2}\left[\frac{(x_{2k} + x_{2j})}{2} - \frac{(x_{1k} + x_{1j})}{2}\right] \qquad (1)$$

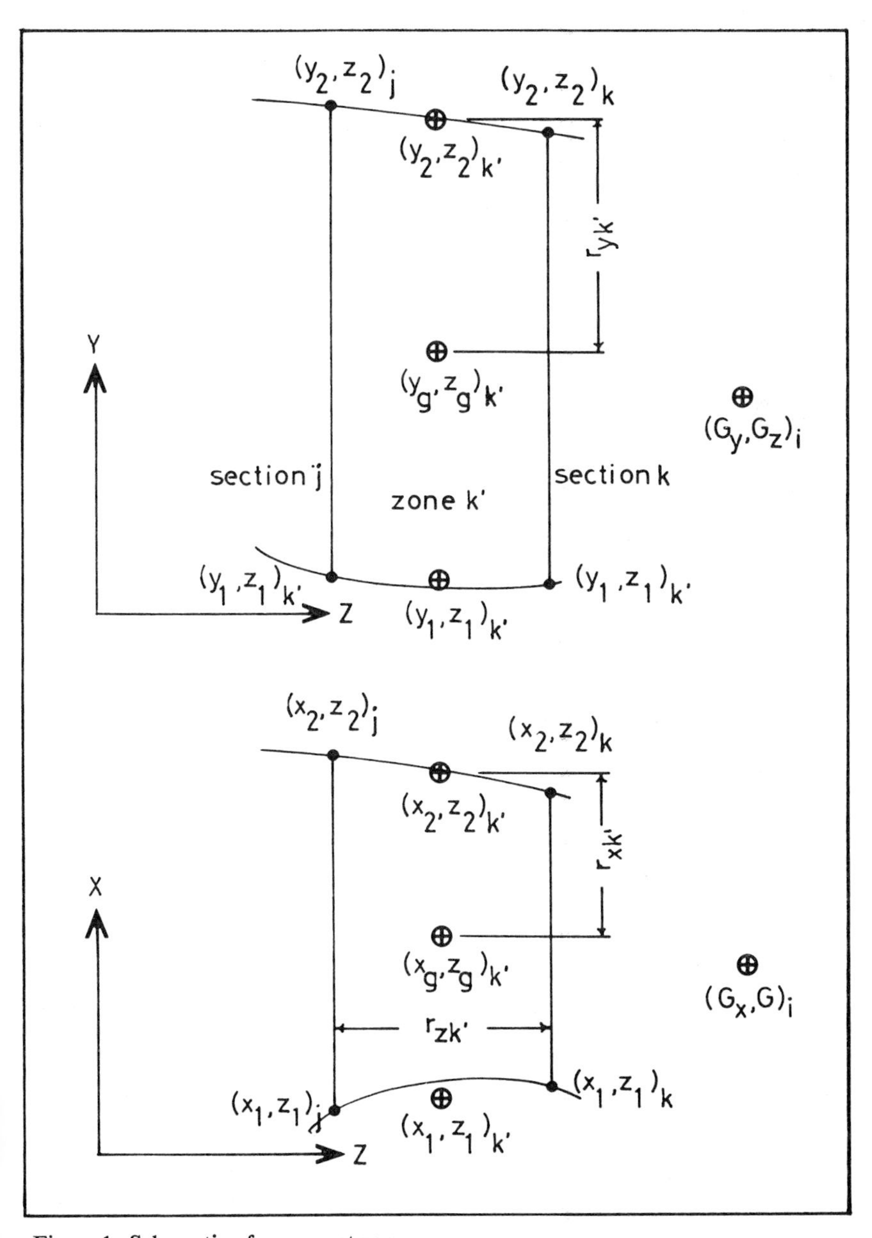

Figure 1. Schematic of a segment zone.

$$r_{xk}' = \frac{1}{2}(x_{2k}' - x_{1k}') \tag{2}$$

and the zone thickness, r_z, by

$$r_{zk}' = z_k - z_j. \tag{3}$$

The centroid of the zone, $(x_g, y_g, z_g)_k'$, is, by construction,

$$x_{gk}' = \frac{x_{2k}' + x_{1k}'}{2} \tag{4}$$

with y_{gk}' and z_{gk}' in similar form. Standard formulas for elliptical plates can now be applied. Zone volume is given by

$$v_k' = \pi r_{xk}' \, r_{yk}' \, r_{zk}'. \tag{5}$$

The first moments of volume, $(t_x, t_y, t_z)_k'$, are given by

$$t_{xk}' = v_k' \, x_{kg}' \tag{6}$$

and the moments of inertia of area, $(I_{xk}'^a, I_{yk}'^a, I_{zk}'^a)$, for the plane through the zone centroid are

$$I_{xk}'^a = \frac{1}{4} \pi \, r_{xk}' \, r_{yk}'^3, \tag{7}$$

$$I_{yk}'^a = \frac{1}{4} \pi \, r_{xk}'^3 \, r_{yk}', \tag{8}$$

$$I_{zk}'^a = I_{xk}'^a + I_{yk}'^a. \tag{9}$$

The properties of the segments can be found by summing across $(n - 1)$ zones and the properties of the whole body by summing across m segments. Segment volume, V_i, and whole body volume, V, are given by

$$V_i = \sum_{2}^{n} v_k' \tag{10}$$

and

$$V = \sum_{1}^{m} \sum_{2}^{n} v_k'. \tag{11}$$

Segmental buoyancy is

$$B_i = V_i \, \rho a_g$$

where ρ is the density of water and a_g is acceleration due to gravity. The segment centroid, G_i, and the whole body centroid, G, are found from the first moments,

$$G_{xi} = \frac{\sum_{2}^{n} t_{xk}'}{V_i}, \tag{12}$$

$$G_x = \frac{\sum_{1}^{m} \sum_{2}^{n} t_{xk}'}{V}. \tag{13}$$

In order to determine the remaining parameters, segment mass, M_i, or segment density, ρ_i, have to be known as

$$\rho_i = \frac{M_i}{V_i}. \tag{14}$$

Weights and the whole body mass center can then be calculated. The mass moments of inertia of the zone, $(I_{xk}'^{g}, I_{yk}'^{g}, I_{zk}'^{g})$, are found from

$$I_{xk}'^{g} = I_{xk}'^{a}\, \rho_i\, r_{zi}, \tag{15}$$

and the mass moments of inertia of the segment about its centroid axes are found by applying the parallel axis theorem,

$$I_{xi}^{g} = \sum_{2}^{n} (I_{xk}'^{g} + \rho_i\, v_k'\, c_k'^{2}), \tag{16}$$

where c is the distance between the segmental and the zone centroids,

$$c_k^{2\prime} = (G_{yi} - y_{gk}')^2 + (G_{zi} - z_{gk}')^2. \tag{17}$$

The parallel axis theorem can also be used to find the moments of inertia about the axes through other points including the whole body center of mass. These reference axes are not necessarily principal axes, however.

PROCEDURE

The above equations were used to write a program (ZONE) in the FORTRAN IVG language. A slightly modified version of Hanavan's program (1964) was also written (DESIGN). The apparatus consisted of a horizontal reaction board (the *Y–Z* plane) and a sheet of plexiglass fixed vertically to the board (the *X–Z* plane). A large mirror at 45° reflected the *Y–Z* plane and the planes were photographed with a 35 mm camera and 200 mm lens. An 11-yr-old, outstanding male swimmer was tested (height, 1.52 m; weight, 40 kg). Surface markings of relevant anatomical landmarks and estimated joint centers (Plagenhoef, 1971, pp. 10–27) were made and the subject placed supine with the feet plantar flexed. The projected photographs were sectioned into 16 segments and zones 2 cm wide (Figure 2) and the coordinates digitized. The reaction board transducer was sensitive to masses of 20 gm. Anthropometric data were also recorded for the DESIGN program.

RESULTS AND DISCUSSION

Because there appears to be no agreement as to the best method of evaluating each of the segmental parameters, the results of the two programs were compared as follows. First, a geometric human figure was constructed and the properties of the solids of revolution from DESIGN used as the criteria for evaluating the accuracy of the ZONE program. Second, because the ZONE program took into account the shape of the segments, it was used as the criterion for evaluating the accuracy of the DESIGN analysis of the subject. The segment masses used in the two programs were from Barter's regression equations (1957). The results for the geometric human figure indicated that overall the errors for the ZONE program were less than five percent. Higher percentage errors were noted for the hand, a sphere, and the foot, a frustrum of a cone placed at a 45° angle.

Selected body segment parameters for the subject are presented in Table 1. The volume of the ellipsoid for the head (DESIGN) was greater

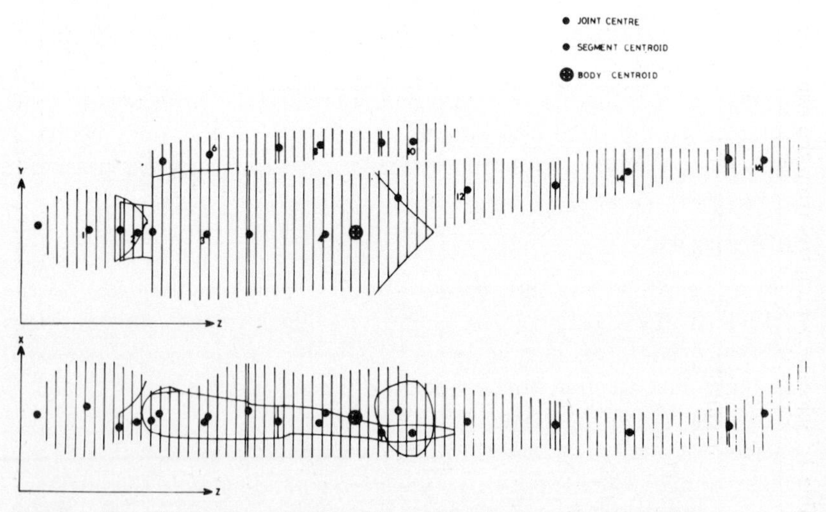

Figure 2. Simplified representation of the subject in the reference position.

Table 1. Selected parameters from the two programs

Segment	No.	Program	Volume 10^{-3} m^3	Mass kg	I_{xx} (10^{-3}) kg · m^2	I_{yy} (10^{-3}) kg · m^2	I_{zz} (10^{-3}) kg · m^2
HEAD	1	ZONE	3.44	2.79	10.05	11.76	9.24
	1	DESIGN	4.55	3.16	17.95	17.95	9.51
NECK	2	ZONE	0.45	0.37	0.40	0.29	0.44
UP TRUNK	3	ZONE	6.23	7.33	55.24	36.27	42.95
	2	DESIGN	6.00	6.72	43.46	28.39	42.12
LO TRUNK	4	ZONE	10.76	12.66	165.49	147.29	66.53
	3	DESIGN	10.80	13.28	154.40	124.40	82.25
UP ARM	5–6	ZONE	1.15	0.90	5.40	5.55	0.75
	6–7	DESIGN	0.95	0.90	5.02	5.02	0.54
LO ARM	7–8	ZONE	0.44	0.69	2.63	2.61	0.30
	8–9	DESIGN	0.50	0.66	2.22	2.22	0.27
HAND	9–10	ZONE	0.19	0.31	0.47	0.41	0.10
	4–5	DESIGN	0.24	0.34	0.21	0.21	0.21
THIGH	11–12	ZONE	3.90	4.13	38.38	39.24	8.24
	10–11	DESIGN	3.19	4.13	42.26	42.26	6.64
LO LEG	13–14	ZONE	2.03	1.70	19.05	18.74	2.04
	12–13	DESIGN	1.49	1.69	18.88	18.88	1.11
FOOT	15–16	ZONE	0.83	0.69	1.54	2.41	1.27
	14–15	DESIGN	0.47	0.71	3.48	3.48	0.23
WHOLE BODY		ZONE	37.98	39.99	5223	4942	408

I_{xx}, I_{yy}, and I_{zz} are the moments of inertia about the reference axes through the segment centroid.

than the total volume for the head and neck (ZONE). The trunk volumes for the two programs were similar but the error for the moments of inertia ranged up to 25 percent. Moments of inertia for the ZONE program tended to be higher, indicating the effect of the distribution of trunk mass. A comparison of arm and leg segment volumes showed errors ranging from 14 to 43 percent, and the DESIGN program underestimated three of the four segments. Interprogram moments of inertia for the X and Y axes were comparatively similar with errors ranging up to 15 percent, but the errors for the Z axis were somewhat greater. The results for the hand and foot indicate the need for redesigning their geometric shape (Hanavan, 1964, p. 69).

The representation of the body segments by the DESIGN program is inadequate. However, the major disadvantage of the ZONE system is the amount of data reduction needed, and a digitizing table would greatly expedite the process. Further refinements to the program, such as coefficients to allow for variations in density for the segment (Hatze, 1975) and comparisons with other criteria, are needed. The basis for such refinements has been presented in this paper.

REFERENCES

Barter, J. T. 1957. Estimation of the mass of body segments. WADC Technical Report, 57–260, Wright-Patterson Air Force Base, Ohio.

Dempster, W. T. 1955. Space requirements of the seated operator. WADC Technical Report 55–159, Wright-Patterson Air Force Base, Ohio.

Hanavan, E. P. 1964. A mathematical model of the human body. AMRL Technical Report, 64–102, Wright-Patterson Air Force Base, Ohio.

Hatze, H. 1975. A new method for the simultaneous measurement of the moment of inertia, the damping co-efficient and the location of the centre of mass of a body segment in situ. J. Biomechan. (in press).

Plagenhoef, S. 1971. Patterns of Human Motion, Prentice-Hall, Englewood Cliffs, N.J.

Weinbach, A. P. 1938. Contour maps, center of gravity, moment of inertia and surface area of the human body. Human Biol. 10: 356–371.

Simulation of human locomotion in space

V. M. Zatziorsky and S. Y. Aleshinsky
Central Institute of Physical Culture, Moscow

Determining moments and forces that arise at the joints of the human support-motion apparatus is a fundamental problem of biomechanics. Since direct measurement of these values proved impossible, no wonder that an idea was put forward by O. Fischer in 1895, who suggested that the inverse problem of dynamics be used for the purpose. Later on research workers tackled the problem time and again; see Aleshinsky and Zatziorsky (1974), and also Beletsky (1974), Chow and Jacobson (1971), Bresler and Frankel (1950), Gurfinkel, Fomin, and Stilkind (1970), Hill (1971), Korenev (1972), Moreinis et al. (1969), Nubar, and Contini (1961); a valuable contribution in this respect was made by Bernstein (1935).

Much more favorable conditions for the problem have appeared thanks to unprecedented progress in computation techniques and development of new experimental methods, such as stereophotogrammetric registration of the subject's coordinates in space (Zatziorsky, 1973), followed by processing on a stereocomparator, and determination of body mass geometry (Sarsaniya et al., 1974) with the aid of radioisotope technique, etc.

The mechanomathematical method of simulation (described below in brief) and employment of the above-mentioned methods allow determination of actual moments and forces at the joints of a moving human body.

The proposed model of the human support-motion apparatus is presented here in the form of a branching 15-link chain (Figure 1). The model is described in mathematical terms and the following assumptions are made:

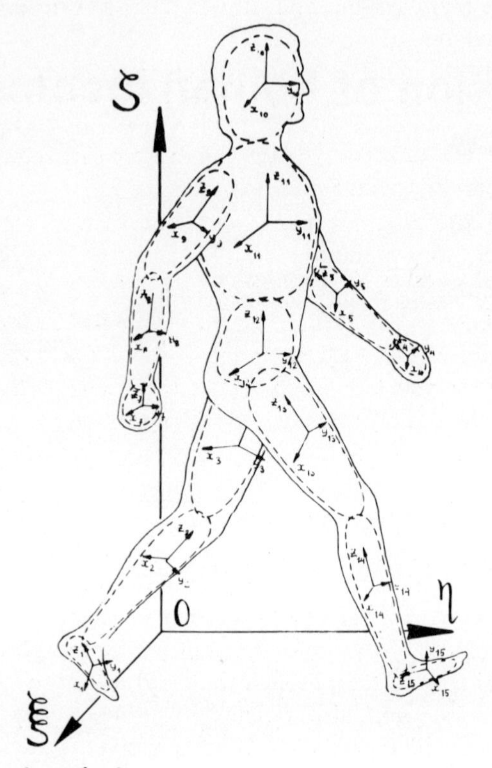

Figure 1. Model schematic view.

1. links are rigid bodies;
2. the parameters of model links, namely, their lengths, masses, primary central moments of inertia and arms of effective forces, coincide with the corresponding parameters of the subject's body segments;
3. links are interconnected by means of ideal spherical hinges;
4. the twisting of segments of extremities along their longitudinal axes is defined by rotation of cross sections of corresponding segments (the secant planes run normally to the segments, longitudinal axes and pass through centers of mass);
5. lines of action of interlinkage forces pass through the centers of rotation of the hinges;
6. locomotion is realized by control moments, acting at the joints.

Locomotion will be considered as relative to immovable rectangular system of Cartesian coordinates $0\xi\eta\zeta$. Let us connect the centers of mass of each link to the systems $k\xi_k\eta_k\zeta_k$ whose axes run respectively parallel to those of the main system, and also to $kx_ky_kz_k$ whose axes run along principal central axes of inertia ($k = 1,2,\ldots, 15$).

The following parameters have been determined through experiments:

1. coordinates of "reference points," fixed on the subject (data from bilateral stereophotogrammetric filming);
2. inertia and gravity properties of body segments (data obtained by means of radioisotope method);
3. $p - 1$ resultant vectors of external forces and $3 \times (p - 1)$ coordinates of the points, to which these forces are applied, when p is the number of links with external forces acting (except gravity). For example, for a two-support phase of walking it is the vector of ground reaction and three coordinates of the point, in which this vector is applied to the foot (in fact, these are data obtained by means of a force plate).

The solution has been based on d'Alembert's principle and also on the theorem of Chasles.

Let us consider an algorithm of the solution of our problem.

HOW TO DETERMINE COORDINATES OF CENTERS OF ROTATION OF HINGES AND CENTERS OF MASSES THROUGH "REFERENCE POINTS"

Coordinates of unknown points are defined from a system of 10 equations that imply geometric connections between "reference point" coordinates. Determination of these coordinates is much easier if no twisting is considered.

HOW TO DETERMINE COSINES OF ANGLES BETWEEN THE AXES OF $k\xi_k\eta_k\zeta_k$ AND $kx_ky_kz_k$ SYSTEMS

These cosines are derived from elementary geometrical considerations.

DETERMINATION OF ANGLES THAT DEFINE LINK POSITION

Selected as link position defining angles were:

θ_k, angle between positive direction of ξ_k axis and link;

ψ_k, angle between negative direction of ζ_k axis and link projection onto plane $\eta_k k \zeta_k$;

ϕ_k, angle between line of intersection of planes $x_k k y_k$ and $\xi_k k \eta_k$ to axis kx_k.

The selection of such angles precludes situation of the so-called "cardan suspension" (e.g., when selecting Eulerian angles for θ_k, ψ_k, ϕ_k and at zero angle nutation).

The θ_k, ψ_k, and ϕ_k angles can be expressed through the data, defined earlier, namely,

$$\theta_k = \arccos\,(\xi_k, z_k),$$

$$\psi_k = \begin{cases} \pi - \arccos\,(\cos\,(\zeta_k, z_k)/\sin\theta_k), & \text{when } \cos\,(\eta_k, z_k) < 0 \\ \pi + \arccos\,(\cos\,(\zeta_k, z_k)/\sin\theta_k), & \text{when } \cos\,(\eta_k, z_k) > 0 \\ \pi, \text{ when } \cos\,(\eta_k, z_k) = 0 & \end{cases}$$

$$\phi_k = \begin{cases} \pi - \arccos\,(\cos\,(\xi_k, x_k)/\sin\theta_k), & \text{when } \cos\,(\xi_k, y_k) < 0 \\ \pi + \arccos\,(\cos\,(\xi_k, x_k)/\sin\theta_k), & \text{when } \cos\,(\xi_k, y_k) > 0 \\ \pi, \text{ when } \cos\,(\xi_k, y_k) > 0. & \text{when } \cos\,(\xi_k, y_k) > 0 \end{cases}$$

HOW TO DETERMINE ACCELERATIONS OF CENTERS OF MASSES AND THE FIRST AND SECOND DERIVATIVES OF THE θ_k, ψ_k, AND ϕ_k ANGLES

By plotting approximation curves on the basis of least squares method and by marking elementary conversions one can obtain a formula for the derivative, utilizing empirical data on n consecutive values of function to the left and to the right of a point the derivative of which we are interested in:

$$f'(t_m) = \frac{\sum_{s=-n}^{n} s \cdot f(t_m + s\Delta t)}{2 \sum_{s=1}^{n} s^2 \Delta t}$$

HOW TO DETERMINE FORCES AT JOINTS

Assume that we have previously determined the forces, applied on the side of the kth link to all but one adjacent links. Then we can deduce equations for determining the unknown force in the absolute system of coordinates:

$$\vec{F} = m_k \ddot{\vec{r}}_k - \Sigma \vec{F}^{k-1} - \vec{F}^{ke}$$

Here m_k is the mass of the kth link; $\ddot{\vec{r}}_k$ is the vector of absolute acceleration of the kth link center of mass; $\Sigma \vec{F}^{k-1}$ is the symbolic designation of a sum of vectors of the known forces, applied to the kth link on the side of adjacent links; $\vec{F}^{ke}$ is the resultant vector of external (relative to the whole body) forces, applied to the kth link (gravity force inclusive).

DETERMINATION OF JOINTS MOMENTS

Let us take as basic the equations that express the theorem on changing the link kinetic moment, plotted in the form of projections onto movable axes x_k, y_k, and z_k.

As in the case of force determination, assume we have previously inferred the moment, active at all but one (from the kth link side) adjacent links. It is possible to derive equations (or, rather, formulas) that enable a component of unknown moment to be deduced. This, however, must be preceded with conversion of all the required forces and moments (in the systems connected with adjacent links), first into the absolute system of coordinates and then, from the absolute, into the x_k, y_k, z_k system. Such conversions are done with the aid of matrices of leading cosines.

Thus

$$\begin{aligned} M_{x_k} = I_{kx} \frac{d\omega_{kx}}{dt} + (I_{kz} - I_{ky})\, \omega_{ky}\, \omega_{kz} - \Sigma M_{x_k}{}^{k-1} \\ - \Sigma M_{x_k} F - \Sigma M_{x_k} F^{ke} \end{aligned}$$

$$\begin{aligned} M_{y_k} = I_{ky} \frac{d\omega_{ky}}{dt} + (I_{kx} - I_{kz})\, \omega_{kx}\, \omega_{kz} - \Sigma M_{y_k}{}^{k-1} \\ - \Sigma M_{y_k} F - \Sigma M_{y_k} F^{ke} \end{aligned}$$

$$\begin{aligned} M_{z_k} = I_{kz} \frac{d\omega_{kz}}{dt} + (I_{ky} - I_{kx})\, \omega_{kx}\, \omega_{ky} - \Sigma M_{z_k}{}^{k-1} \\ - \Sigma M_{z_k} F - \Sigma M_{z_k} F^{ke}. \end{aligned}$$

Here I_{kx}, I_{ky}, I_{kz} are principal central moments of inertia of the kth link; ω_{kx}, ω_{ky}, ω_{kz} are projections of angular velocity vector of link onto x_k,

y_k, z_k axis; $\omega_{kx} = \dot{\psi}_k \sin \theta_k \cos \phi_k - \dot{\theta}_k \sin \psi_k$, $\omega_{ky} = \dot{\psi}_k \sin \theta_k \sin \psi_k + \dot{\theta}_k \cos \psi_k$, $\omega_{kz} = - \dot{\psi}_k \cos \theta_k - \psi_k$; $\Sigma M_{xk}{}^{k-1}$, $\Sigma M_{yk}{}^{k-1}$, $\Sigma M_{zk}{}^{k-1}$ are the designations of sums of projections onto x_k, y_k, z_k axis of known moments, applied to the *k*th link from the side of adjacent links; $\Sigma M_{xk} F$, $\Sigma M_{yk} F$, $\Sigma M_{zk} F$ are the designations of sums of projections onto x_k, y_k, z_k axis of force moments at all adjacent joints; $\Sigma M_{xk} F^{ke}$, $\Sigma M_{yk} F^{ke}$, $\Sigma M_{zk} F^{ke}$ are the designations of sums of projections of moments of external (relative to the whole body) forces, applied to the *k*th link.

Note that repetitive conversions from the absolute system of coordinates into movable coordinates and vice versa are the cost we must pay for employment of equations on movable axes.

Let us consider the above description for the case of walking. While moving along the links, corresponding to upper extremities, determine the forces and moments in shoulder joints (this can be done, since for wrists $\Sigma \vec{F}^{k-1}$ and $\Sigma \vec{M}^{k-1}$ are equal to zero). Summing, then, the actions of arms and head upon the upper part of the trunk, determine the effect of the latter on the pelvis. Summing up the actions of the pelvis and free leg (for a single support phase) or the supporting leg, for which the floor reaction is being measured (for a two-support phase), one can derive not only forces and moments in the hip, knee, and ankle joints of a supporting leg, but also the reaction of floor and coordinates of the point of its application.

Figure 2 shows some results, obtained with the aid of the above-described technique—moments taken in respect to transverse, longtudinal, and vertical axes and developed in the ankle, knee, and hip joints of a swinging leg in running for a short distance.

In conclusion it should be noted that the human support-motion apparatus is, in fact, a very complicated system, which has many degrees of freedom. Having this in mind, the present authors consider the above-described method only an instrument for an approximate solution of the problem of determination of dynamic characteristics of man's three-dimensional movement.

REFERENCES

Aleshinsky, S. Yu., and V. M. Zatziorsky. 1974. Mechano-mathematical models of human locomotion. Biomechanics of Physical Exercises, Riga. 1: 60–123 (in Russian).

Beletsky, V. V. 1974. Dynamics of two-leg walking. Preprint No. 32 Institute of Applied Mathematics of the USSR Academy of Sciences, 1–65 (in Russian).

Bernstein, N. A. 1935. Studies of locomotion biodynamics. Viem, M. (in Russian).

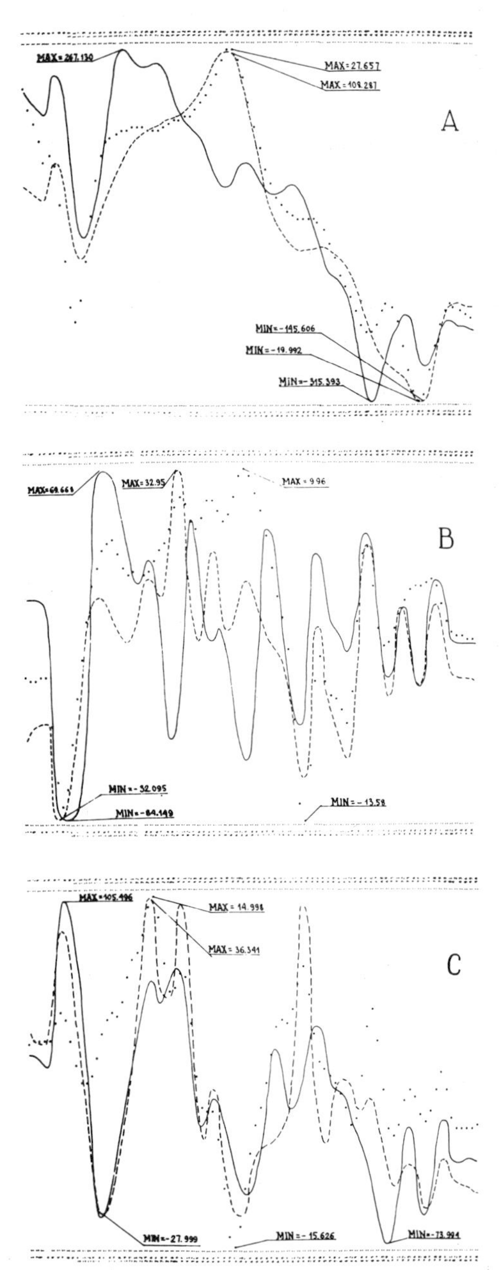

Figure 2. Moments developed at the joints of a swinging leg when running over a short distance (*abscissa,* time through 0.005-sec intervals; *ordinate,* moment in Newton-meters). *A,* relative to transverse axis; *B,* relative to longitudinal axis; *C,* relative to vertical axis. *Dotted line,* ankle; *dashed line,* knee; *solid line,* hip.

Bresler, B., and J. P. Frankel. 1950. The forces and moments in the leg during level walking. TASME 72, No. 1: 27–36.

Chow, C. K., and D. H. Jacobson. 1971. Studies of human locomotion via optimal programming. Math. Biosci. 10, No. 3/4: 239–306.

Gurfinkel, V. S., S. V. Fomin, and T. I. Stilkind. 1970. Determination of joints moments in locomotion. Biophysics 16, issue 2: 380–383 (in Russian).

Hill, I. C. 1971. A dynamic model of the human postural control system. IEEE Syst. Man and Cybern. Group. 104–109.

Korenev, G. V. 1972. About human movements, reaching a specified goal. Automation and Telemechanics, No. 6: 131–142. (in Russian).

Moreinis, I. Sh., et al. 1969. Mathematical simulation of walking and electromiography. *In:* Coll. of Trans. (Central Research Institute for Prosthesis Engineering), pp. 109–118 (in Russian).

Nubar, Y., and R. Contini. 1961. A minimal principle in biomechanics. Bull. Math. Biophys. 23, No. 4: 377–390.

Sarsaniya, S. K., et al. 1974. Radio-isotope method in determining the geometry of masses of sportsmen bodies. *In:* Coll. Trans. of the First All-Union Scientific Conference on Sport Biomechanics. Part I. Gtsolifk, M. (in Russian).

Zatziorsky, V. M., et al. 1973. Stereophotogrammetric methods for investigation of sport movements. Theory and Practice of Physical Culture. M., No. 12 (in Russian).

Instrumentation and methodology

Introductory paper

Recent advances in instrumentation and methodology of biomechanical studies

P. R. Cavanagh
The Pennsylvania State University, University Park

In this review paper I hope to achieve three objectives: first, I want to examine recent progress in certain areas of instrumentation and methodology that are common to most areas of biomechanical study. The second objective, which is a corollary to the first, is to prognosticate about future trends. This should be a direct result of the identification of areas where there has been stagnation or where progress has been slow. The last objective is to reflect somewhat philosophically upon the purpose and justification for biomechanical experiments. Since I am not an engineer I can afford to be irreverent about the topic of instrumentation and I hope at least to be controversial.

MEASUREMENT OF THE DISPLACEMENT OF BODY SEGMENTS

An important principle, sometimes referred to as Kelvin's first rule of instrumentation (Geddes and Baker, 1968), states that the measuring instrument must not alter the event being measured in any way. Although it is rather an extreme example, one would imagine that the experiment shown in Figure 1 illustrates a near perfect contravention of this principle. The position of this experimenter, who incidentally was Fick, almost 70 years ago, in relation to Kelvin's rule was an intriguing one and is exactly the same as our position today. He had an excellent technique for the determination of the center of gravity location which obeyed Kelvin's rule, but he still searched for another method because of the laborious nature of the better technique—which is, of course, cinematography.

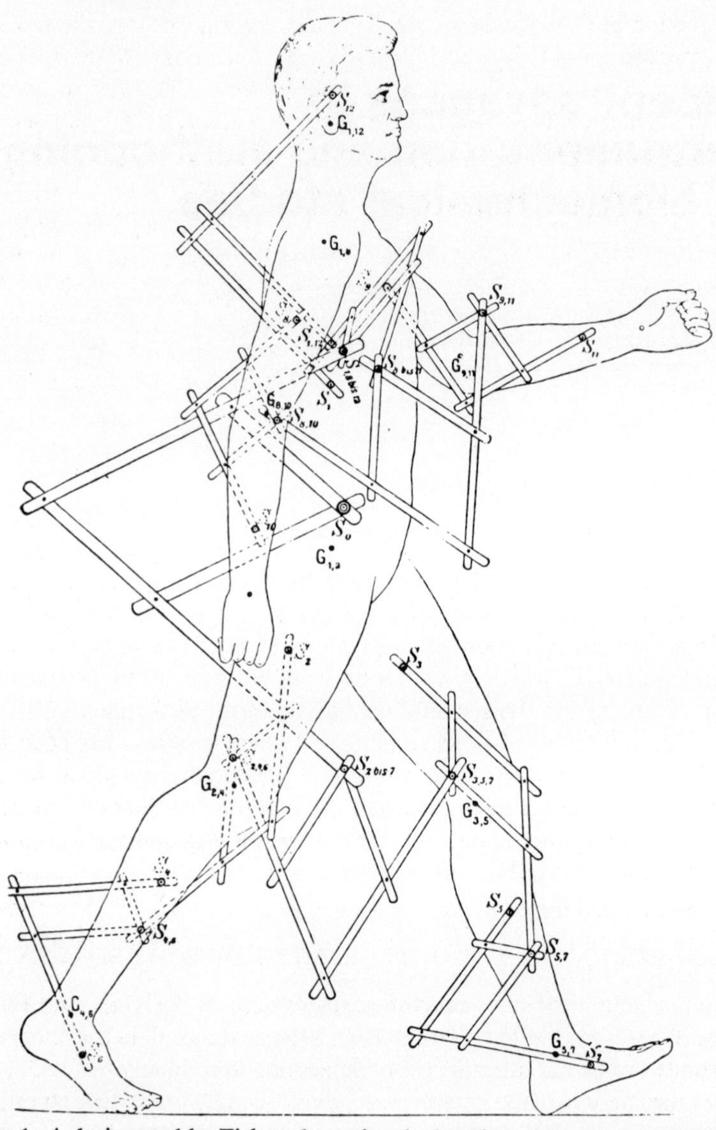

Figure 1. A device used by Fick to determine the location of the whole body center of gravity during movement.

High-speed cinematography is perhaps the most frequently used research tool in biomechanics and there is no question that it can be a valuable and precise measuring instrument. There is also no question that it is expensive, definitely an "off-line" system with slow feedback to the experimenter and, finally, an extremely laborious technique for the extraction of displacement data. This last point is perhaps the factor that has proved most limiting of all in biomechanical experiments. We have never seen, for instance, an experiment in which the vertical oscillation of the center of gravity during running or walking at a given speed has been presented as the mean pattern from fifty cycles with stated confidence limits. Because of the limitations of our methodology we are permitting ourselves experimental laxity that workers in other fields would not tolerate. Taking an analogy from neurophysiology, it is hard to imagine a paper that is based upon single evoked response rather than a computed average transient from many stimulii. It is not surprising then that many investigators, from Marey onward, have devoted their attention toward alternative methods for the direct recording of movement.

OPTOELECTRONIC DEVICES

The major effort in research and development for techniques to replace the cine camera has been in the area of optoelectronic devices. Four main categories can be identified: (1) polarized light goniometry, (2) automatic scanning of television, (3) laser-based systems, and (4) devices using the lateral photo effect of diodes.

Polarized Light Goniometry

First reported by Grieve (1969) and subsequently by Reed and Reynolds (1969), and Mitchelson (1973), the technique of polarized light goniometry, which has become known as Polgon, is confined to the measurement of angles. The area in which the subject moves is illuminated by plane polarized light (Figure 2) produced by light from a D.C.-energized source passing through a polarizing filter and being focused by a standard projector lens, not shown in the figure. Each detector consists of two matched photocells covered by a slip of polarizing filter, but with the axes of the filters at 90 degrees to each other. The plane of polarization of the light source is rotated at constant angular velocity, each complete revolution generating a reference pulse from a photocell placed near the rotating disk. The time from this reference pulse to the instant when both cells are equally illuminated can be converted to an angular measurement, since the angular velocity of the disk is known. The specifications for linearity given by the investigators range from 0.16 to 1.6 percent of FSD when the planes of the revolving disk and the slips covering the detector are parallel. Since its appearance six years ago, very few studies have taken advantage of this important new technique for angular measurement. However, since the devices are now available commer-

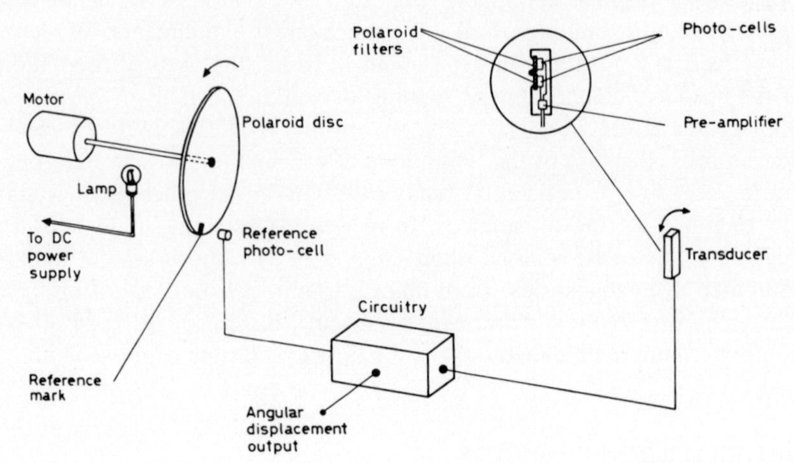

Figure 2. Schematic diagram of a POLGON (reproduced from Mitchelson, 1973).

cially, perhaps an indication of their usefulness will be forthcoming from extended field tests in the near future.

Automatic Scanning of TV Image

Automatic image analysis has been a rapidly growing field in recent years, enabling workers in such diverse areas as cell pathology and weapons systems to automate previously tedious analysis. Winter, Greenlaw, and Hobson (1972) described a television and computer system that has found ready application in a clinical setting for the kinematic analysis of gait (Winter et al., 1974). A maximum-contrast video tape recording of a subject wearing 4-cm-diameter markers is made at a rate of 60 fields per second from a camera on a moving tracking cart. The portion of each field in which the subject image exists is subsequently digitized and results in a 96×96 binary matrix. Resolution is improved by averaging the X and Y coordinates of all the 1's in the display and absolute location in two-dimensional space is achieved by making the coordinates of the limb markers relative to fixed external markers. Although automatic scanning of cine film has been attempted (Kasvand, Milner, and Rapley, 1971), the advantage of videotape is its extremely fast turnaround time, which makes it ideal for a clinical environment. However, technical developments in television that would bring to this medium the same resolution, linearity, and data acquisition rates as cine film, have been slow to develop and extremely expensive to acquire.

Laser-Based Systems

Mitchelson (1975) reports on the design concept of a laser-based device that he calls CODA (Cartesian Optoelectronic Dynamic Anthropometer). The positions of small gallium arsenide injection lasers attached to the moving subject are sensed by detectors consisting of arrays of photodiodes covered by a gray code optical mask. Cylindrical optics are used in the three detectors from which a theoretical resolution of 1:4,000 is anticipated for coordinates in three dimensions. Since no actual device exists yet, it is impossible to estimate the performance but if the design specifications can be met the instrument would be extremely useful in biomechanics research.

The Lateral Photoeffect of Diodes

Adapting a technique that had been developed in nuclear physics Lindholm (1974) described a new optoelectronic device for the measurement of displacement that he called SELSPOT. While the technical details in Lindholm's paper were lacking, Woltring (1974) presents a comprehensive discussion of various forms of real-time measurement of light spot position and it is from his work that most of the following discussion is taken.

The hydraulic analogy of the lateral photoeffect, redrawn from Woltring, in Figure 3A, indicates that, dependent upon the lateral position of the falling water, different flow rates will be detected by the two observers. By comparing notes they will be able to determine the position of the incident water. This is fundamentally different from the operation of a normal diode where total current flow across the p-n junction would be measured, whereas in this case the difference between current leaving opposite ends of the same layer is of interest. The simplified presentation of an actual one-dimensional diode (Figure 3B) indicates the reverse biasing of the junction and the addition of a sheet of resistive material on the bottom of the n-type layer. Incident photons create an electron–hole pair at the surface of the p-type material. The electron is swept across the reverse biased junction and appears as current flow in the terminal shown. The difference in current flow from the two contacts on the resistive sheet is nonlinearly related to the position of incident light.

The actual devices used are able to detect both the X and Y coordinates of many incident light spots, using time multiplexing of the illumination pulse to identify multiple sources. A standard optical system focuses the light spots onto the surface of a tetralateral photodiode and

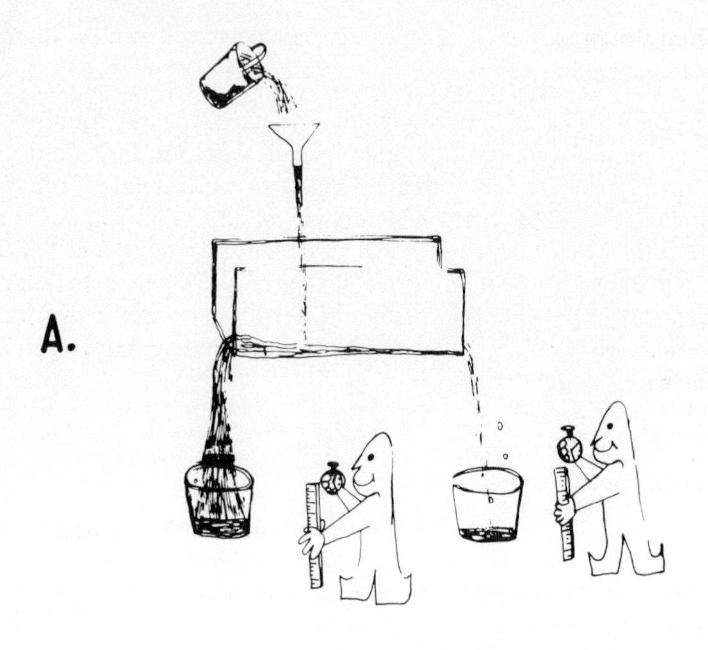

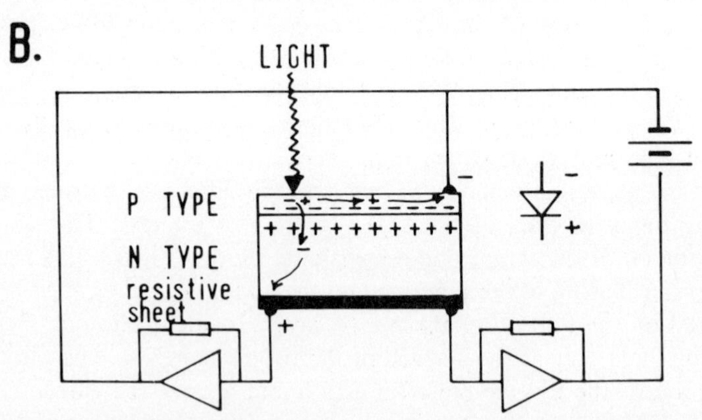

Figure 3. *A*, a hydraulic analogy of the lateral photoeffect of diodes (modified from Woltring, 1974). *B*, a schematic diagram of the photodiode in operation. See text for an explanation of both figures.

analog signals proportional to position are then available for direct use or for digitization. A resolution of between 1:500 and 1:1,000 is currently attainable.

From this brief survey of optoelectronic devices it is clear that the area is an expanding one with exciting new development occurring. In this kind of atmosphere, purchasing decisions are often difficult to make because of the rapidity with which specifications may change and because of the fear of buying a prototype device. It is essential at this early stage for the manufacturers of each device to provide comprehensive and accurate specifications for their systems so that experimenters can assess the performance and suitability of the various systems. It seems very likely that some form of optoelectronic system will eventually replace the cine camera especially in lab studies where the attachment of small sensors will not be restrictive. Exactly which type of device will eventually prove to be the most versatile is still very much open to question.

DETERMINATION OF ACCELERATION

A continuing methodological problem in biomechanics research is the determination of linear and angular acceleration of body segments from displacement-time data. The movements that we measure are often of interest only in that they provide a pathway for the calculation of accelerations and subsequently forces or torques. This pathway, although frequently used, has a very dubious theoretical basis.

The area of numerical differentiation and the accompanying smoothing processes remain perhaps the weakest links in our chain of tools for application to the analysis of human movement. It would be hard to find a numerical analyst who would condone the process of double numerical differentiation with the small amount of information that we can provide regarding the signal characteristics and the properties of the noise that is always present. The situation is somewhat analogous to requesting an electrical engineer to design a receiver for a certain radio signal without telling him the transmission frequency. I believe that this is another area where progress has been lacking and also that we are using techniques without a theoretical basis in a manner that would not be tolerated in other fields of science.

There are two basic questions that must be posed when displacement-time data from any source are to be used for subsequent numerical analysis. These are: (1) How fast should the sampling rate be? and (2) What techniques of smoothing and differentiation should be used? It would not be unfair to say that the answers to both of these questions have been obtained on a completely arbitrary basis by most investigators in biomechanics. A survey of the literature reveals an interesting variation in sampling rates as shown in Table 1. The range is from 16 samples per second used by Ariel (1974) to calculate forces at the knee joint during squats to more than 190 frames per second used by Bernstein (1967) in the study of ground reaction forces in walking.

The representation of continuous data by discrete samples is governed by Shannon's sampling theorem, which specifies what is known as the Nyquist rate, the minimum frequency at which sampling can occur if the original signal is to be reconstructed from the discrete samples. To determine this minimum frequency it is necessary to know the bandwidth that contains most of the significant frequency components of the signal. An attempt must be made at the definition of various kinds of movements in frequency spectra terms so that a more informed choice of data acquisition rates can be made. Isolated examples do exist in the literature; for instance Winter et al. (1974) indicated that 99.7 percent of the signal power of the toe marker during normal gait was below the eighth harmonic in a cycle of approximately 1.4 sec. More data of this nature would build a much firmer foundation for frame rate decisions.

The second area of data smoothing and differentiation is, if anything, even more arbitrary than the choice of frame rate. Three methods in current use involve (a) fitting a single least squares polynomial to all available data points (Plagenhoef, 1971) and taking the analytical de-

Table 1. A survey of frame rates used in various studies

Investigator	Purpose	Frame speed (exposures per second)
Elftman (1939)	Running kinetics	33*
Bernstein (1967)	Biodynamics of locomotion	60–190
Bresler and Frankel (1950)	Walking kinetics	40
Morrison (1970)	Locomotion kinetics	50
Plagenhoef (1971)	Generalized body motion	60
Ramey (1973)	Angular momentum (long jump)	50
Miller (1970)	Angular momentum (diving)	33*
Ariel (1974)	Forces at knee joint	16*

*Denotes that the data collection rate was faster but not every frame was used for analysis.

rivatives of the curve obtained, (b) a piecewise fit such as a cubic spline technique, and (c) a weighted moving average or digital filtering method (Lanczos, 1957; Wilcock and Kirsner, 1969; Pezzack, Norman, and Winter, 1975). It is important with the single polynomial fit to realize that the choice of degree for an approximating polynomial immediately imposes limits that may not be justified on the form of the acceleration-time curve. As elementary as this point might seem, it has been ignored by many investigators. In general, it would seem that the moving average and the piecewise fit approaches place fewer restrictions on the displacement-time data, without using polynomials of unrealistically high degree. There is still a danger of oversmoothing as the methods of Wilcock and Kirsner (1969) can be shown to demonstrate. The nine-point second-derivative filter, for example, has a 3-dB point at approximately 3–4 Hz at a frame rate of 50 frames per second.

A further aspect that needs attention in the area of smoothing and differentiating is the use of a priori information concerning the characteristics and statistical properties of the measurement error or noise. The mathematics of communications theory shows how data on the variance and covariance of the noise can be incorporated into a digital filter to improve the prediction of the smoothing process (Morrison, 1969). This, in turn, will improve the prediction of the acceleration and lead to better estimates of force.

BODY SEGMENT PARAMETERS

There continues to be an acute need for better methods of determining body segment parameters in living human subjects. The work of Clauser, McConville, and Young (1969) provided a considerable improvement in prediction techniques by using multiple regression analysis relating certain body segment parameters to anthropometric measurements in a group of 13 male cadavers. Although the resulting estimates are somewhat more "personalized" than those from earlier techniques (for example, Drillis and Contini, 1966), the method is still indirect and does not contain any data that could be used to estimate inertial parameters. Chandler et al. (1975) expanded our knowledge of the inertial properties of body segments considerably in a recent study of six male cadavers, but no methods were presented to extend the results of living subjects. It has been suggested (Goulet, Cuzzi, and Herron, 1974) that stereometric photogrammetry will provide a means for the determination of inertial parameters in vivo. In reconstructive surgery or prosthetic fitting where an accurate determination of shape is needed, stereophotogrammetry can provide both an accurate representation of surface contours and input data to computer controlled machine tools. However, the assumptions required to extend measurements of volume to estimates of mass and mass moment of inertia are substantial, and the estimate must still be dependent upon data obtained from cadavers. It is therefore unlikely that stereophotogrammetry will provide these estimates with the required accuracy.

One of the more promising techniques that has appeared in recent years for the determination of body segment parameters is the use of gamma mass spectrometry (Casper, 1971; Brooks and Jacobs, 1975). The first generation of this device, shown in Figure 4, is similar in concept to radiation density gauging. High-energy gamma rays from a cobalt 60 source undergo Compton scattering as they pass through the tissue specimen, in this case, a leg of lamb. A mass distribution map is built up in the computer memory, and on completion of the whole scan, a variety of parameters such as total mass, center of mass, and moment of inertia about any axis can be determined. For a complete scan, the radiation dose is of the order of one milliroentgen, which is only approximately 1 percent of the dosage to which a patient is exposed during a normal diagnostic x-ray. Of all the techniques for the determination of body segment parameters that have been used, it is likely that the gamma scanner will, in the long term, prove to give the most accurate results. It is vital that this device be further developed and validated for use on live human subjects.

I want to bring this paper to a conclusion by a retrospective look at the relatively young discipline of biomechanics and to make the suggestion that it is about time that we realized that it has, in fact, come of age. In reviewing some 218 papers from the volumes containing the proceedings of the first four meetings of this group, the number of subjects used in the papers totals 2,260 or an average of 10.37 subjects per paper. While I am not suggesting that there is merit simply in volume, I do believe that this represents a trend indicating a discipline that is somewhat inward looking and still concerned with developing methodology—perhaps even for methodology's sake.

The question of relevance must be faced. If the work being done in biomechanics is to be worthwhile, then there must be a wider application than to one or two subjects or to the further refinement of technique. A good example is now being set in gait studies—a traditional mainstay of biomechanics research. The techniques and methodologies that have been developed over a number of years in laboratory studies are now being used in a routine clinical setting by several centers for the quantitative evaluation of rehabilitation programs. Murray et al. (1975) have published papers concerned with the analysis of the gait of large samples

Figure 4. A prototype of the gamma mass scanner shown in use for the determination of the inertial properties of a leg of lamb.

of patients undergoing rehabilitation, in an attempt to provide some quantitative evidence to direct treatment and rehabilitation.

It is vital that we realize that instrumentation and methodology are not ends in themselves. With equipment no more complicated than steel balls, planks of wood, and water clocks, Gallileo performed experiments that were fundamental to our view of mechanics. In the face of a contribution that represented so much with so little experimental sophistication, it is a challenge to us to contribute a little with so much.

REFERENCES

Ariel, B. G. 1974. Biomechanical analysis of the knee joint during deep knee bends with heavy load. *In:* R. C. Nelson and C. A. Morehouse (eds.), Biomechanics IV, pp. 44–52. University Park Press, Baltimore.

Bernstein, N. A. 1967. Coordination and Regulation of Movements. Pergamon Press, Oxford.

Bresler, B., and Frankel, J. 1950. The forces and moments in the leg during level walking. Trans. ASME 72: 27–36.

Brooks, C. B., and A. M. Jacobs. 1975. The gamma mass scanning technique for inertial anthropometric measurement. Med. Sci. in Sports 7: 290–294.

Casper, R. M. 1971. On the use of gamma ray images for the determination of human body segment mass parameters. Unpublished master's thesis, Penn State University.

Chandler, R. F., C. E. Clauser, J. T. McConville, H. M. Reynolds, and J. W. Young. 1975. Investigation of inertial properties of the human body. DOT Report. HS-801-403. Washington, D.C.

Clauser, C. E., J. T. McConville, and J. W. Young. 1969. Weight, volume and centre of mass of segments of the human body. AMRL-TR-69-70. Wright Patterson AFB, Ohio.

Drillis, R., and R. Contini. 1966. Body segment parameters. Tech. Rept. 1166.03. School of Engineering and Science. N.Y.U.

Elftman, H. 1939. The function of muscles in locomotion. Am. J. Physiol. 125: 357–366.

Geddes, L. A., and L. E. Baker. 1968. Principles of Applied Biomedical Instrumentation. John Wiley and Sons, New York.

Goulet, D. V., J. R. Cuzzi, and R. E. Herron. 1974. A parametric description of the human body using biostereometric techniques. *In:* Biostereometrics 74, Am. Soc. of Photogrammetry. Falls Church, Va.

Grieve, D. W. 1969. A device called Polgon for the measurement of parts of the body relative to a fixed external axis. J. Physiol. 201: 70.

Kasvand, T., M. Milner, and L. F. Rapley. 1971. A computer based system for the analysis of some aspects of human locomotion. *In:* Human Locomotor Engineering, pp. 297–306. Inst. of Mech. Eng., London.

Lanczos, C. 1957. Applied Analysis. Pitmans, New York.

Lindholm, L. E. 1974. An optoelectronic instrument for remote on-line movement monitoring. *In:* R. C. Nelson and C. A. Morehouse (eds.), Biomechanics IV, pp. 510–512. University Park Press, Baltimore.

Miller, D. I. 1970. A computer simulation of the airborne phase of diving. Unpublished doctoral dissertation. Penn State University.

Mitchelson, D. L. 1973. An opto-electronic technique for analysis of angular movement. *In:* S. Cerquilini, A. Venerando, and J. Wartenweiler (eds.), Biomechanics III, pp. 181–184. University Park Press, Baltimore.

Mitchelson, D. L. 1975. Recording of movement without photography. *In:* Techniques for the Analysis of Human Movement, pp. 59–65. Lepus Books, London.
Morrison, J. B. 1970. The mechanics of the knee joint in relation to normal walking. J. Biomech. 3, 1: 51–61.
Morrison, N. 1969. An Introduction to Sequential Smoothing and Prediction. McGraw-Hill, New York.
Murray, M. P., B. J. Brewer, D. R. Gore, and R. C. Zuege. 1975. Kinesiology after McKee Farrar total hip replacement. J. Bone Joint Surg. 57-A, No. 3: 337–342.
Pezzack, J. C., R. W. Norman, and D. A. Winter. 1975. An assessment of derivative determining techniques used for motion analysis. J. Biomech. (in press).
Plagenhoef, S. 1971. Patterns of Human Motion. Prentice-Hall, Englewood Cliffs, N.J.
Ramey, M. R. 1973. Use of force plates for long jump studies. *In:* S. Cerquilini, A. Venerando, and J. Wartenweiler (eds.), Biomechanics III, pp. 370–380. University Park Press, Baltimore.
Reed, D. J., and P. J. Reynolds. 1969. A joint angle detector. J. Appl. Physiol. 27, 5: 745–748.
Wilcock, A. H., and R. L. G. Kirsner. 1969. A digital filter for biological data. Med. Biol. Eng. 7: 653–660.
Winter, D. A., R. K. Greenlaw, and D. A. Hobson. 1972. Television-computer analysis of kinematics of human gait. Comput. Biomed. Res. 5: 498–504.
Winter, D. A., A. D. Quanbury, D. A. Hobson, H. G. Sidwall, G. Reimer, B. G. Trenholm, T. Steinke, and H. Shlosser. 1974. Kinematics of normal locomotion—A statistical study based on T.V. data. J. Biomech. 7: 479–486.
Woltring, H. J. 1974. New possibilities for human motion studies by real-time light spot position measurement. Biotelemetry 1: 132–146.

Instrumentation

Mechanical parameters describing the standing posture, based on the foot-ground pressure pattern

M. Arcan
Tel-Aviv University, Tel-Aviv

M. Brull
Tel-Aviv University, Tel-Aviv

A. Simkin
Hadassah University Hospital, Jerusalem

Knowledge of the pressure distribution between the feet and the ground may be very useful for the biomechanical analysis of the standing posture as well as for locomotion studies. The purpose of this investigation is to introduce and measure a number of mechanical parameters that can serve to characterize the standing posture. The foot–ground pressure (FGP) pattern will play a basic role in this work. The standing posture serves both for working and as an intermediate phase between all other postures and locomotion. The only contact area during standing is between the ground and the feet. The forces acting through this small area (less than 500 cm^2 or about 3 percent of the total body surface) must react, control, and stabilize the gravity and inertial forces of all body segments.

The FGP pattern for the standing posture may be found useful in different domains: (a) In mechanics of postures, it may help to find parameters that define balance achieved between the conflicting demands from this posture: stability and alertness (ability to quickly change in direction and motion). (b) In foot orthopedics, it may help to achieve a better redistribution of the forces and thus relieve pains and correct deformities through orthopaedic appliances or inserts and also to evaluate these appliances. (c) In rehabilitation problems and (d) In mechanics of sporting activities.

Morton and Fuller (1952) and Carlsöö (1964) have investigated the weight distribution on the feet but, because they were limited to measurements of the total forefoot and heel loads, they did not have sufficiently detailed information.

Elftman (1934) and Pliquett and Helm (1966) have attempted to determine the pressure distribution on the foot using a rubber mat with pyramidal projections on which the subject stands. This gives, however, only a qualitative measure. Other investigators (Lerein and Serck-Hanssen, 1973) have used small pressure transducers or strain gauges placed on elastic strips (Stokes, Stott, and Hutton 1974). These measurements give more detailed data but unfortunately require many sensors, which in turn increases equipment complexity and cost.

Thus, for a variety of reasons, it has been impossible in the past to define geometrical and mechanical parameters that could suitably describe posture or gait and could be measured with suitable accuracy.

METHOD AND MATERIALS

In a previous paper Brull and Arcan (1974) presented a method and an instrument for measuring the pressure distribution between the human foot and the ground.

The proposed technique will be briefly outlined here in order to make the present paper self-contained. The new method is an optical one that uses optical interference to measure local pressure simultaneously over the entire contact area.

The complete apparatus consists of (I) Standing or walking platform, (II) The Imaging system, (III) Loading device.

(I) The platform consists of an optical sandwich hereafter referred to as the optical interference sandwich; its constitutive elements are presented in Figure 1.

(II) The imaging system consists of an intense light source, a mirror or half a mirror mounted at 45° to the plane of the platform and a suitable recording system (i.e., camera for standing measurements, high-resolution motion picture or TV system with a proper recorder for locomotion studies).

(III) The loading device consists of a thin leather or rubber sole on which are mounted a number of specially shaped solids used to "dis-

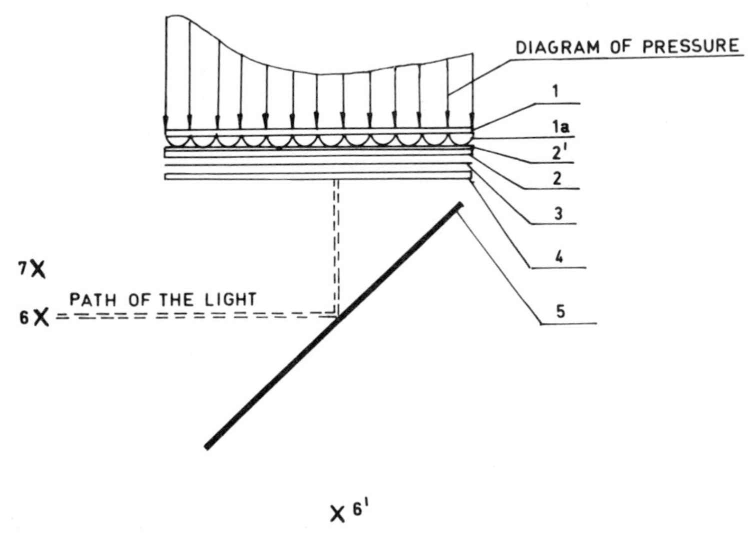

Figure 1. Schematic diagram of contact pressure measurement apparatus. (1) Flexible sheet with contact solids to apply the pressure. (1a) Contact solids. (2) Optically sensitive (photoelastic) sheet. (2′) Reflective layer. (3) Optical filter (polarizer). (4) Sheet of rigid transparent material (glass). (5) Mirror or semi-mirror (6 or 6′). Possible positions of the light source. (7) Photographic or cine-photo equipment.

cretize" the load, which yield a circular interference fringe. The normal force is thus determined by measuring the fringe diameter and using a suitable calibration curve.

The sequence of data recording and processing is as follows: pattern imprinting, photographic recording, coding of the circles, measurements of diameters and coordinates, computer calculations which mean using the calibration data, and finding the weight distribution between foot segments, geometrical-mechanical parameters of these segments (individual pressure centers), and so on.

It should be noted that the interference pattern obtained indicates also the shape of the true contact area. It is also of interest to note that visual examination of this foot–ground pressure pattern (FGP) immediately gives a picture of the pressure distribution that could be used, without further processing, if only a qualitative evaluation were required. Thus the instrument can be used for routine clinical and qualitative use as well as for detailed quantitative investigation.

Measurements were carried out on five subjects, all male students ranging from 20 to 28 yr of age. All of the subjects consider themselves free from orthopedic diseases except No. 5 (D.S.), who knows that he is slightly flat-footed. The subjects were instructed to stand comfortably but did not make any special effort to distribute their weight evenly between their feet. The measurements were made on the subjects standing barefoot and repeated with the subject wearing his own shoes.

RESULTS AND DISCUSSION

Table 1 is a summary of the measured results for the nine parameters describing the foot–ground pressure (FGP) pattern that were defined in a previous paper (Arcan and Brull, 1975).

An examination of the results for barefoot subjects show that for subjects 1 to 4, 45–65 percent of the weight is carried on the heels while 30–47 percent is carried on the forefoot for this particular group. The most variable parameter is k_m, indicating middlefoot load variations from less than 1 percent of body weight to about 9 percent. The flatfooted patient no. 5 exhibits startling deviations with an FGP indicating that he is carrying nearly 60 percent of the weight on his forefoot while the heel region is underloaded, carrying only 26 percent; the middlefoot region in this subject carries 15 percent of the weight, which is almost twice as much as any other case in this group.

When the same measurements are repeated with the subject wearing his own shoes, some changes are seen to occur. Thus for subjects no. 1 to 4, k_h again ranges from 0.40 to 0.61 and k_f from 0.39 to 0.52 while k_m is close to zero in all cases except No. 4 who was wearing very light flexible shoes. Comparing the results for each subject between the barefoot and the "shoed" condition a significant feature is noted, namely that in the "normal" subjects (no. 1 to 4) the middlefoot load, in the barefoot case, tends to shift to the forefoot. The effect of wearing shoes for the flatfooted subject (no. 5) is a spectacular shift of the middlefoot load to the heel while the forefoot load remains essentially constant. Thus it would appear that the flatfoot condition causes, in this case, unloading of the heel region and a shift of the resultant load forward.

It should be noted that all the results above are concerned with parameters that describe the foot loads and make no direct connection to characteristics of the standing posture. It therefore seems desirable to define additional parameters that will serve to describe the standing posture. This can be done by utilizing the detailed data furnished by the

Table 1. Weight distribution between foot segments

Subject	Barefoot									With shoes									Remarks
	Heel			Midfoot			Forefoot			Heel			Midfoot			Forefoot			
	k_{hL}	k_{hR}	k_h	k_{mL}	k_{mR}	k_m	k_{fL}	k_{fR}	k_f	k_{hL}	k_{hR}	k_h	k_{mL}	k_{mR}	k_m	k_{fL}	k_{fR}	k_f	
1 (S.S.)	60.6	61.1	60.8	0.9	0.5	0.7	38.5	38.4	38.5	62.0	58.1	60.0	0.1	0.2	0.1	37.9	41.7	39.9	
2 (M.B.)	62.0	67.0	64.5	8.2	3.6	5.8	29.9	29.5	29.7	58.1	64.5	61.3	0	0	0	41.9	34.0	38.7	
3 (A.G.)	43.9	46.2	45.0	6.4	8.9	7.6	49.8	44.8	47.4	41.7	40.5	41.1	0	0.1	0.1	58.3	59.4	58.8	
4 (H.B.)	38.4	53.3	44.9	12.5	4.0	8.8	49.1	42.7	46.3	38.4	54.2	46.0	3.8	0.5	2.2	57.8	45.3	51.8	Flexible shoes
5 (D.S.)	30.4	18.3	25.7	16.4	12.7	15.0	68.9	53.3	59.3	46.3	37.7	42.4	0.4	0.8	0.6	53.3	61.5	57.0	Flat footed subject

k, ratio of partial load to total load (percent).
The indexes L, R, h, m, f refer to left, right, heel, midfoot and forefoot respectively.
Absence of L or R indexes indicates total body parameters.

measurements of local pressures. Referring to Figure 2, we wish to introduce the following points. Let C denote the center of pressure of the FGP of both feet, and let C_R and C_L denote the centers of pressure for the right and left foot, respectively. It will be shown below that the determination of first moments of the individual FGP is of importance and we therefore introduce the points C_{FR} and C_{HR}, which are, respectively, the centers of pressure of the forefoot and heel loads on the right foot. C_{FL} and C_{HL} are corresponding points of the left foot. Finally, C_F and C_H denote, respectively, the centers of pressure of both forefoot and both heels when together.

It should be noted that C_F must be located on the segment $C_{FR}C_{FL}$. It should also be noted that C_L need not be located on the segment $C_{FL}C_{HL}$ because C_{FL} and C_{HL} do not include the middlefoot loads while C_L does. In fact, the distance from C_L to the line segment C_{FL}–C_{HL}

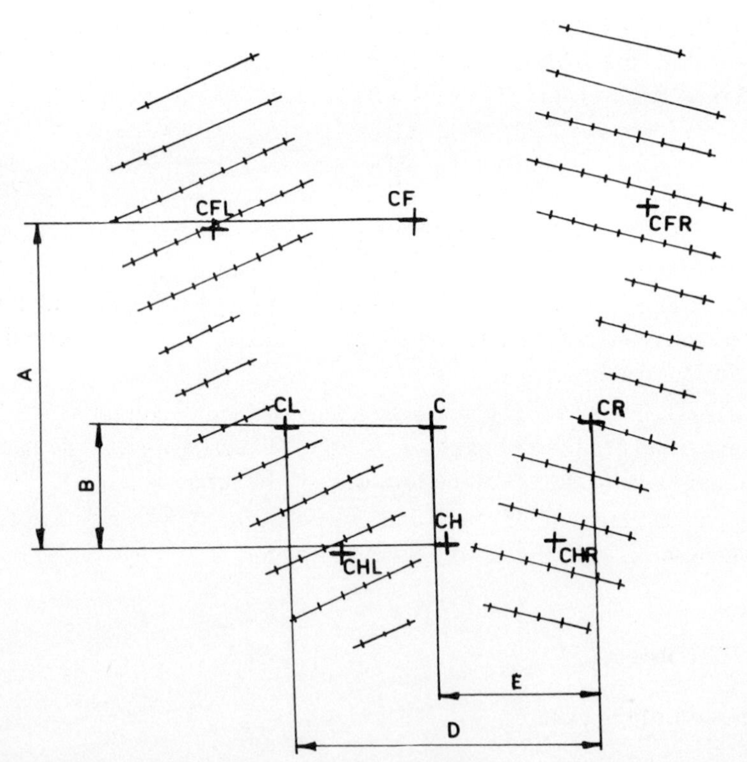

Figure 2. Scheme of pressure centers. C, total body; CL, CR, left, right foot; CF, forefeet; CH, heels; CFL, CFR, left, right forefoot; CHL, CHR, left, right heel.

indicates the eccentricity of the middlefoot load. (Similar remarks hold for C_{FR}, C_{HR}, and C_R.)

Distances A and B are measured in the sagittal plane and are simply the projections in this plane of the lines C_F–C_H and CC_H, respectively.

It is clear that A is the measure of the range of center of gravity positions that are allowable without major changes in external forces. In addition, B indicates how close to this boundary the subject is. Thus the ratio B/A (hereafter called sagittal stance ratio) is a measure of the stability of the stance. Clearly, values in the neighborhood of 0 or 1 indicate that small motions require the introduction of new external forces. In order to bring into play the height of the center of gravity, we make use of previous extensive measurements (Drillis and Contini, 1966) and we assume that the center of gravity height H is equal to 57 percent of the total height. We then introduce the ratio H/A as an additional posture parameter that should be related to stability or to the ability to initiate changes in posture. In Table 2, we present for the previous subjects the values of A, B, H, B/A, and H/A. It is immediately clear that the distances A and B do not exhibit any particular pattern (even for the flatfooted subject) but that the sagittal stance ratio B/A varies from a relatively low value 0.39 for subject no. 1 and generally increases as the arches get lower to a value of 0.64 for the flatfooted subject no. 5. For the measurements with shoes, the values of B/A are slightly higher for subjects no. 1 to 4 but lower for no. 5 and the differences generally decrease, indicating the correcting effects of shoes. Thus the sagittal stance ratio B/A seems to reflect load shifts due to orthopedic conditions but extensive clinical studies should be conducted before any definitive conclusions can be drawn.

The ratio H/A does not exhibit any particular variation in the present group of subjects but this is probably because this ratio will likely be of most significance in connection with the initiation of motion.

The parameters in Tables 1 and 2 were all obtained from the FGP of which some examples are shown in Figure 3*A–C*.

CONCLUSIONS

Mechanics of the Foot

Measurements of the proposed nine parameters which describe the foot load distribution (k_{hL}, k_{hR}, k_h, k_{mL}, k_{mR}, k_m, k_{fL}, k_{fR}, k_f) indicate that

Table 2. Geometrical-mechanical parameters of the FGP

					Barefoot								With shoes							
Subject	Age	Weight (kg)	Height (cm)	H (cm)	A (cm)	B (cm)	D (cm)	E (cm)	B/A	E/D	H/A	H/D	A (cm)	B (cm)	D (cm)	E (cm)	B/A	E/D	H/A	H/D
1. (S.S.)	24	76	184	104.9	14.6	5.7	12.4	6.5	0.388	0.522	7.18	8.46	12.4	5.0	13.9	6.5	0.400	0.471	8.46	7.55
2. (M.B.)	26	59	170	96.9	12.6	4.2	11.9	6.1	0.330	0.509	7.69	8.14	14.3	5.6	15.3	9.2	0.387	0.603	6.78	6.33
3. (A.G.)	28	76	177	100.9	13.7	7.0	12.7	6.6	0.513	0.520	7.36	7.94	12.8	7.6	13.1	6.8	0.589	0.517	7.88	7.70
4. (A.B.)	23	70	185	105.4	15.3	7.7	13.9	7.8	0.502	0.565	6.89	7.58	15.4	8.1	15.4	8.1	0.523	0.521	6.84	6.84
5. (D.S.)	21	70.5	166	94.6	12.6	8.1	15.2	9.3	0.640	0.610	7.51	6.22	12.4	7.1	14.6	7.9	0.570	0.540	7.63	6.48

For symbols A, B, D, E see Figure 2.

H, height of the whole body gravity center.

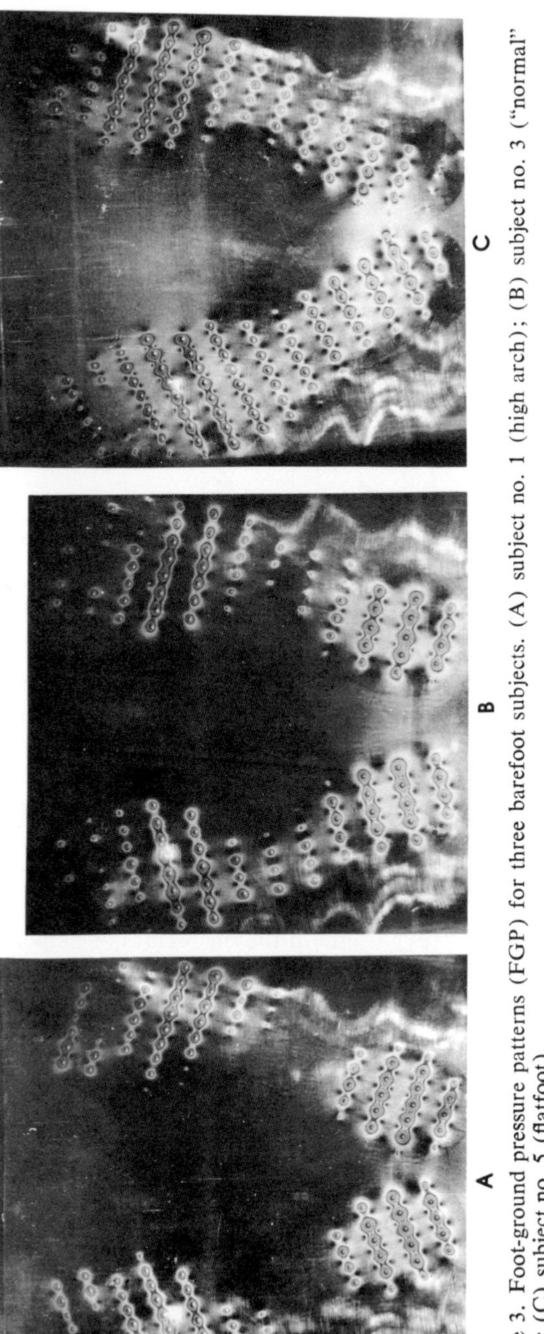

Figure 3. Foot-ground pressure patterns (FGP) for three barefoot subjects. (A) subject no. 1 (high arch); (B) subject no. 3 ("normal" arch); (C) subject no. 5 (flatfoot).

these parameters are suitable measures of the mechanical properties of the foot. In particular the parameter k_m is a good indicator of "flatfootedness" ($k_m > 10$ percent may well be a suitable definition). More generally it is noted that for the present group of subjects, as the portion of weight k_m taken by the midfoot gets larger, the heel load decreases, so that the middlefoot load can be thought of as taking up part of the heel load.

When shoes are worn, this midfoot load is, for the "normal" subjects, shifted to the forefoot. We note that in the case of the "flatfooted" subject, the shoe increased the heel load and led to a more "normal" force distribution. This result indicates that the proposed method will be useful in the evaluation of normal or orthopedic shoes as well as orthopedic inserts and appliances.

Mechanics of the Standing Posture

The results also indicate that the position of the points C_{FL}, C_{FR}, C_{HL}, and C_{HR} define a region in which the projection of the center of gravity must lie to insure equilibrium without additional external forces. The sagittal stance ratio B/A appears to be a good parameter to quantify the forward–backward stability of the standing posture, while the position of the center of pressure is a relatively insensitive quantity as shown in a previous paper (Arcan and Brull, 1975).

The quantities involving center of gravity height, namely H/A and H/D, are probably of some significance in locomotion and in initiation of motion, but the information available at this stage does not allow us to evaluate their usefulness.

It is likely that the mechanical and geometric parameters defined in this paper will also prove useful in the mechanics of human locomotion. We anticipate that the variations of these parameters with time and body position will permit systematic investigation of gait problems.

REFERENCES

Arcan, M., and M. A. Brull. 1975. A fundamental characteristic of the human body and foot: The foot-ground pressure pattern. Paper presented at The Israel Conference on Biomechanics, Haifa, January 8.

Brull, M. A., and M. Arcan. 1974. A method and instrument for recording the pressure distribution between the foot and the ground. Paper presented at The World Congress of Prosthetics and Orthotics, Montreux, October 8–12.

Carlsöö, S. 1964. Influence of frontal and dorsal loads on muscle activity and on the weight distribution in the feet. Acta Orthop. Scand. 34: 299–309.

Drillis, R., and R. Contini. 1966. Body segment parameters. Technical Report no. 1166.03. New York University, School of Engineering and Science, University Heights, New York.

Elftman, H. 1934. A cinematic study of the distribution of pressure in the human foot. Anat. Rec. 59: 481–492.

Lereim, P., and F. Serck-Hanssen. 1973. A method of recording pressure distribution under the sole of the foot. Bull. Prosthetics Res., Fall: 118–125.

Morton, D. J., and D. D. Fuller. 1952. Human Locomotion and Body Form. Williams and Wilkins, Baltimore.

Pliquett, F., and W. Helm. 1966. Die Druckverteilung unter der Fussohle Wahrend des Abrolevorganges (The pressure distribution under the sole by locomotion). Orthop., 102: 285–295.

Stokes, I. A. F., J. R. R. Stott, and W. C. Hutton. 1974. Force distributions under the foot—a dynamic measuring system. Biomed. Eng. 9: 140–143.

On the new force plate study

T. Matake
Nagasaki University, Nagasaki

It is well known that the dynamic methods of research on the gait include observation of the body motion by photograph, measurement of locomotion of center of gravity, use of electromyography, and the measurement of floor reaction by the force plate. All of these have contributed to the analysis of dynamics of gait. Among them, the force-plate studies (Elfman, 1939; Cunningham, 1950; Saunders, Inman, and Eberhart, 1953; Gotoh, 1959; Mizui, 1961; Matsuki, 1966; Suzuki, 1972) have offered the best method to measure dynamic forces during walking. Their force plates are supported by three or four columns connected by wire resistance strain gauges, and are able to record three components (vertical, antero-posterior, and transverse direction force) of ground reaction force on the oscillograph paper.

In the case of three columns, namely three load cells, the applied load and its coordinate (x, y) can be determined by the three values of load cells. However, one needs a device to combine these three different forces in the force plate, including the case of four columns. Furthermore, except for the vertical direction force, it might be inaccurate to mix torsion simultaneously in measuring antero-posterior and transverse forces.

On the other hand, an ordinary bridge usually contains one strain gauge, but a bridge with a resultant gauge, which combines the forces of two or four strain gauges, may not indicate the correct value on the strain meter or the oscillograph. In this case, it is necessary to calibrate the force plate values to determine whether they correspond to the variation of resistance of the resultant gauge, following the principle of bridge connection.

Thus, there are many problems in the use of the force plate. It is the purpose of this paper to describe a method for measuring step force and a new force plate that can measure compression, antero-posterior and transverse force, and also torsion simultaneously. Furthermore, six diagrams of resultant forces are defined and their characteristics and applications discussed.

METHODS

The new force plate is supported by four columns and eight poles to restrain the level motion with 12 ballbearings to reduce the friction to these columns and poles, on which one pair of strain gauges are mounted (Figure 1).

This force plate is considered as having the same feeling as the floor elasticity. As this instrument indicates the same value for any point of contact on the footboard, it is set in the middle and on the same level of the wooden floor, covered by a vinyl sheet. Therefore, it is possible to avoid the restriction in movement caused by requiring the subjects to step on a specific point.

Consequently, it is important to connect the strain gauges correctly. Assuming the rigid footboard and the symmetry of the design of the columns, the correct compressive force is obtained by the connection of the strain gauges on the diagonal columns, since the resultant strain of one pair of diagonal columns is the same as another pair of columns (Figure 1).

Since it is difficult to step exactly on the center of the footboard, motion of translation and rotation are normally present (Figure 1); thus each strain gauge on the eight poles must measure both deformations, after which they are separated. Although the case of four poles is a statically indeterminate problem, the separation and the correct measurement are achieved by considering the connection of strain gauges the same as four columns.

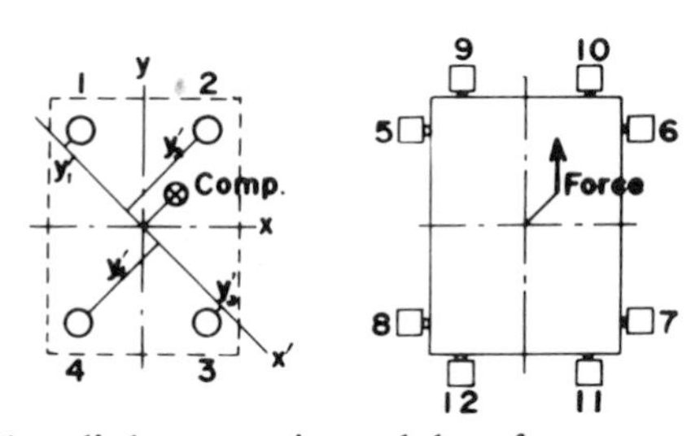

Figure 1. Footboard applied compression and shear force.

The resultant force of two arbitrary forces is plotted in the x–y plane. The tip of the resultant vector appears as a spot on the Braun tube of x–y synchroscope and its locus draws a loop diagram in one step.

RESULTS

The six resultant forces are combined in sets of two and isolated among four elements. The combinations and the names of the resultant diagrams are noted in Table 1. For example, Figures 2–7 are α-ζ diagrams of normal walking. In these figures the location of the spot indicates some vibration and high-speed running from the shock landing on the footboard, but it becomes clear during foot flat in the slow motion. Point M is near the middle and it returns to the origin 0 at toe off.

Table 1. Combinations of forces

	Comp.	Ant.-post. force	Trans. force	Torsion
Comp.		α	β	ξ
Ant.-post. force			γ	η
Trans. force				ζ

DISCUSSION

It is recognized that the step force is not applied at a precise point and the center of the contact area of the foot sole is moving continually during walking. However, assuming it to be a fixed point, Figures 2–7 provide a reasonable approximation of the characteristics of the magnitude and direction of instantaneous vector of the step force. For example, its vector in three dimensions can possibly be easily constructed with the α, β, and γ diagram.

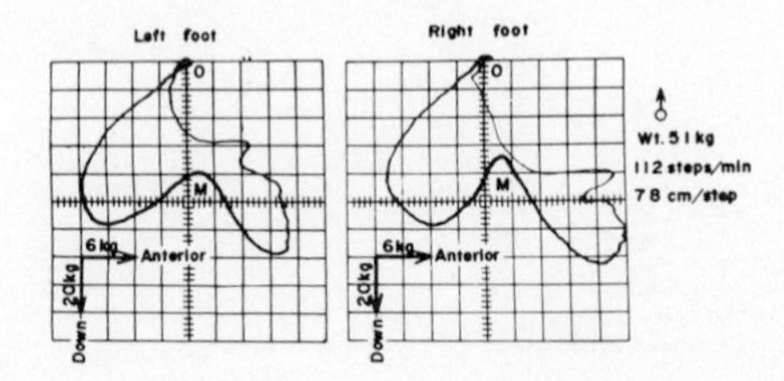

Figure 2. The α diagram.

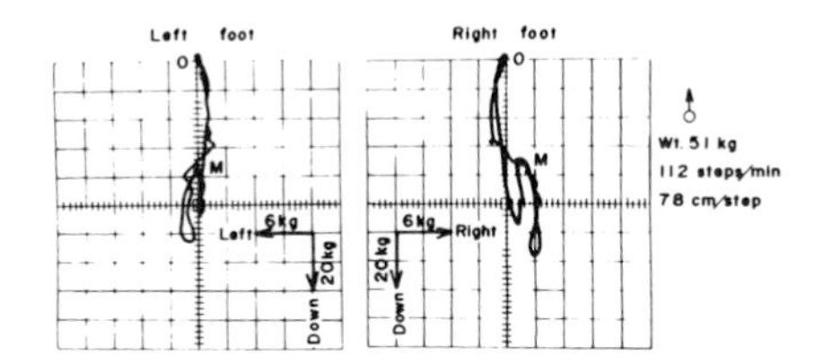

Figure 3. The β diagram.

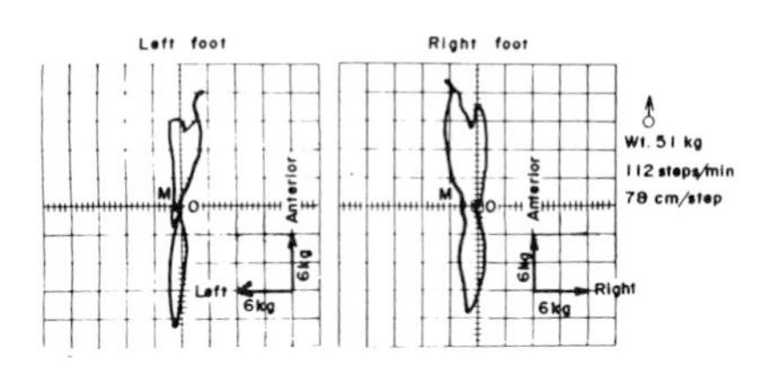

Figure 4. The γ diagram.

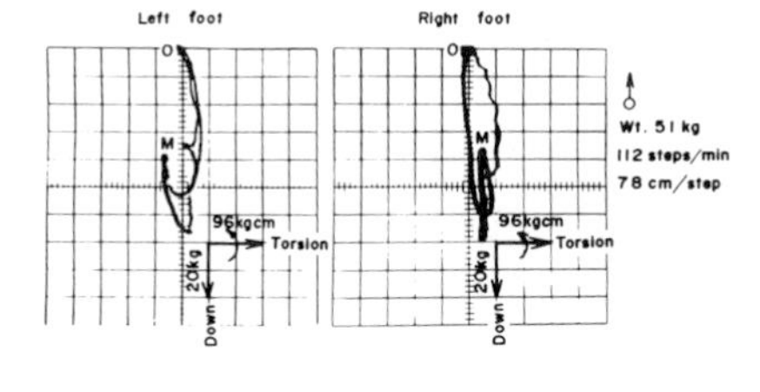

Figure 5. The ξ diagram.

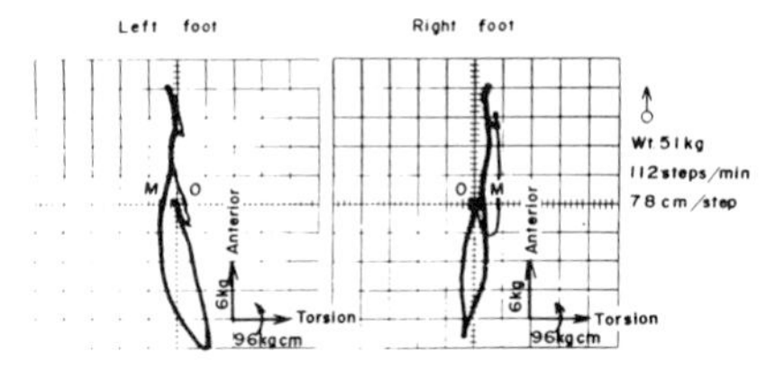

Figure 6. The η diagram.

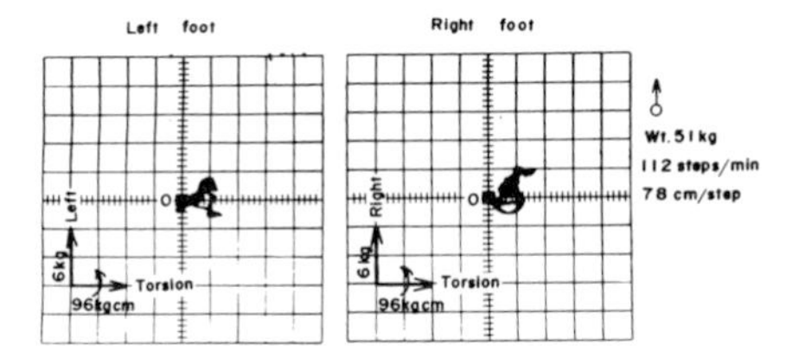

Figure 7. The ζ diagram.

Among these diagrams, the α diagram has the most specific character; that is, it is the largest and has an inverse heart shape, taking the axis of compression down. From this figure, the braking and driving period and the resultant force of body can be clearly seen and gait abnormalities determined.

In order to research the effect of walking speed on the α diagram, two cadences were examined (96 and 112 step/min, Figures 8 and 9).

If the dimension of the α diagram is defined as a, m, and b as in Figure 10, using the ratios a/b, a/m, and b/m, named the step ratios, the character of the α diagram is magnified. If the step ratios are constant, its shape is similar and the decrease of a/b and a/m makes the diagram flatter, as seen in the case of fast walking (Figures 8 and 9).

As an example of abnormal walking, Figure 11 indicates an α diagram of the right foot of a subject whose ankle joint is restrained. In this case the heart shape becomes flat compared with normal walking and the ratio a/m is near unity, because it is impossible to use the motion of the ankle joint. These values and shapes can be adopted as an aid in diagnosis of pathological walking and its rehabilitation.

Torsion of the body is important for walking, but it is small compared with weight. Then ξ, η, and ζ diagrams have less meaning, but must contribute to analysis of motion of the body.

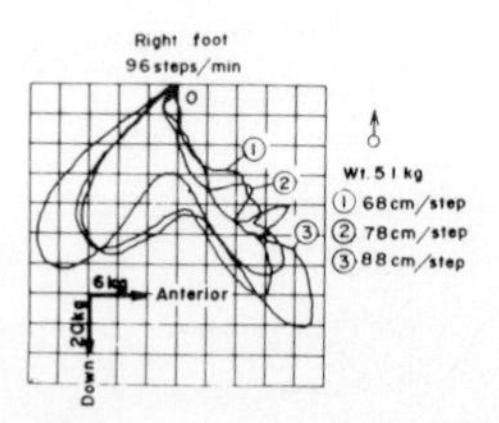

Figure 8. Speed effect on α diagram, cadence 96.

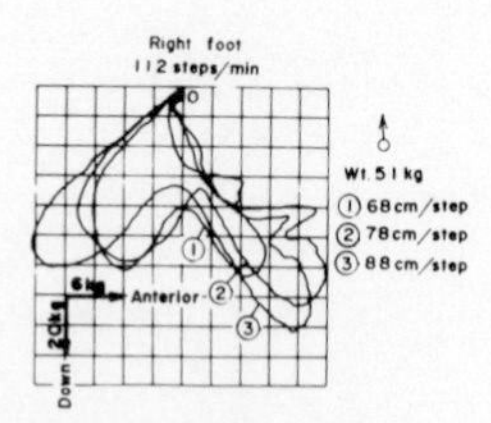

Figure 9. Speed effect on α diagram, cadence 112.

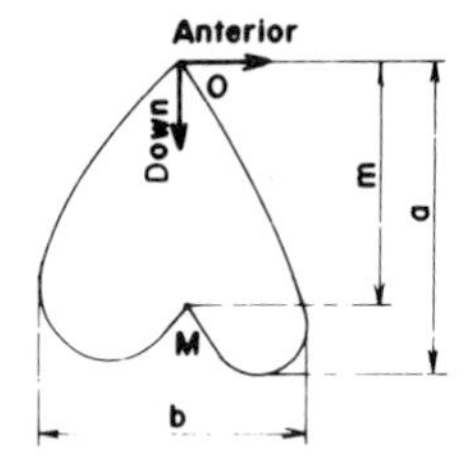

Figure 10. Standard shape of α diagram.

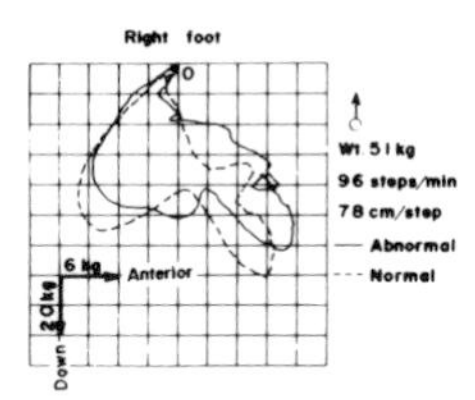

Figure 11. Abnormal walking by sprint.

CONCLUSIONS

The new force plate makes possible the measurement of three components and the torsion of step force simultaneously and to draw the resultant force of two components on the Barun tube. These possibilities of measuring the correct force and combining them depend on the separation bending from torsion and the research on the connection of strain gauges.

The α diagram is the most important among the resultant diagrams and will be surely valid for the diagnosis of pathological walking and its rehabilitation. Other diagrams will be used effectively to make artificial legs and for research on human motion.

REFERENCES

Cunningham, H. 1950. Components of floor reactions during walking. Advisory Cong. on Artificial Limbs, Berkeley, pp. 3–14.

Elfman, H. 1939. The force exerted by the ground in walking. Arbeitsphysiology 10: 485–491.

Gotoh, T. 1959. The dynamic studies on human gait. J. Japan. Orthoped. Assoc. 33: 775–792.

Matsuki, Y. 1966. Individual characteristics in the human locomotion. Central Japan J. Orthoped. Traum. Surg. 9: 751–766 (in Japanese).

Mizui, K. 1961. Gait studies of the normal and the patients with congenital dislocation of the hip by wire resistance strain gauge. Central Japan J. Orthoped. Traum. Surg. 4: 88–106. (in Japanese).
Saunders, M., V. T. Inman, and H. D. Eberhart. 1953. The major determinants in normal and pathological gait. J. Bone Joint Surg. 35-A: 543–558.
Suzuki, K. 1972. Force plate study on the artificial limb gait. J. Japan. Orthoped. Assoc. 46: 503–516.

Time-dependent method for measuring force distribution using a flexible mat as a capacitor

K. Nicol and E. M. Hennig
Johann Wolfgang Goethe-Universität, Frankfurt/Main

Biomechanical measurements, in addition to serving as a basis of operation for scientists in the areas of medicine, sport science, and work science, could also be successfully utilized by doctors, physical therapists, trainers, manufacturers of machinery and furniture and last but not least by patients and by athletes for equipment-supported motor learning (using technical devices in the feedback loop).

In reality, however, biomechanical measurements have been conducted primarily by the scientists. Moreover, they generally limit themselves to the quantitative interpretation of photographs and films. The important measurements that make instant information possible—especially the dynamographic measurements—are seldom carried out.

There are a number of reasons for this. Certainly the instruments capable of performing these measurements are too expensive for most of the above-mentioned groups. Moreover, the usable methods, especially for dynamography, are derived from industrial measurements. Forces are measured at rather few points; large-area forces must be transferred to the measuring point by sufficiently stiff materials, which are difficult to handle. Either these apparatuses cannot follow the movement or they must be adapted by elaborate equipment. The entire apparatus must be conveniently placed, possibly even sunk into the ground. If measurements on a different part of the body are required, a different construction is frequently necessary. Thus the financial, spatial, and personnel requirements for biomechanical measurements are considerable.

For these reasons we looked for measuring methods that take into consideration the circumstances described above. A possible solution will be described here which deals with the determination of kinematic and dynamic characteristics by assessment of the effect of geometrical arrangements on an electrical air capacitor and of forces on a capacitor with a compressible dielectric.

In this manner we obtained measuring instruments which: (a) are inexpensive (average price for transducer and transformer is about $100), (b) when necessary can adjust themselves to the movement, since they can be in the form of flexible mats, and (c) can be easily adapted to new problems (sometimes by simply cutting the mat to the desired form with scissors).

However, one must realize that with these methods, the precision reached by use of certain other methods will not be attained. In addition, at this time only force measurements with one component are possible, although an extension to three components is possible in principle. However, in biomechanics a high degree of accuracy is often not necessary; sometimes a yes–no response is sufficient. Moreover, sufficient information can also be frequently obtained from measurements with just one component.

The transducer and the measuring system that will be described here in detail are conceived for these cases.

AIR CAPACITOR MEASUREMENT

The transducers that function by influencing an air capacitor will only be mentioned briefly; they are described in more detail elsewhere. These are instruments for the determination of: (a) time characteristics of stationary movements, for example the repetition frequency while tapping, (b) time and space characteristics during walking–separate for each leg, (c) joint angle with a comparatively unimpeding arrangement (two 4 X4-cm plates)–a method that is particularly convenient for motor learning studies, and (d) vertical forces by use of a modified bathroom scale. Here the distance from the base and the cover plate is determined capacitively.

CAPACITIVE MEASUREMENTS WITH A COMPRESSIBLE DIELECTRIC

The second type of transducer functions according to the following principle. When a condensor plate of area A, filled with a compressible dielectric, is charged with power F, the plate distance d and the dielectric constant $\mathcal{E}$ and thus the capacity C is changed

$$C = \int_A \frac{\mathcal{E}}{d} (p) \, dA$$

When $\mathcal{E}/d$ is linearly dependent on pressure p,

$$\frac{\mathcal{E}}{d} = R + S \cdot p$$

one can obtain

$$C = \int (R + S \cdot p) \, dA = \int R \, dA + \int S \cdot p \, dA = C_0 + S \cdot F$$

i.e., the variation in capacity $C - C_0$ is proportional to the total force F. $C - C_0$ is independent of the size of the charged surface, the pressure distribution within the surface, and the position of the surface on the transducer. The following instruments were developed according to this principle.

PORTABLE MEASURING SYSTEM FOR FOOT PRESSURE

An insole, which is covered with two conducting foils, serves as the transducer. Each surface element produces—even under minimal pressure—a maximum variation in capacity; the total signal is proportional to the foot sole area under pressure. Smaller receivers indicate whether a defined part of the sole of the foot is under pressure. A harder foam rubber makes force measurements possible. The changes in capacity are transformed by small portable oscillators into changes of sound pitch and are controlled by the subject by ear phones. In the case of a onesided handicap (hemiplegia), the signal from the healthy side serves as the ideal value. Figure 1 shows the entire measuring system. Applications can be made in therapy of pes equinus, control of partially burdening of the foot during a habilitation program, symmetrization of the gait, and as a training aid for walking.

LARGE-AREA TRANSDUCER FOR GROUND CONTACT INTERVALS

A 1 $\times$ 20-m mat was used to measure flight and stance intervals during running and jumping. It consists of a 5-mm-thick foam rubber surface

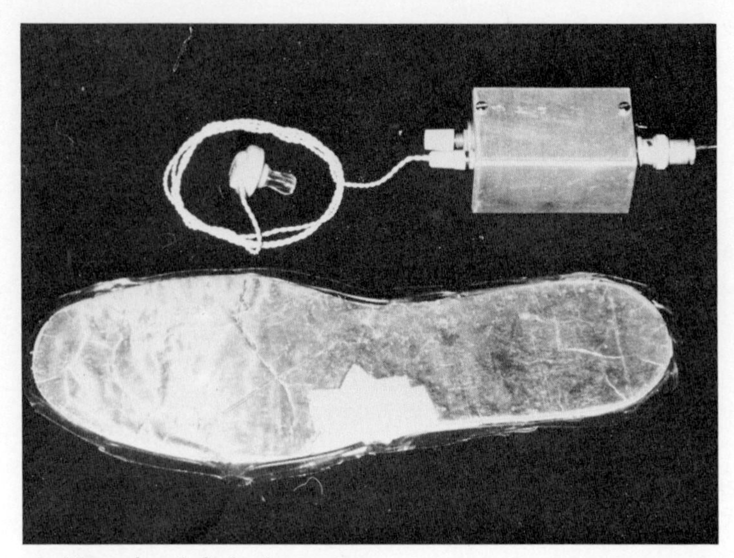

Figure 1. Measuring chain for foot pressure.

that is covered with metallic paper on both sides and welded into plastic foil. The capacity of the spot acted upon increases by a factor of three, the capacity of the entire mat changes by 0.3 percent. The capacity/voltage transducer described below offers sufficient resolution so that one can obtain the signal reproduced in Figure 2. It was recorded during a series of jumps on a box and mat, that is, a high jump test. Comparison with the vertical force measured with a piezoelectrical measuring platform resulted in an accuracy of ± 5 msec.

MEASURING PLATFORM FOR VERTICAL FORCES

This platform consists of two 40 × 60-cm aluminum plates, between which pieces of synthetics are fastened at several spots. Auxiliary electrodes are arranged between the plates in order to raise the linearity between force and capacity (the linearity is 2 percent full scale) and in order to reduce the position dependence of the sensivivity (±5 percent parallel to the longer side and ±1 percent parallel to the shorter side).

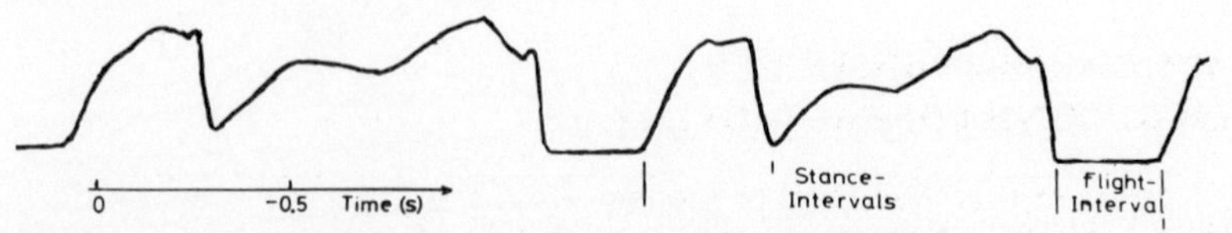

Figure 2. Signal of the large area transducer for ground contact.

CAPACITY/VOLTAGE TRANSDUCERS

For purposes of storage, evaluation, and representation, the variations in capacity must be transformed into voltage. For this, in addition to the systems for acoustical representation and the voltage divider (see below) and in addition to a loaded impulse-integrator that is advantageous for large signals, a phase-locked loop system is used. It is especially applied when high resolution and linearity are required. Its output voltage is linear for the variation in capacity, if this remains below 10 percent. Its response time is less than 1 msec. The system can be used with capacities in a range from 10 pF up to 30 nF.

MEASURING SYSTEM FOR PRESSURE DISTRIBUTION ON A SURFACE

The most developed measuring system (Figure 3) determines the pressure which a body exerts on a (perhaps soft or uneven) base in relation to place and time. The transducer consists of a flexible mat covered with 256 capacitor plates on each side.

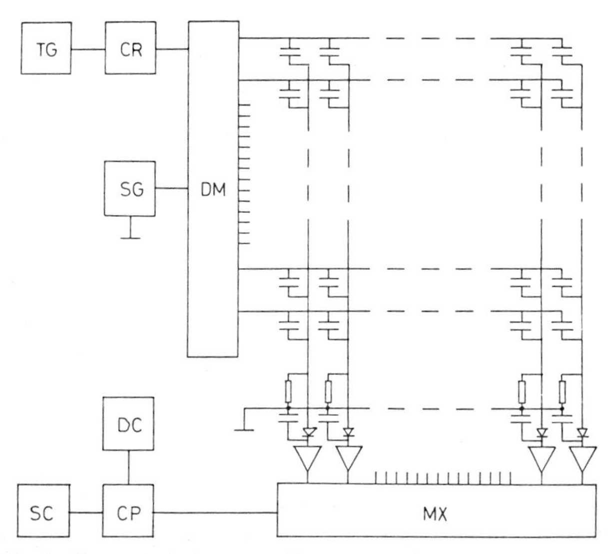

Figure 3. Block diagram of the measuring system for pressure distribution on a surface (for abbreviations see text).

The capacity of the force receivers is recorded via multiplexers. The upper condensor plates are joined in rows, the lower ones in columns; an alternating current is intermittently connected to the rows by means of a switch ("Demultiplexer"). A second switch ("Multiplexer") successively connects all the columns with a measuring resistor while the Demultiplexer is in a constant position. Thus each measuring capacitor is an element of a voltage divider; the voltage which then appears on the resistor depends on the acting force. The advantages of this circuit are that only 32 channels are required instead of 256 and that one can mount merely 32 continuous metal strips instead of 512 single condensor plates.

By loading the input lines one can prevent alternating current from reaching the output not only through the measuring condensor just switched on, but also through a chain of condensor circuits connected in parallel.

Figure 4 shows the construction of the transducer. A strip system is fastened to both sides of a 3-mm-thick mat made of a special synthetic which corresponds very well to the requirements of Equation 2. The condensors are 25 × 12 mm, the entire mat measures 48 × 24 cm. The mat can be bent on both axes to a half-circle, which makes both measurements on soft surfaces (for example jumping mat, upholstered chair, bed) and on uneven surfaces (for example, artificial limb pressure points) possible.

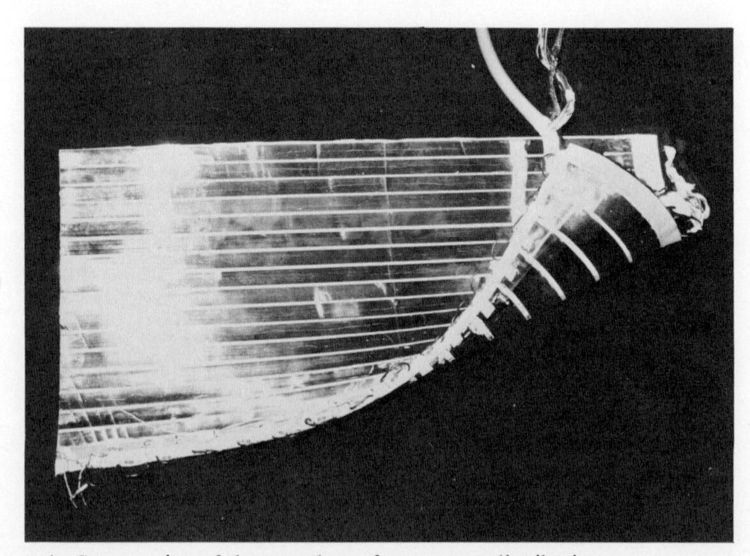

Figure 4. Construction of the transducer for pressure distribution.

The electronics necessary to supply the measuring condensors with alternating current consist of four simple ICs (see Figure 3), the signal generator SG, the tact generator TG, the demultiplexer DM, and a counter CR for addressing it. The mat is also a low-cost device because of its simple construction. Unfortunately this inexpensiveness does not yet extend to the arrangements for measuring, storage, and representation. At the moment a computer is necessary for these tasks. However, further development is aimed at solving these tasks with simple standard equipment.

The output voltage of each matrix row is rectified, smoothed out, amplified, and then connected to the Multiplexer MX of the computer CP; a total of 256 measuring values occur during a single measuring process. Since there are 4,000 available memory words, 16 single measurements can be carried out and stored within 1 sec at the minimum. After this the measuring values are stored on disc DC. This requires a few tenths of a second, then the measurement may go on. The representation of the results on the computer screen SC, which has been divided

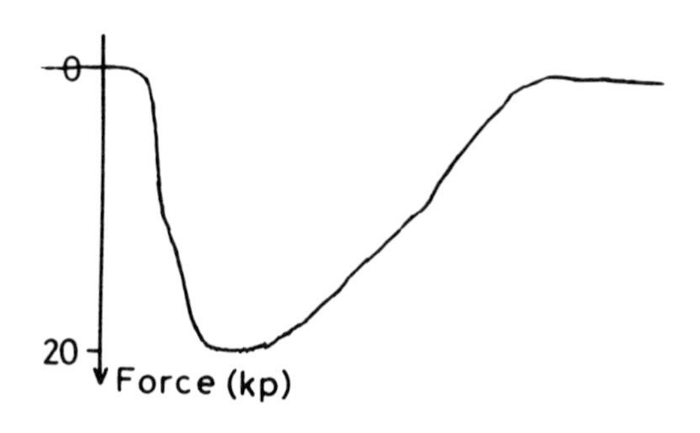

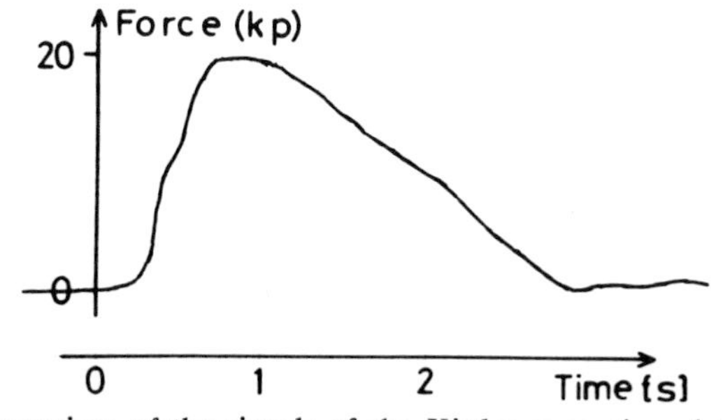

Figure 5. Comparison of the signals of the Kistler-measuring platform above and one condensor element.

into 16×16 fields occurs in two ways. In the first place, the light intensity of each field represents the force that is exerted on a measuring condensor during a single measurement. The 16 single measurements of a series can be shown on the screen with a maximum frequency of 1 picture per second. Secondly the force–time diagram of all measuring positions can be presented simultaneously on the screen.

In order to demonstrate the measuring accuracy, a mat was laid on a Kistler-measuring platform, a condensor element was put under pressure. Figure 5 shows the signals of the platform and of the pressured element. The comparison results in only unimportant differences up to a pressure of 10 kp/cm.

Digital speedometry at high sampling rate

W. Gutewort and M. Sust
Friedrich Schiller University, Jena

Speed measurement is a central problem in biomechanical analysis. Analog methods are still prevailing, with a speed-versus-time diagram being plotted, known as speedography. The present state of digital measuring techniques with their inherent gain in accuracy and time resolution, in particular the automation of data acquisition and processing by means of digital computers, makes numerical speed measurement with a very rapid sequence of discrete samplings appear a reasonable concept.

ANALOG AND DIGITAL MEASUREMENTS COMPARED WITH REGARD TO INFORMATION THEORY

According to the "fundamental law of digital measurement," the dynamic behavior (measuring time T) is inversely proportional to accuracy (a):

$$a = \frac{1}{Tf} \quad (1)$$

A comparison of analog and digital techniques yields the following results (Woschni, 1964):

Analog	Digital
Number of discriminable grades m:	
$m = 1 + \frac{1}{2F}$	$m = \frac{1}{a} + 1 = 1 + fT$ (2)

Frequency cutoff f_g
(dynamic behavior):

$$f_g = \frac{1}{2\, t_E} \qquad\qquad f_g = \frac{1}{2\, T} \tag{3}$$

Channel capacity C_t
(Maximum number of
bits/s transferred from
measuring instrument):

$$C_t = 2\, f_g\, ld\, m \qquad\qquad C_t = \frac{1}{T}\, ld\, (1 + Tf)$$

$$= 2\, f_g\, ld\left(1 + \frac{1}{2\, F}\right) \qquad\qquad 2\, f_g\, ld\left(1 + \frac{f}{2\, f_g}\right) \tag{4}$$

where f = pulse rate
F = minimum error of an analog measurement
t_E = build-up time
ld = binary logarithm

We can conclude that "digital techniques are definitely superior where frequency cutoff is relatively low and the number of amplitude steps relatively large, while analog techniques are preferable at relatively high frequency cutoffs with relatively few amplitude steps." (Woschni, 1965.)

DIGITAL SPEED MEASUREMENTS WITH HIGHER RESOLVING POWER

In view of the above relations, digital measurements were employed only for establishing mean speeds of longer sections of a motion path or of individual phases of a motion by way of short-time, electronic measurements. For that purpose, path signals are generated by sequences of light barriers (Ikai, 1967), aperture discs and impellers switched electro-optically, or by Hall effect switches (Neukomm, 1973), by reflex targets, contact transmitters in the athlete's shoe, pneumatic transmitters in swimming, or non-contacting proximity switches employing high-frequency probes (Gutewort, 1974). The path signals (up to 10^{-2} m) effect address indexing for storing defined digital time signals. If the measuring time T can be kept sufficiently short [to satisfy equations (1)–(4)] so as to achieve an adequately high sampling rate, and if digital encoding accepting high pulse rates can be implemented, it is feasible and reasonable to conceive digital speedometric methods based on a pulse technique.

THE DIGITIZER SYSTEM

Numerical measurements of approximate instantaneous speeds can be made with the IGR Incremental-Code Shaft-Angle Digitizer of VEB Carl Zeiss Jena (Figure 1). It has a glass disc bearing an incremental, radial grid of alternating light-transmittent and opaque segments, which is scanned by silicon phototransistors through a counter-grid. The phototransistor signals produced upon rotation of the radial grid are processed by a pulse-shaping circuit and an electronic signal-processing system, which delivers a train of square-wave pulses at a rate of up to 2500 pulses/ revolution. If, as in speedography, a pulling cord converts a translatory movement into the rotation of a pulley, the pulley diameter determines the linear distance between two pulses. It may be between 10^{-2} and 10^{-5} m. Speeds up to 6000 rev/min ($\leq$ 100 kHz) are possible.

Apart from its high-distance resolution and operating speed, the IGR digitizer system has the big advantage that the inertia of its rotating parts is as low as 20 g · cm^2, and its torque is not greater than 5 g · cm. The minimum braking force required to stop the entire system is 0.016 kg. If a motor with a friction clutch is employed running opposite to the pulling direction, the cord is kept taut and can easily be reeled back. The tensile force required in this case increases to about 0.25 kg.

THE STORAGE SYSTEM

The pulse train delivered by the IGR, representing the distance information in a digital code, is entered into a ferrite core memory or a multi-

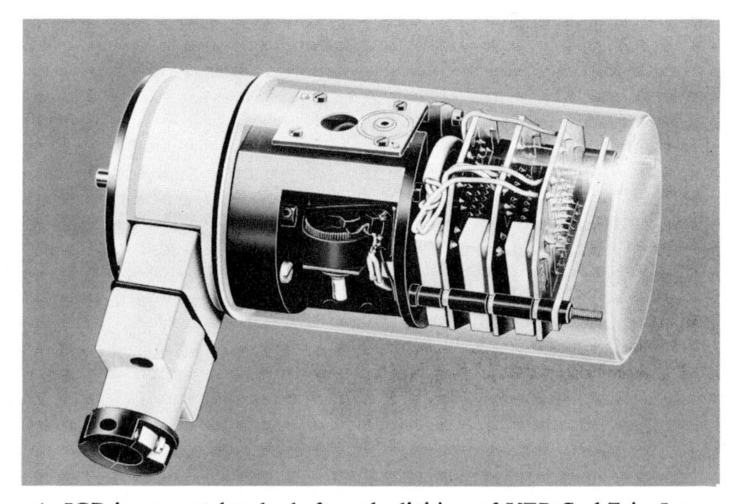

Figure 1. IGR incremental-code shaft-angle digitizer of VEB Carl Zeiss Jena.

channel analyzer. External timing pulses at 0.1- or 0.01-sec intervals effect address indexing, so that the pulses arriving during an interval are stored in one and the same channel. The fill level of each channel represents a distance covered within a certain time and thus has the dimension of a speed. Results are presented automatically, either in analog form on the screen of a storage oscilloscope, or by a numerical record in clear print and on punched tape.

Figure 2 shows the block diagram of the digital measuring system.

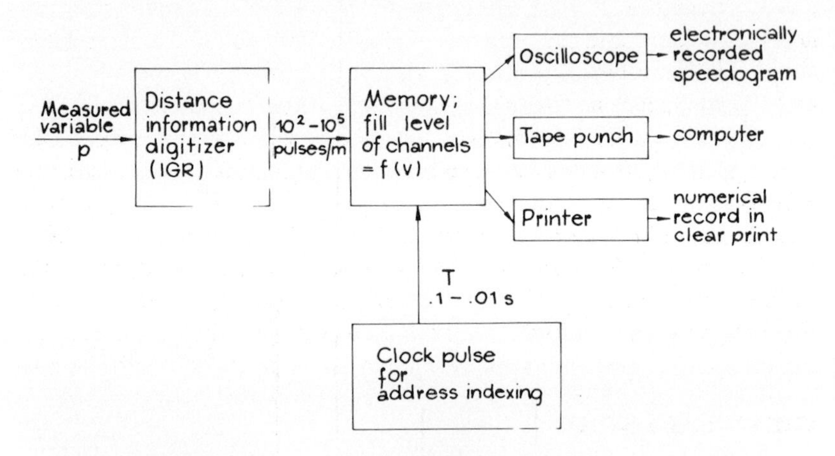

Figure 2. Block diagram and photograph of the digital measuring system.

Figure 3 outlines the performance level that can be reached with the system. A fall movement was used as a defined test motion. The quadratic distance–time equation $p = a/2\ t^2$ and the linear speed–time equation $x = a \cdot t$ are rendered well. Acceleration is less than gravitational, owing to the resistive forces of the test setup, which increase with the square of speed.

From the above, the possibility of continuous numerical measurements of distances and speeds in athletic movements and their direct processing by digital computers appear practicable.

EXAMPLES

If the pulling cord is aligned as closely as possible with the putting direction of a shot-putter and fastened to his or her middle finger, the speeds will be as given in Table 1. Figure 4 is a graphical representation of these results. Figure 5 shows the distance, speed, and acceleration behavior during the start of a 100-m sprinter, and the comparison of step lengths measured with the IGR and the tape measure, respectively. The unexpectedly large variations of speed and acceleration are, primarily, artefacts produced by longitudinal and transverse vibrations of the pulling cord, although the fact that these variations fit in well with the configuration of steps indicates that there are other causes, too.

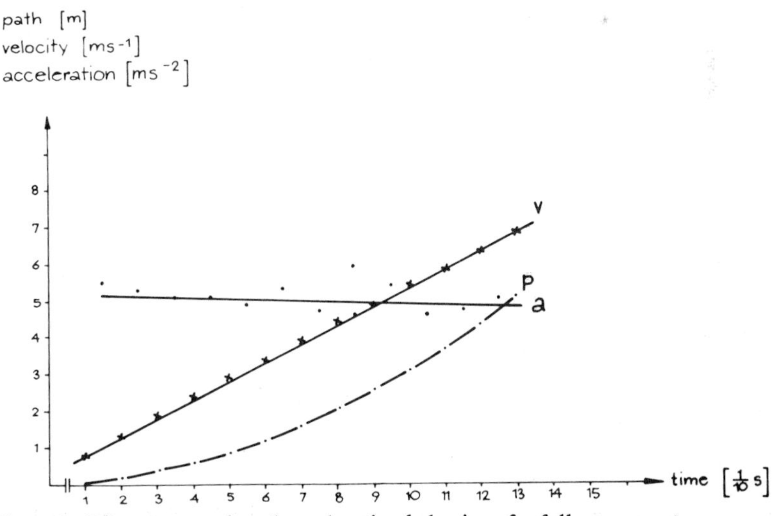

Figure 3. Distance, speed, and acceleration behavior of a fall movement.

Table 1. Putting speeds (in m/sec) of putting a 4-kg shot 11.30 m. Measuring interval 1/100 sec. Digital distance encoding by an IGR at 1000 pulses/rev (corresponding to $\Delta\, p$ between two pulses = 0.25 mm)

sec	m/sec	sec	m/sec	sec	m/sec
0.01	2.10	0.19	1.63	0.37	2.70
0.02	1.86	0.20	1.75	0.38	3.03
0.03	2.21	0.21	1.86	0.39	2.80
0.04	1.98	0.22	1.86	0.40	3.50
0.05	2.10	0.23	1.98	0.41	3.26
0.06	2.10	0.24	2.10	0.42	4.19
0.07	1.86	0.25	2.10	0.43	4.19
0.08	2.10	0.26	2.33	0.44	4.89
0.09	2.10	0.27	2.33	0.45	5.83
0.10	1.86	0.28	2.21	0.46	6.64
0.11	1.98	0.29	2.56	0.47	7.57
0.12	1.98	0.30	2.33	0.48	8.62
0.13	1.75	0.31	2.56	0.49	9.20
0.14	1.86	0.32	2.56	0.50	9.78
0.15	1.75	0.33	2.45		
0.16	1.63	0.34	2.80		
0.17	1.75	0.35	2.70		
0.18	1.63	0.36	2.91		

The braking interactions in the amortization phase and the accelerating forces in pushing off decrease markedly as speed is increased. Obviously, the method presented here can be employed only for the quantification of movements not longer than a few meters. Still there are quite a number of quantification tasks in sports that can be solved that way, such as the speed of the starting steps of a sprinter or of a player in

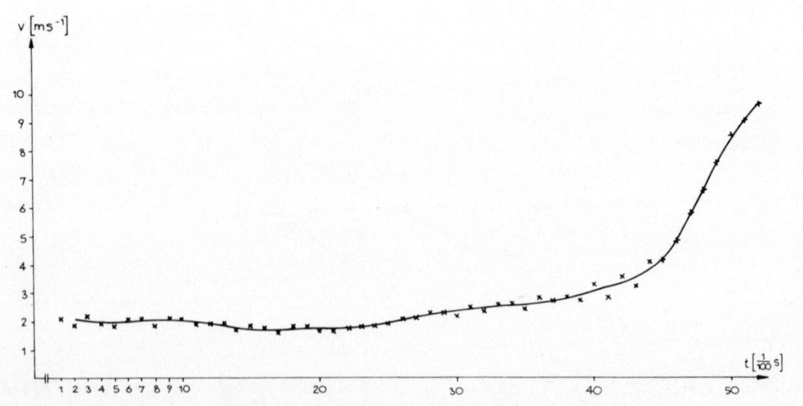

Figure 4. Speed-time diagram of putting a shot. Distance of putting, 11.30 m. Measuring time $T = 1/100$ sec. $\Delta p = 0.25$ mm.

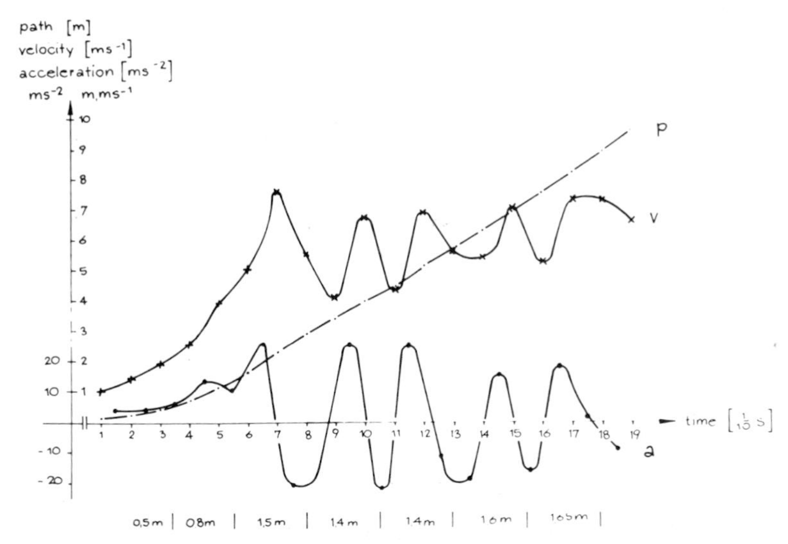

Figure 5. Distance, speed, and acceleration behavior during the starting steps of a 100-m sprinter. Step lengths measured by IGR and tape measure.

a ball game, the speed behavior in throwing, shot-putting, and jumping, the speed distribution in a cycle of motions of swimmers, rowers, and others, and speed tests in the speed-oriented strength training.

Frequently the only point of interest is to determine maximum speed, such as the initial ballistic speed of a shot leaving the hand, the speed of a jumper at the moment of leaving the ground, etc. In these cases, the DC voltage generated by a tachogenerator can be measured directly with a maximum-reading digital voltmeter. For example, if the exact take-off speed and take-off height of a shot are known, a table of values can be compiled that permits estimating the putting angle and the ballistic-technical potential inherent in the respective putting technique. In Figure 6, crosses mark the putting distances, whereas the dots in the circle indicate the maximum possible distance at a certain speed and height of shot take-off. In experiments 2 and 6, the putting angle was next to optimum, while experiments 3, 4, and notably 7 indicate considerable reserve potential for improvement.

CONCLUSION

Notwithstanding the gap between the relatively high error influences from the mechanical parts and the precise operation of the electronic digitizing system, the method reported is, in our experience, a suitable means of

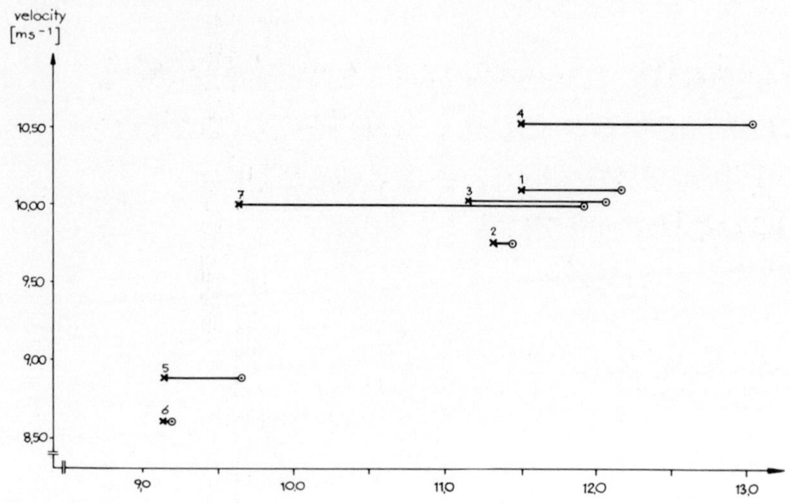

Figure 6. Potential and actual distance of putting a shot.

providing immediate, numerical feedback information on performance-determining kinematic factors in sports training; as a method of analysis, it is compatible with digital computation techniques.

REFERENCES

Gutewort, W. 1974. Determination of biomechanical parameters of athletic movements utilizing a noncontact measurement system. *In:* R. C. Nelson and C. A. Morehouse (eds.), Biomechanics IV, pp. 131–136. University Park Press, Baltimore.

Woschni, E.-G. 1964. Meßdynamik, eine Einführung in die Theorie dynamischer Messungen (Measurement Dynamic, an introduction in the Theory of Dynamic Measurements). Hirzel-Verlag, Leipzig.

Woschni, E.-G. 1965. Informationstheoretischer Vergleich der analogen und digitalen Messung (Analog and Digital Measurement Compared with Regard to Information Theory). Messen, Steuern, Regeln—technisch-wissenschaftliche Zeitschrift für Automatisierungstechnik 8: 367–370.

Velocity measurement without contact on body surface points by means of the acoustical Doppler-effect

E. M. Hennig and K. Nicol
Johann Wolfgang Goethe-Universität, Frankfurt/Main

When an acoustical transmitter is moved relative to a stationary receiver, a frequency is recorded on the receiver that is dependent on the speed of the transmitter. This phenomenon is called the acoustical Doppler-effect. For movements of the sound source in the direction of the receiver a shift in frequency results that leads to an increase in the reception frequency. A decrease is registered for movements away from the receiver. If we limit ourselves to transmitter speeds that are small compared to the signal's speed of propagation, the shift in frequency is approximately proportional to the transmitter speed.

$$\Delta f = f_0 \cdot \frac{v}{c} \cos \theta \qquad (1)$$

$$v = \frac{c \cdot \Delta f}{f_0 \cos \theta} \qquad (2)$$

where f_0 = transmitted frequency
Δf = shift in frequency
v = speed of transmitter
c = speed of sound
θ = angle between the direction of movement and the connecting line from the transmitter to the receiver.

For velocities up to 10 m/sec, the relative error of equation (2) by the approximation is smaller than 3.0×10^{-2}. By the shift in frequency Δf between transmitter and receiver, it is possible to determine the speed of the transmitter. Thus the velocities of a number of body surface points during a movement process can be measured both continuously and simultaneously from a measuring station, provided that there are sound sources of different frequencies on the body surface points. This method is an excellent aid for biomechanical movement analysis and for training purposes.

ACCURACY OF MEASUREMENT

As can be seen in equation (2), the accuracy of a speed recording under constant angle depends mainly upon the measurability of the speed of sound and the stability of the transmitting frequency.

The speed of sound depends upon the temperature, the pressure, and the humidity of the gaseous medium that transmits the sound. Movements of the sound carrier change the speed of sound, too. While influences from atmospheric pressure and humidity are extremely low for air, the sound velocity increases with rising temperature (Bergmann, 1954). According to the Laplacian equation for the spreading speed of pressure variations in ideal gases and according to the thermic equation of state, an increase in the speed of sound of 0.6 m/sec is calculated for the transporting medium air, when the temperature increases by 1°C. This corresponds to a relative change of the sound velocity in the range of 1.8×10^{-3}/°C. The sound velocity also changes according to the strength and the direction of the air flow. Under homogeneous air flow conditions, the speed of sound results from the vectorial addition of sound velocity and air flow velocity. The error rate by wind speeds of less than 3.5 m/sec remains below $\pm$ 1 percent. After consideration of temperature and wind speed, the error due to external influences is also $< \pm 0.5$ percent for the speed of sound during unfavorable conditions (rapid temperature changes, sudden turbulence).

A second important factor for the accuracy of measurement is the constancy of the transmitting frequency during the measurements. The frequency stability reached by use of piezoquartzes is $\pm 1.5 \times 10^{-4}$. This corresponds to an absolute error of $\pm$ 5.5 cm/sec in the determination of the transmitter velocity. The total error thus consists of a relative error of 0.5 percent resulting from the uncertainty of the speed of sound and an absolute error of $\pm$ 5.5 cm/sec resulting from the instability of the transmitter oscillator.

INTERFERENCE PHENOMENA

In measurements carried out within narrow spaces, interference phenomena from strong reflections are found. They can lead to a weakening of the signal (destructive interference) and additionally can falsify the measurement result.

Beats on the receiver alter the measuring frequency. In this manner non-sine signals result for frequencies that lie between the useful frequency and the transmitting frequency. However, falsification of the measurement results occurs only in the case of very strong reflections, since in addition to reflection losses, also a large weakening of the interfering signal because of the longer track appears. Errors from reflection could be avoided simply by covering the reflecting areas with sound absorbing materials. Likewise, an increase in the useful signal in relation to the interfering signal can be attained by use of an acoustical concave mirror on the receiver. In the measurements to date, even without additional appliances, the influence of interference has proved to be very small.

TRANSMITTER AND RECEIVER

Piezoelectrical crystals were used as ultrasonic transducers for transmitter and receiver. They show a distinct resonance behavior and a high electroacoustical efficiency of 10 percent. The transducers have a mass of less than 10 g and are quite convenient for attachment to parts of the body and hardly impede movement. The supply unit with battery and oscillator has only a mass of 50 g. Figure 1 shows three different pairs of ultrasonic transducers.

For the transmitter the quartz is both the transducer as well as the frequency-determining component of the oscillator. In this manner it is guaranteed that the vibrating quartz swings with its own resonance and thus exhibits a high degree of efficiency and a very high frequency stability. Quartzes with low resonance frequencies between 25 and 40 kHz were used, because the absorption of ultrasonic waves through the air increases very quickly with rising frequency. The conversion of sound pressure into corresponding electrical charge variations takes place through the quartz in the receiver. The receiving quartz is synchronized to the transmitting quartz, so that both have the same resonance frequency. The electronic processing results in the following manner: By a frequency–voltage conversion the change in frequency is reproduced in the form of a voltage signal.

MEASUREMENT RESULTS

Two different measuring methods were used for the investigations. For low speeds of up to 3 m/sec a testing apparatus was built in which the acoustical transmitter was placed on the sliding carriage of an air-

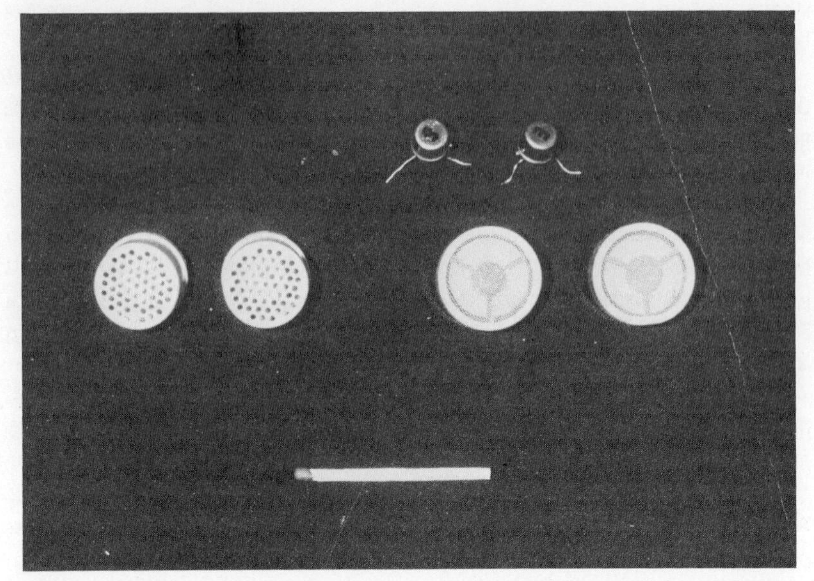

Figure 1. Pairs of ultrasonic transducers as transmitting and receiving elements.

cushioned track. A photo–diode combination is fastened to the carriage. Through optical contact with reflecting marks (400/m) on the track it delivers a series of impulses whose frequency is a measurement for the speed of the carriage. The impulse frequency is converted into a voltage signal that is proportional to the speed. With this testing arrangement, an average error rate of 0.9 percent for the acoustical speed measurement was found.

For high speeds, a photographic light impulse procedure (chronophotographic method) was used (Baumann, 1974). The acoustical transmitter was fastened with two electroluminescence diodes onto an aluminium plate. The diodes were pulsed with a frequency of 40 Hz and had a fixed distance in order to have a standard for each phase of the movement. With speeds up to 14 m/sec an average difference of less than 2 percent between the speeds of acoustical and light track measurement was found.

ONE-CHANNEL SPEED MEASURING INSTRUMENT

The measuring instrument consists of a portable, light-weight transmitter and a stationary receiver. The transmitter is fastened to the body of the test person. The use of the instrument is restrained by the weakening of the transmitted signal. A decrease in the reception signal occurs for strong angle deviations from the receiver direction and for long distances between transmitter and receiver.

Because of the low frequency (25 kHz) of the quartzes used, the directional radiation pattern was broad enough to ensure reception even with rotational movements. For turns of about $\pm$ 25° the sound pressure was found to be reduced by half.

The weakening of the transmitter signal is considerably dependent on ultrasonic frequency, temperature, and humidity (Kneser, 1933). For low frequencies, low temperature, and a relative humidity of less than 50 percent a range of more than 50 m can be obtained. For unfavorable conditions the range is limited to 20 m. Figure 2 shows the attenuation of an ultrasonic wave with the distance.

The one-channel speed measuring instrument was used for recording of different movements in a gymnasium. The distance between test person and measuring station was 15 m (temperature 22°C, relative humidity 60%). Figure 3 shows the speeds of three different movements.

MULTIPLE-CHANNEL SPEED MEASURING INSTRUMENT

At the moment multiple-channel speed measuring instruments are being developed that allow the simultaneous recording of the speeds of various

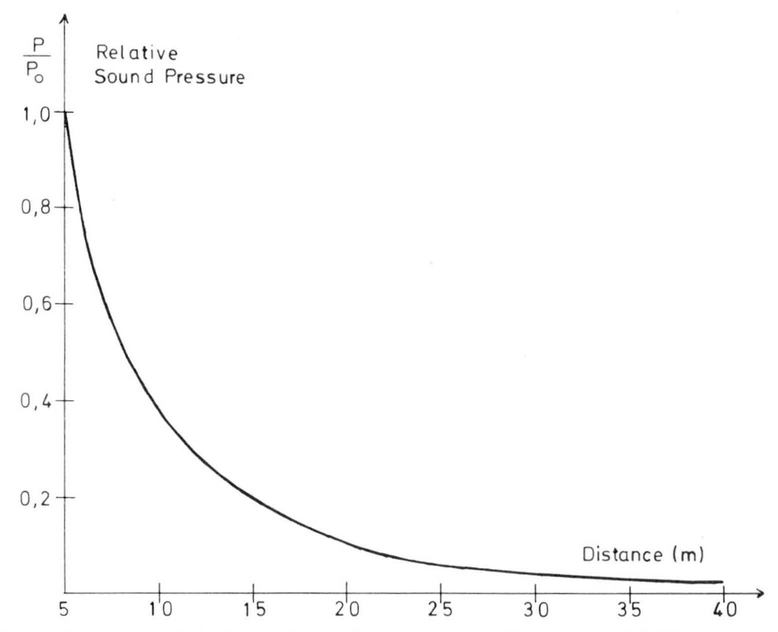

Figure 2. Attenuation of an ultrasonic wave in air (frequency, 25 kHz; temperature, 10°C; relative humidity, 60 percent).

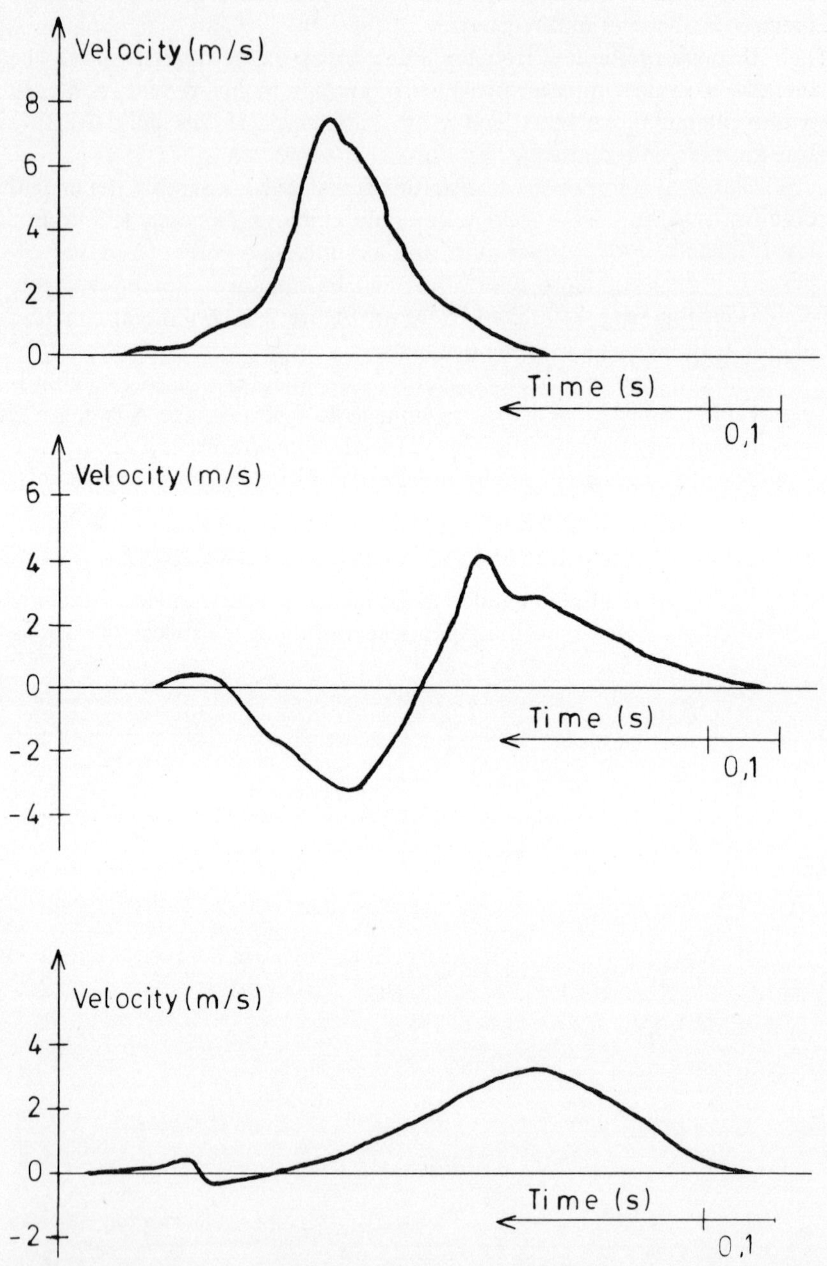

Figure 3. Velocities of different movements. *Top,* kick (measuring point: ankle joint); *middle,* rotational movement of an arm (measuring point, wrist): *bottom,* walking (measuring point: ankle joint).

parts of the body during a movement. A reciprocal influence is not expected if one uses piezo quartzes of different resonance frequencies. Their distinct resonance behavior makes these ultrasonic transducers excellent acoustical transmitters and receivers. The bandwidth required for one channel is 2.5 kHz for measuring speeds up to 100 km/hr. Because of the small bandwidth required, the favorable range of low frequencies can be maintained even with multiple-channel measuring instrument.

DEVELOPMENT OF THE SPEED MEASURING METHOD

A three-dimensional speed recording is possible with three receivers placed in different positions, which record the different shifts in frequency for one moved transmitting element.

A further development consists in differentiation and integration of the velocity signal. Thus distances and accelerations can be determined by only one mathematical operation. It is planned to record simultaneously distances, velocities, and accelerations for movements by electronic differentiation and integration of the velocity signal.

REFERENCES

Baumann, W. 1974. New chronophotographic methods for three-dimensional movement analysis. *In:* R. C. Nelson and C. A. Morehouse (eds.), Biomechanics IV, pp. 463–468. University Park Press, Baltimore.

Bergmann, L. 1954. Der Ultraschall (Ultrasonics), pp. 523–537. S. Hirzel-Verlag, Stuttgart.

Kneser, H. O. 1933. The interpretation of the anomalous sound-absorption in air and oxygen in terms of molecular collisions. J. Acoust. Soc. Am. 5: 122–126.

LED drivers: useful tools in biomechanics

P. R. Francis
Iowa State University, Ames

R. Gabel
University of Iowa, Iowa City

For a number of years individuals engaged in biomechanics research have recognized that the technique of chronocyclography has great potential for analyses of human movement. Light sources which repeatedly emit light pulses at some constant rate are attached to selected anatomical landmarks on a subject and a "long exposure" photograph is made of the subject in motion in a darkened environment. The resulting photograph shows broken traces indicating the loci of the points to which the light sources were attached. Average velocities between adjacent points of light emission can then be calculated from the displacements of the segments of the trace and the constant time intervals between the segments. Prior and Cooper (1968) used six-volt light bulbs that were pulsed five times per second; however, Howie (1971) reported that tungsten filament light bulbs require significant periods of time to heat and cool when they are turned on and off. Thus segments of light traces produced in this manner consist of elongated images. From such images it is difficult to distinguish those portions of the trace corresponding to the time intervals when current is flowing through the bulb filaments. Therefore the use of light bulbs may not be appropriate for recording activities involving relatively high velocities.

Hochmuth (1967) described an infrared chronocyclography technique that employed a rotating disc in front of a camera and Baumann (1974) described stereoscopic chronocyclography using pulse-operated infrared emitting diodes.

Chronocyclography apparently has some advantages over conventional 16 mm cinematography for many kinematic analyses. Several hundred data points can be recorded on a single photograph, thus reducing both the bulk of material to be handled and the overall cost of data acquisition. Further, a discrete light pulse can be recorded simultaneously by several different cameras, thus guaranteeing the temporal synchronization of three-dimensional data.

In an effort to overcome a number of the problems associated with existing chronocyclography techniques, a series of devices, named LED drivers, have been developed at the University of Iowa. These devices permit a large number of anatomical landmarks to be examined simultaneously in such a way that the light trace for each landmark is readily identifiable.

EQUIPMENT

Light emitting diodes (LEDs) were used as light sources for a number of reasons. LEDs are small and robust and are available in a variety of colors. Using appropriate circuitry, certain LEDs can be made to emit light pulses that will register adequate photographic images on suitable film even when viewed from a relatively oblique angle.

The elapsed time from the onset of current flow until the emitted light reaches maximum intensity is typically less than 50 nsec ($= 50 \times 10^{-9}$ sec). Similarly, the elapsed time from the cessation of current flow until the cessation of light emission is of a similar order of magnitude. Therefore, for all practical purposes, both switching-on and switching-off are instantaneous. The time interval for which an LED emits light (on-time) and the time interval between successive light pulses (off-time) can be determined with great accuracy using an electronic counter or oscilloscope. These values can be readily adjusted over a wide range of time intervals.

For most practical applications, an on-time of 0.5 of a millisecond (5×10^{-4} sec) will produce an adequate image that appears as a discrete point on a 35 mm color transparency. Pulse rates as high as 200 Hz have been obtained, but this frequency is generally in excess of values required for normal kinematic analyses of human movement.

Figure 1 shows, schematically, an arrangement that permits 18 LEDs to be driven from a single power source. The primary frequency from the pulse generator is progressively subdivided to produce two additional secondary frequencies. The zero detector ensures that all LEDs pulse as soon as current flows through the pulse generator. In the

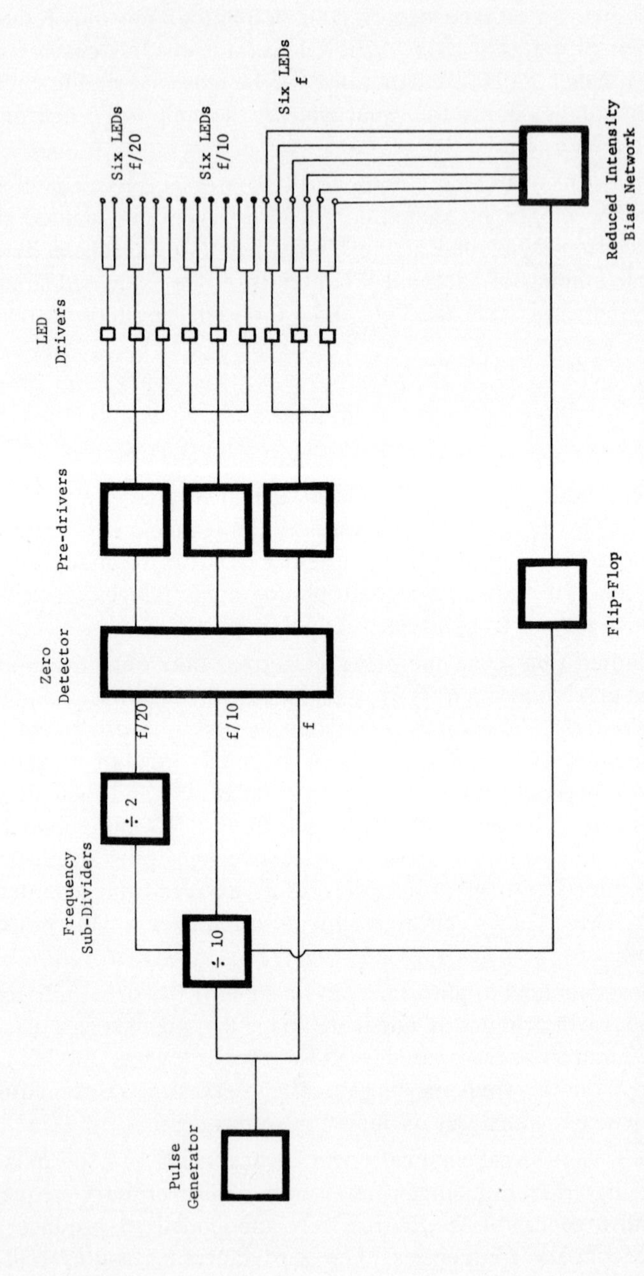

Figure 1. Schematic diagram of LED driver circuit. The primary frequency (f), and hence the secondary frequencies, may be varied over a wide range of values. Individual LEDs may be chosen from any of the colors currently available.

example shown, the six LEDs driven at the primary frequency will pulse synchronously. In addition, all six LEDs driven at 0.1 of the primary frequency will pulse simultaneously on each tenth pulse of the LEDs driven at the primary frequency. The six LEDs driven at 0.05 of the primary frequency will pulse simultaneously on each twentieth pulse of the LEDs driven at the primary frequency.

Identification of the locus of any landmark is ensured by using combinations of pulse frequencies of paired-LEDs. For example, Figure 2 shows LED traces of a basketball free-throw in which the joint axes of the wrist, elbow, and shoulder of the throwing arm have been marked with red LEDs pulsing at a suitable primary frequency. An additional

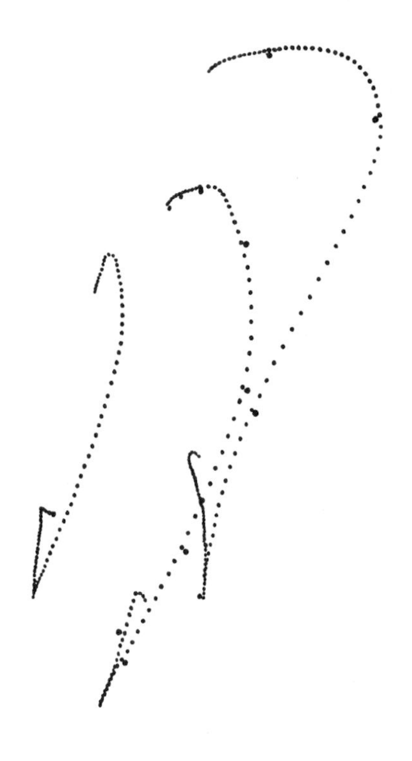

Figure 2. LED traces of a basketball free-throw. A secondary frequency of f/10 identifies the elbow joint and a further secondary frequency of f/20 identifies the wrist joint. A linear scale, consisting of LEDs placed 5 cm apart, is included in the plane of the motion.

LED pulsing at 0.1 of the primary frequency was placed close to the LED at the elbow joint, and an additional LED pulsing at 0.05 of the primary frequency was placed close to the LED at the wrist joint.

Distinctive identification of the loci of any four landmarks is facilitated by using LEDs that emit light in the red, orange, yellow, and green regions of the visible spectrum, respectively. Using a combination of pulse frequencies and colors, an almost unlimited number of anatomical landmarks can be readily identified.

The reduced intensity bias network (Figure 1) may be used as an aid in the analysis of photographic data. If this device is included in the circuit, the current flow through the LEDs is reduced to some low constant value between the successive higher levels of current flow that produce the high-intensity light pulses. Thus, an image produced during the movement of an LED pulsing in this manner consists of consecutive distinct points linked sequentially by a continuous faint light trace. This "streaking" effect is particularly useful when loci pass through the same region of space repeatedly.

A flip-flop may be incorporated to provide a further aid in analysis. When this component "gates" the bias network "off," the LEDs operating at the primary frequency emit five consecutive, discrete pulses of light. The flip-flop then gates the bias network "on" so that "streaking" occurs between the next five pulses of light. The bias network is again gated "off" and "streaking" ceases for a further five pulses. This process continues for as long as the LEDs are pulsed.

Figure 3 shows LED traces of a golf swing in which the joint axes of the shoulder and wrist, and the head of the golf club, have been marked with red LEDs pulsing synchronously at some suitable primary frequency. "Streaking" has been interrupted at regular intervals to facilitate interpretation of the traces. It is much easier to select a pulse on one locus and to determine the location of the synchronous pulse on another locus using this technique than it is in the absence of "streaking," as in Figure 2.

LED drivers have been adapted for use in a variety of studies. The device described in Figure 1 is a rechargeable unit measuring $18 \times 10 \times 7$ cm and can be attached to the subject by means of a Velcro belt. A much larger semiportable unit allows independent selection of several parameters over a wide range of values. These parameters are: the on-time, the off-time (which determines the pulse frequency), the ratio of the primary pulse frequency to secondary frequency in paired-LEDs, and the intensity of emitted light. A subminiature device with six pairs of LEDs pulsing at a wide range of primary and secondary frequencies has also been developed. This instrument incorporates two "event marker"

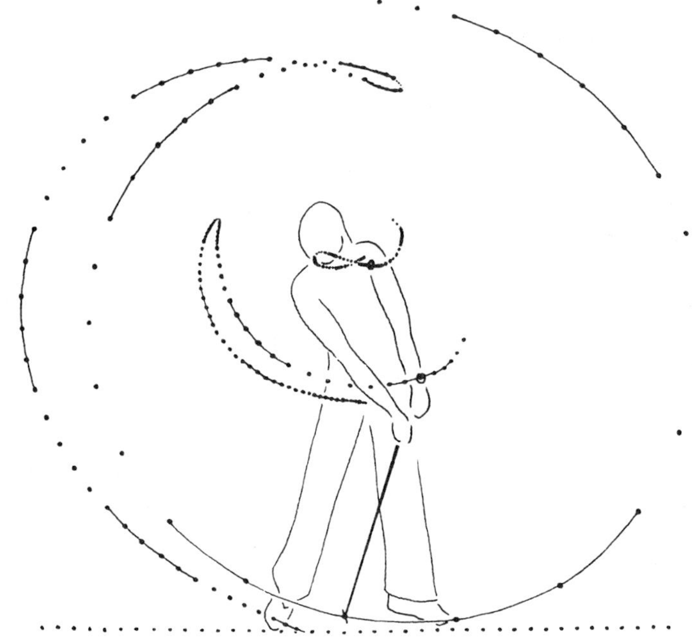

Figure 3. LED traces of a golf swing performed by a novice golfer. The primary frequency is 40 Hz.

LEDs which can be switched by some external mechanical means. The subminiature device can be carried in a coat pocket.

Film

Kodachrome II Professional Film (A.S.A. 40) produces transparencies with sharply defined images. This film is therefore recommended when a high degree of accuracy in analysis is required. High Speed Ectachrome (A.S.A. 160) is cheaper and can be processed by the experimenter in about one and one-half hours using the Kodak E-4 kit.

Polaroid Land film may also be used so that data is available with a minimum of delay. However, Polaroid film is relatively insensitive to certain colors emitted by LEDs.

PROCEDURE

Appropriate LEDs, or pairs of LEDs, are attached firmly to specifically designated points on the subject. The LEDs should be orientated such that optimal levels of emitted light will enter all cameras during the execution of the activity being examined. For single LEDs, 22-gauge wires are used to connect the LEDs to the LED driver, but 18-gauge wire may be more suitable for a "common ground" connection to a number of LEDs. Teflon-covered wire has proved to withstand consistent vigorous usage. The LED driver is connected to a console switching unit that is controlled by a single operator.

Suitable 35 mm cameras are positioned at accurately determined locations appropriate to the study being undertaken. The operator reduces ambient light conditions to low-intensity blue light. When the subject is ready to perform, the switching console is used to open the camera shutters and subsequently drive the LEDs for the duration of the subject's performance. The camera shutters are then closed and the film is automatically advanced in preparation for a further performance.

A linear scale, consisting of a series of LEDs placed known distances apart, is included in the field of view of all cameras. Current flow through these stationary LEDs is sufficiently low that their photographic images appear as discrete points. Complete identification of each photographic record is made by including an appropriate digitally coded display in the field of view of the camera.

ANALYSIS

Two instruments are currently being used for the reduction of LED driver data. Soderberg et al. (1976) have employed a computerized graphic tablet system similar to that described by Owen and Adrian (1974). The present investigators are using a modified Vanguard Motion Analyzer that permits digitization in the plane of the film. The latter instrument is used in an attempt to reduce errors caused by distortion during projection of the photographic image.

ACKNOWLEDGMENT

Peter R. Francis is indebted to the personnel of the Biomechanics Laboratory, Department of Orthopaedics, University Hospitals, Iowa City, Iowa, for their co-operation and encouragement in the completion of this project.

REFERENCES

Baumann, W. 1974. New chronophotographic methods for three-dimensional movement analysis. *In:* R. C. Nelson and C. A. Morehouse (eds.), Biomechanics IV, pp. 463–468. University Park Press, Baltimore.

Hochmuth, G. 1967. Biomechanik sportlicher Bewegungen. Wilhelm Limpert-Verlag GmbH.
Howie, A. 1971. Light spur photography. N. Zeal. J. Health, Phys. Ed. Rec. 4: 47–48.
Owen, M. G., and M. Adrian. 1974. Versatile uses of a computerized graphic tablet system. *In:* R. C. Nelson and C. A. Morehouse (eds.), Biomechanics IV, pp. 491–495. University Park Press, Baltimore.
Prior, T., and J. M. Cooper. 1968. Light tracing used as a tool in analysis of human movement. Res. Quart. 39: 815–821.
Soderberg, G., R. Reiss, R. C. Johnston, and R. Gabel. 1976. Kinematic and kinetic changes during gait as a result of hip disease. *In:* P. V. Komi (ed.), Biomechanics V-A, pp. 437–443. University Park Press, Baltimore.

Methods

Comparison of three methods for determining the angular momentum of the human body

B. D. Wilson and J. G. Hay
University of Iowa, Iowa City

Several computational techniques have been proposed for calculating the parameters of angular motion. However, a comparison of these computational techniques in terms of possible sources of error, complexity, and versatility does not appear to have been attempted. It is the purpcse of this paper to make such a comparison with respect to the methods of Miller (1970), Ramey (1973), and Hay and Wilson (1975) for calculating the angular momentum of a human body.

The Miller method requires a quasi-rigid body position during a period in which no external forces act on the body. The moment of inertia of the whole body about its transverse axis I_T is calculated from cinematographic data using a segmental model and the parallel axes theorem. The angular velocity (ω) of the body is determined from its angular displacement during a number of film frames, and the relationship

$$H_T = I_T \, \omega$$

is then used to calculate H_T, the angular momentum of the body about the transverse axis.

In the Ramey method coordinated film and force platform records are used to determine torque-time curves and hence the change in angular momentum experienced by the body. The following relationship is used to calculate the change in angular momentum produced:

$$\Delta \vec{H} = \int \left(\vec{F} \times \vec{r} \right) dt$$

where $\vec{F}$ is the force exerted by the subject on the force platform and $\vec{r}$ is the position vector of the center of gravity (c.g.) of the body with respect to the point of application of $\vec{F}$.

In the Hay–Wilson method a segmental model is used with cinematographic data to determine the average H_T between consecutive film frames by the following formula:

$$H_T = \sum_{i=1}^{14} (I_{Ti}\,\omega_i + m_i\,r_i^2\,\omega_i^*)$$

where ω_i is the angular velocity of the segment about a transverse axis through the c.g. of the segment, m_i is the mass of the segment, r_i is the distance from the segment c.g. to the whole body c.g., and ω_i^* is the angular velocity of the segment c.g. about the whole body c.g.

METHODS

The subjects in this study were eight male college gymnasts. Subject age, height, and weight are given in Table 1. Each subject performed three trials of a standing front somersault from a force platform. Coordinated film and force data were recorded for subsequent analysis.

The force platform used was 0.76 m square and utilized strain gauges mounted on four proving rings to sense vertical forces and strain gauges mounted on four square-sectioned deformable beams to sense horizontal forces. In this study the horizontal force in the direction of motion was the only horizontal force considered. The platform had a frequency response of 115 Hz in the vertical direction and 50 Hz in the horizontal direction. The highest-frequency component of the standing front somersault was observed to be of the order of 8 Hz.

The trials were filmed using a Locam 16 mm high-speed motion picture camera mounted on a fixed tripod of height 1.25 m and at a distance of 5.31 m from the plane of motion. A 10 mm lens was used with the variable shutter of the camera set at 120° and the frame rate at 100 frames/sec.

For each subject, that trial in which a quasi-rigid position was maintained for the longest time was analyzed with the aid of a Vanguard Motion Analyzer linked on-line to a digitizer, teletype, and paper tape output. The coordinates for the endpoints of a 14-segment model of the human body (Whitsett, 1963) were recorded for every second frame until the end of the flight phase.

A Red Lake Milli-Mite timing light generator operating at 10 Hz coupled to the recorder and to the timing light in the Locam camera

Table 1. Subject data

	Subject no.							
	1	2	3	4	5	6	7	8
Age (yr)	23.8	21.5	20.3	20.1	21.2	28.5	21.8	19.6
Height (m)	1.67	1.82	1.72	1.63	1.96	1.80	1.72	1.83
Weight (kg)	57.94	76.89	66.91	69.06	88.68	76.77	65.32	77.57

provided a time base for both the film and force data. To synchronize the data from these sources, coordinate data for each subject were smoothed using a cubic spline as a function of time. This process yielded coordinate data coincident with force data for each 0.02-sec interval throughout the motion.

For all methods of calculation the positions of the segment centers of gravity were determined using Dempster's cadaver data while the segment masses were determined from Dempster's living subjects (muscular) data.

For the Miller and Hay-Wilson methods for calculation of I_T, the values used for the segment moments of inertia were those derived by Miller and Nelson (1973, p. 85) with the upper and lower trunk considered as one segment. For each subject these segment moment of inertia values were adjusted to allow for individual differences in segment mass and length. This adjustment consisted first of determining the proportionality constant, C, from the Miller and Nelson data, using the relationship

$$C = \frac{I}{ml^2}$$

where I is the moment of inertia, m is the mass and l is the length of the segment. The adjusted value of I was then computed using

$$I_{\text{adj.}} = C\, m' l'^2$$

where m' and l' are the appropriate segment mass and length terms for the subjects used in this study.

In the calculation of angular momentum about the transverse axis the following correction for "out of the plane" motion was made:

$$I_T = I_{yy} + (I_{zz} - I_{yy}) \left(\frac{l_p}{l_a}\right)^2$$

This equation is based on a Mohr's circle representation of I, where I_{yy} is I about the segment longitudinal axis, I_{zz} is I about the segment transverse axis, l_p is the projected length and l_a is the actual length of the segment.

RESULTS

The angular momenta calculated for each subject by the three methods are given in Table 2. A typical force record and derived torque-time

Table 2. Angular momentum (N-m-sec)

Method	Subject no.							
	1	2	3	4	5	6	7	8
Miller	−47.04	−58.41	−34.99	−32.93	−73.50	−55.66	−39.10	−63.31
Ramey	− 9.41	−26.46	−41.36	−24.40	−83.10	−62.92	−55.37	−46.55
Hay–Wilson	−34.50	−53.80	−29.79	−28.13	−63.50	−44.49	−36.69	−54.10

curve used in the calculation of H by the Ramey method are shown in Figure 1.

While there was relatively close agreement between the values determined using the Miller and Hay–Wilson techniques, these values did not agree nearly so well with those obtained using the Ramey technique. It is of some interest to note that the subjects for whom the angular mo-

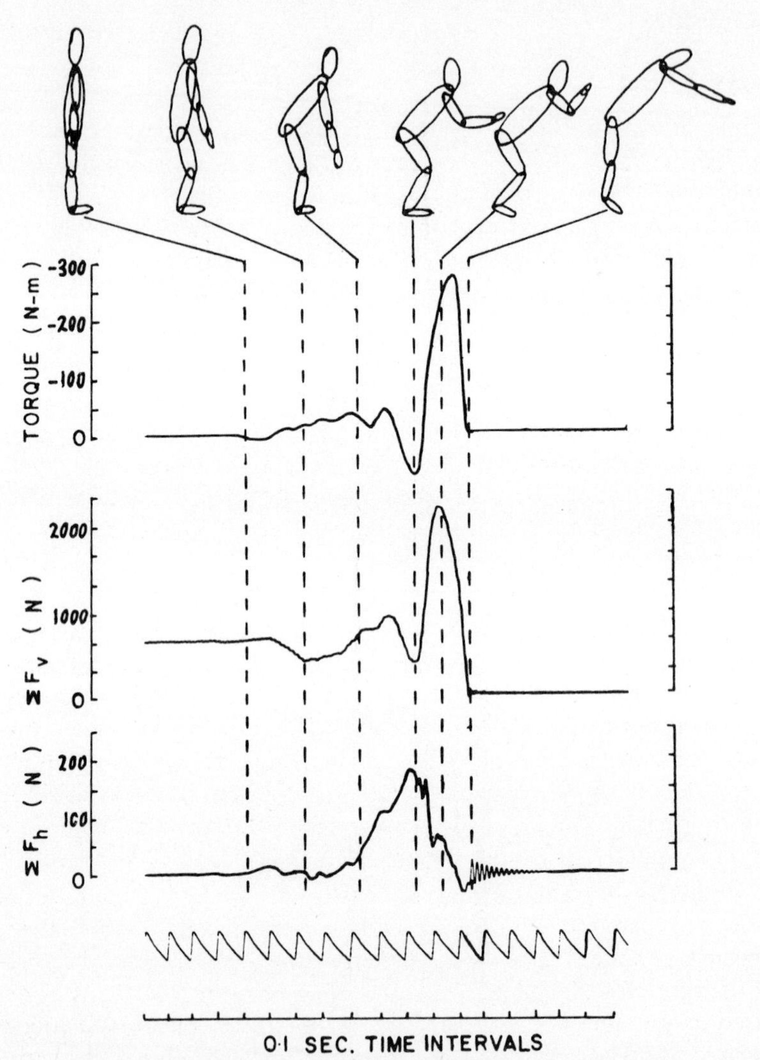

Figure 1. Direct recordings of the horizontal force and vertical force sum (F_h and ΣF_v, respectively), and a derived torque-time curve for subject 4. Body positions are indicated at intervals throughout the motion.

mentum values were in least agreement were subjects one and five—the subjects at the extremes in height and weight.

An analysis of variance with repeated measures design revealed that the obtained H values were not significantly different between methods of calculation.

DISCUSSION

Sources of Error

Inherent in each method of calculation are several sources of possible error. (Note: in this study it has been assumed that measurement error is much smaller than errors attributable to the use of inappropriate input parameters—i.e., segment proportions, masses, and moments of inertia—or to errors arising from the use of invalid assumptions.)

In the Miller technique the accuracy of determination of ω depends to a large extent on the rigor with which the term quasi-rigid is applied. In this study the body was considered to be in a quasi-rigid position if its joint angles changed by less than 5 degrees over 20 frames.

The Miller (and the Hay–Wilson) method of calculation requires the use of segment moments of inertia as input data. It is possible, therefore, that a systematic error in the H value may arise if the moment of inertia values used are inappropriate. In this regard, the trunk segment, which contributes approximately 30 percent to the total I_T when the body is in a tucked position, is probably of greatest importance.

In the Ramey technique the segmental masses and proportions used have a considerable influence on the computed H values. For example, when segment masses calculated from Barter's (1957) regression formula and proportions calculated from Clauser's (1969) mean data were substituted for those used in this study, H increased by approximately 30 percent. Ramey showed a similar effect when he reported that a shift in the x coordinate of the c.g. of approximately 0.07 m was sufficient to alter a calculated H value from approximately $-$ 5.0 to 10 N-m-sec.

In the Hay–Wilson method the H value recorded was the mean of a series of separate determinations of H. Several possible combinations of segment parameter data were tested using this method and the coordinate data for all eight subjects. The resulting H values increased or decreased by no more than 10 percent, suggesting that the computational technique was relatively insensitive to these changes. However, the variation about the obtained means for H increased considerably. Since H should be constant during the flight phase, the segment parameter data used were those which gave the least variation (< 10 percent) in the value of H.

Complexity

The least complex of the three methods tested is obviously the Miller method. In its simplest form the technique requires only the calculation of I_T for the one body position and the determination of ω from the viewing of two film frames.

The Hay–Wilson method is an extension of the segmental methods in widespread use. These procedures can readily be extended to obtain whole body angular momentum although the computational procedures (and in particular the required computer programming) are somewhat involved and tedious.

The most complex of the methods is the Ramey method. The computational technique is not involved, but in addition to requiring more data collection, accurate coordination of the force platform and film records is required.

Versatility

The Hay-Wilson technique appears to be the most versatile; it is applicable to all phases of human motions with the limitation that the motion is essentially planar.

The Miller technique has severe limitations in that few motions may be considered to have a quasi-rigid body position for sufficient time to accurately determine I_T and ω. The technique seems most suited to classroom applications where an understanding of the principles of motion may be more important than the accuracy of the results.

The major limitation of the Ramey method is the requirement that the initial angular momentum be known. The requirement that the performance be recorded by a force platform also prevents the analysis of many events other than in a laboratory situation, a factor which may have considerable influence on the quality of the performance.

CONCLUSION

It seems unlikely that any of the calculation procedures can be relied upon to determine H to a high degree of accuracy. Where possible, the H value determined by one method should probably be checked with that determined by another method, most simply the Miller method.

ACKNOWLEDGMENTS

The authors wish to thank Dr. J. R. Glover for generously giving of his time and making available the facilities of the Department of Mechanics and Hydraulics for the design, construction, and testing of the force platform used in this study.

REFERENCES

Barter, J. T. 1957. Estimation of the mass of body segments. Technical Documentary Report, WADC-TR-57-260. Wright-Patterson Air Force Base, Ohio.

Clauser, C. E., J. T. McConville, and J. W. Young. 1969. Weight, volume and center of mass of segments of the human body. Technical Documentary Report, AMRL-TR-69-70. Wright-Patterson Air Force Base, Ohio.

Hay, J. G., and B. D. Wilson. 1975. A computational technique to determine the angular displacement, velocity and momentum of a human body. Paper presented at the ACSM Annual Meeting, New Orleans, Louisiana, May 22.

Miller, D. I. 1970. A computer simulation model of the airborne phase of diving. Ph.D. dissertation, The Pennsylvania State University.

Miller, D. I., and R. C. Nelson. 1973. Biomechanics of Sport; A Research Approach. Lea and Febiger, Philadelphia.

Ramey, M. R. 1973. The use of angular momentum in the study of long jump take-offs. In R. C. Nelson and C. A. Morehouse (eds.), Biomechanics IV, pp. 144–149. University Park Press, Baltimore.

Whitsett, C. E. 1963. Some dynamic response characteristics of weightless man. Technical Documentary Report, AMRL-TDR-63-18. Wright-Patterson Air Force Base, Ohio.

Behavior in quasistatic balance

B. M. Nigg and P. A. Neukomm
Swiss Federal Institute of Technology, Zürich

Investigation and analysis of human balance behavior have been objects of scientific research for a long time (Vierordt, 1864; Leitensdorfer, 1897; Bass, 1939; Liebert, 1941; Corti, 1959, and others). Results of different balance tests were soon used as an indicator for analysis of psychological and physiological pathological symptoms (Oseretzky, 1931). In such research there are two principal aspects that offer difficulties. On one side there is the difficulty of defining human balance and on the other side there are technological problems, above all using force measurements, to be solved.

In this paper a new method of measuring human balance behavior together with application of this method on normal children and adults between 6 and 25 yr and on psychomotorically troubled children will be presented (Table 1).

METHODS

For measuring the reaction forces on the floor, the subject was standing on a force platform (Kistler system) while assuming the positions described in Table 2.

The test duration was 10 sec and each test was repeated five times. The three-axial output of the platform was amplified by a charge amplifier. With this measuring system, the force amplitude was analyzed.

With a new instrument, the "Impulsmessgerät" (Neukomm, 1974), the force functions were rectified and integrated (Figure 1).

The measurement used in this research was the rectified and integrated force–time function, hereafter called "rectified impulse," since the units of this variable are the units of impulse.

Table 1. Symbols and accuracy

Symbol	Comment	Accuracy (%)
$F_x(t)$	Force component, back and forward	1
$F_y(t)$	Force component, laterally	1
$F_z(t)$	Force component, vertically	1
$J_x(t)$	"Rectified impulse," back and forward	2
$J_y(t)$	"Rectified impulse," laterally	2
$J_z(t)$	"Rectified impulse," vertically	2
$J(t)$	$J(t) = J_x(t) + J_y(t) + J_z(t)$	
t	Time	
A	Age	1

Table 2. Standing position during test

Symbol	Position	Eyes
A1	Bipedal (on both legs)	Open
A2	Bipedal	Closed
B1	Monopedal (on one leg)	Open
B2	Monopedal	Closed

The schematic description of this measuring method is shown in Figure 2. The integrated force–time function, whose parts are all positive (rectified force), is the function used in the following. In this function there is not only the force amplitude included, but also the duration of this amplitude. It therefore seems that this measurement is a good description of the human balance behavior.

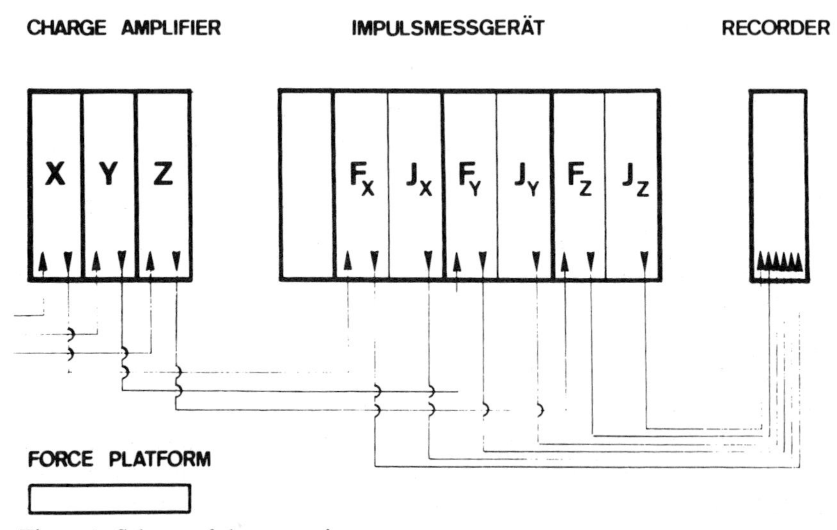

Figure 1. Scheme of the measuring system.

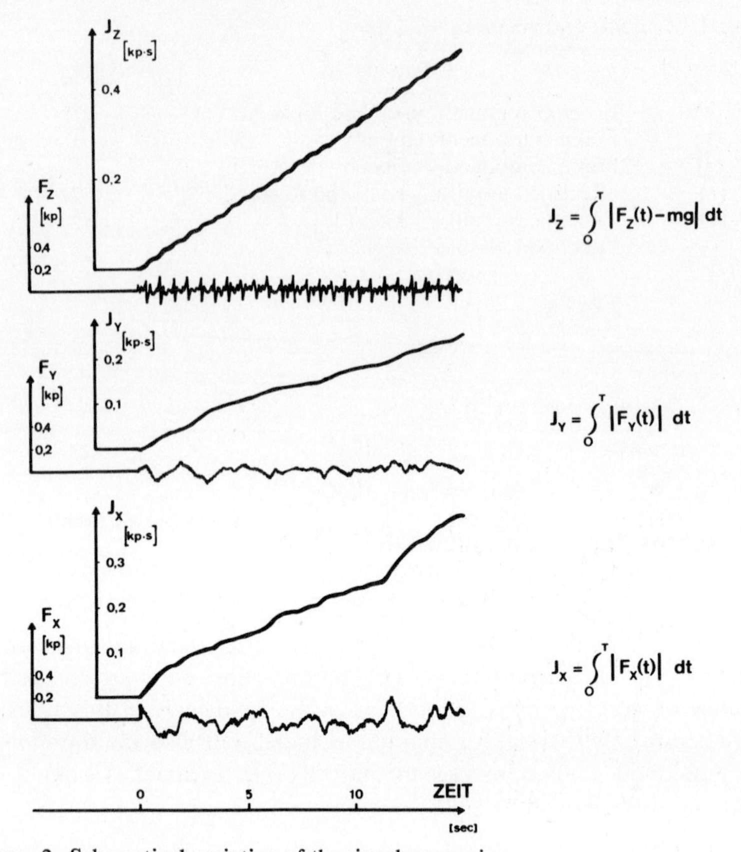

Figure 2. Schematic description of the signal processing.

The "rectified impulse" values after 10 sec were used for the evaluation. These values are equivalent to the increase of the "rectified impulse" curve.

From the five measurements the three with the minimal $J = J_x + J_y + J_z$ were used for the evaluation. In that way disturbing external influences were limited to a minimum and the reliability by the test-retest method was quite good.

Using the described method, 804 subjects between the ages of 6 and 25 were tested. Included among these 804 subjects were 48 with psychomotor troubles.

Figure 3 shows a typical curve of a test subject while standing on both legs (bipedal) with open eyes. The vertical force-time curve shows as a dominant contribution the heart action. The vertical rectified impulse results are closely correlated with the cardiac output. It seems that

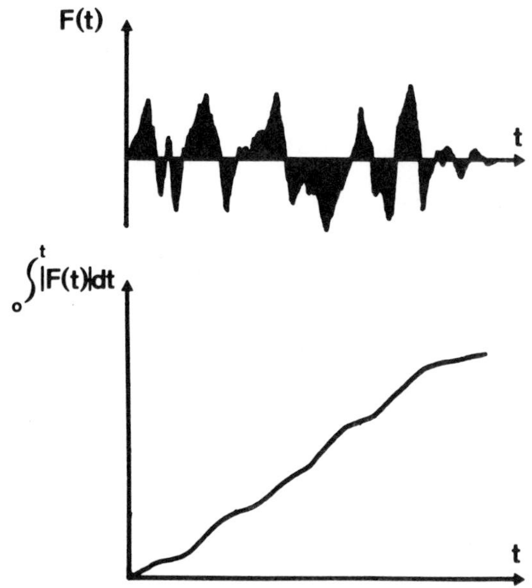

Figure 3. Example of a measuring curve; bipedal with open eyes.

the vertical component does not contribute to the maintenance of balance. The meaning of the horizontal components is a subject of further research.

The use of force to maintain the quasistatic balance depends on the balance ability and on the body weight. Although there is an improvement of the balance ability with age, it must not be forgotten that there is also an increase in body weight. It is therefore expected, that the "rectified impulse" values somewhere have a maximum. This relationship is shown in Figure 4.

The diagram shows that this maximum depends on the sex. The girls have the highest use of force at the age of 14 yr, the boys at 18 yr. The result of an age difference in development between girls and boys corresponds to other anthropometric data. Furthermore, the mean values are higher in males than in females with the differences being statistically significant from the age of 14 yr and older.

It seems that "rectified impulse" related to body weight would be a better measurement for the behavior in the quasistatic balance because test subjects with higher body weight need higher forces to correct small balance troubles.

Figure 5 shows that the "rectified impulse" related to body weight improves with increase of age, which signifies that the microvibrations

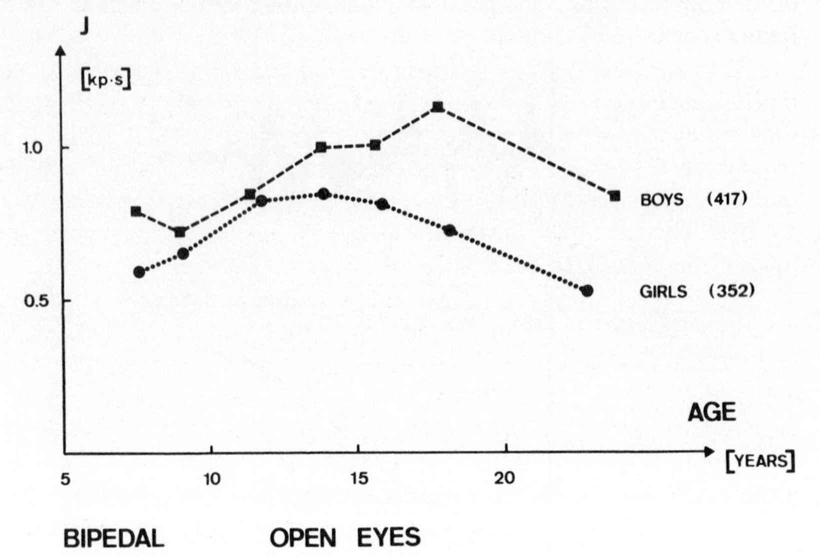

Figure 4. "Rectified impulse" as a function of age.

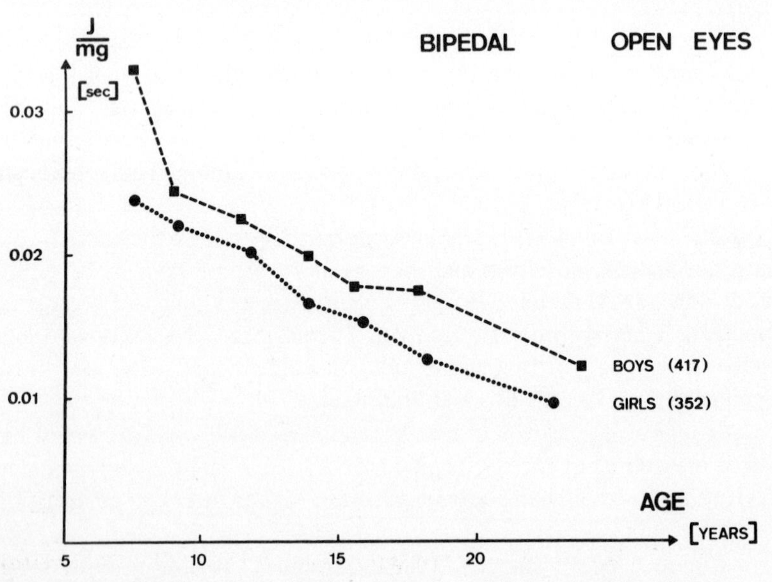

Figure 5. "Rectified impulse" related to the body weight as a function of age.

of the body decrease. One possible explanation could be that the use of force becomes more economical.

It is obvious that forces measured on the force platform in the bipedal test are not the only forces used to maintain balance. The measured result is a summation of different microvibrations in the body. In the monopedal test, where the test subject stands on one leg, the forces measured on the force platform are above all produced by balance corrections. The test results from the monopedal test are approximately ten times higher than from the bipedal test.

The regulation of balance depends above all on information coming from the vestibular system. Furthermore, the visual, the exteroceptive, and the acoustic systems can supply information for this regulation. If any of this information is absent, it is expected that the behavior in balance becomes worse as indicated by higher "rectified impulse" values.

Figure 6 shows the difference between open and closed eyes. All mean values with closed eyes are higher than with open eyes but these differences were not statistically significant. Apparently not all test subjects use the eyes for regulation of balance.

Furthermore it is interesting that nearly 70 percent of the difference is caused by the vertical component, the component having a strong correlation with the cardiac output. It could be that this difference between open and closed eyes is produced by a psychological influence.

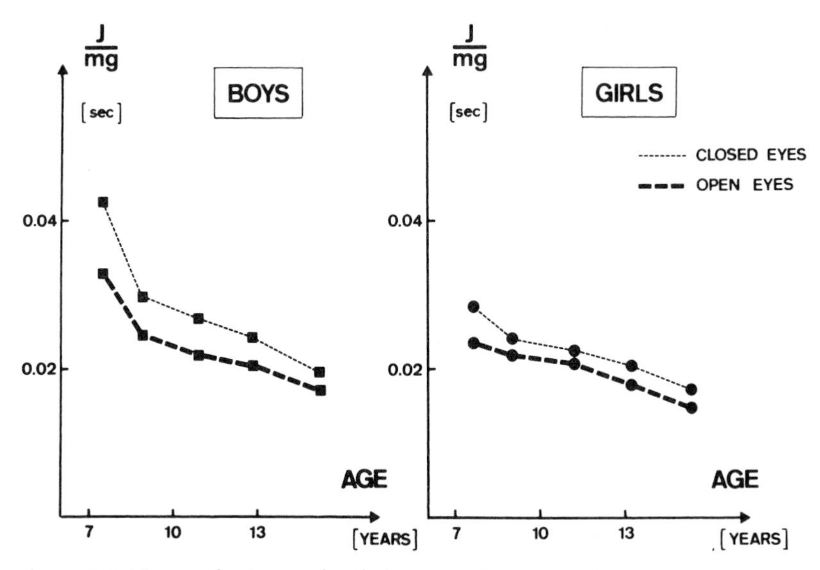

Figure 6. Difference in the quasistatic balance with open and closed eyes.

Another use of this method is the diagnosis of psychomotoric troubles, which are often related to brain damage. A possible origin could be a lack of oxygen during birth. The diagnosis of these troubles is quite difficult and their presence often produces social and school difficulties. It is possible to cure these troubles with special therapy, but it should start at about 5 yr of age before the child enters school.

Some of the measured components show a significant difference between normal and psychomotorically troubled children (Figure 9). In the example described the average difference between these two groups is 34 percent. This signifies that this method can be used for practical diagnosis of psychomotoric troubles, particularly since the test is simple and short.

Another application of this measuring method is in the field of psychological stress analysis. In typical stress situations the measured values amount to more than 15 percent of the normal values. Typical stress situations were measured with 110 medical students immediately before an examination and three weeks later. Immediately before the

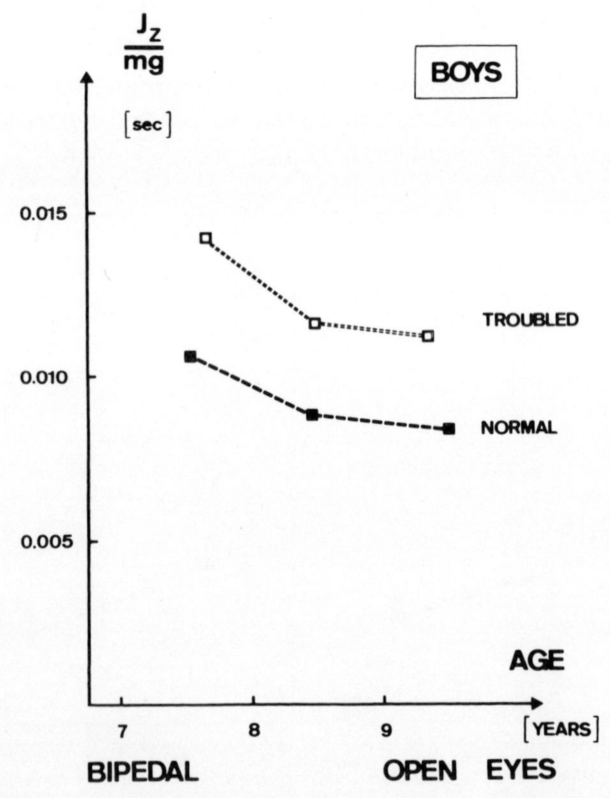

Figure 7. Difference between normal and psychomotorically troubled children.

examination the mean values were 43 percent higher in males and 55 percent higher in females than they were in the normal pre-test situation. (Figure 8).

Analogous measurements were made with the Swiss National Gymnastics Team (boys and girls) in several normal situations and immediately before a competition. The measured vertical component was on the average 71 percent higher than in a normal situation. (Figure 9).

Therefore it is obvious that this new method can be applied in various fields of research as an indicator for psychological and physiological parameters.

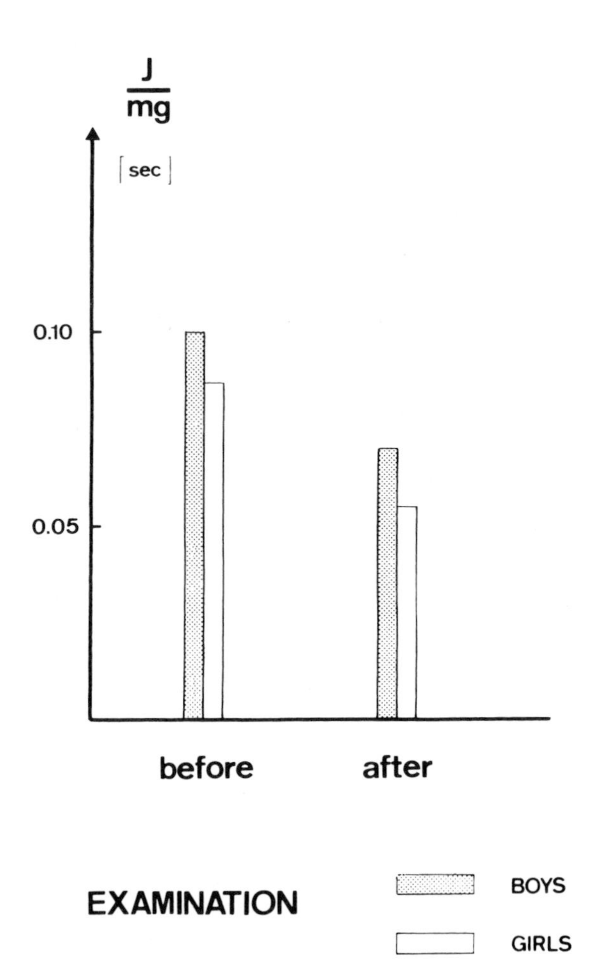

Figure 8. Difference of the related "rectified impulse" before an examination and in a normal situation.

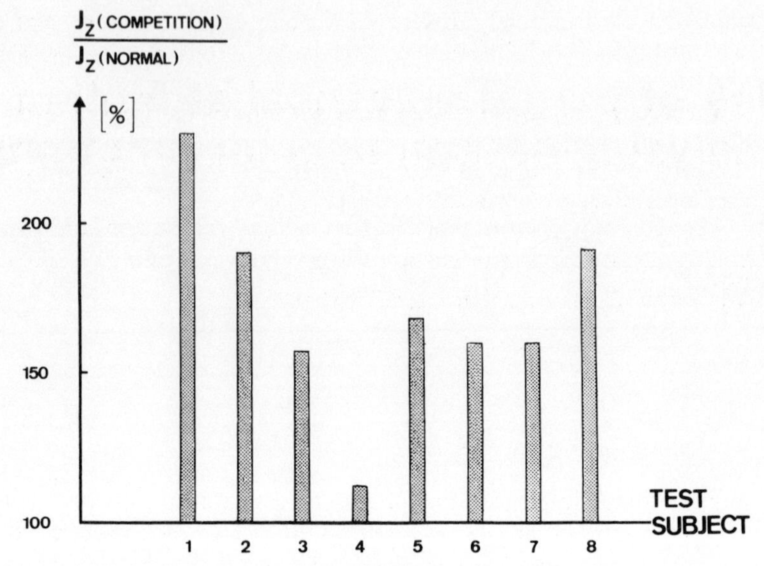

Figure 9. Difference of the "rectified impulse" in gymnastics (boys) before a start (free choice exercise) and in a normal situation.

REFERENCES

Bass, R. I. 1939. An analysis of components of the tests of semi-circular canal function and of static and dynamic balance. Res. Quart. 10: 33–52.

Corti, U. A. 1959. Erschuetterungsmessungen am Lebenden (Vibration measurements in vivo). Schweiz. Medizinische Wochenschrift. 89, 22: 575–581.

Leitensdorfer. 1897. Das militaerische Training usw (The Military Training). Stuttgart.

Liebert, H. 1941. Ueber Schwankungen beim Stehen (Postural sway in standing). Arbeitsphysiologie 11: 151–157.

Neukomm, P. A. 1974. Impulsmessgeraet, Analogrechen- und Auswertelektronik zur Mehrkomponenten-Messplattform (Electronic system for analog data processing and evaluation of force platform output). Interner Bericht am Laboratorium fuer Biomechanik der ETH Zuerich.

Oseretzky, N. 1931. Psychomotorik. Methoden zur Untersuchung der Motorik (Methods for analysis of movement). Z. Angew. Psych. Beiheft 57.

Rohracher, H. 1969. Die Mikrovibration. Verlag Huber, Bern.

Vierordt, K. 1864. Psychologie des Menschen (Human psychology). Tuebingen.

Use of paired electrical stimuli in analyzing neuromuscular functions

G. Rau
Research Institute for Human Engineering, Meckenheim

In addition to applications in the medical field as a diagnostic and therapeutic tool, electrostimulation is going to be utilized more and more in the field of human engineering. Procedures well known from control theory have been applied in complex movement analysis. Accordingly, the rules governing CNS movement control can be studied by analyzing the correcting output as responses to input impulses which disturb ongoing complex movements. These test impulses can be applied either mechanically or by eliciting muscle twitch contractions by electrostimulation.

A twitch contraction of a muscle can be elicited by a single electrostimulus applied either to the muscle itself (direct stimulation) or to its nerve (indirect stimulation). The practical problem with single shock stimulation using surface electrodes is the limited range of gradation in the mechanical response when the stimulus intensity is varied. In nerve stimulation, from the threshold up to maximum response this range is very small. In muscular stimulation the range is larger but is limited by the pain threshold at high stimulation intensities.

Another way to increase the mechanical response besides increasing intensity is by using pairs of electrical shocks as described in this study on "isometric" contraction of the extensor carpi ulnaris muscle in direct stimulation and of the flexor digitorum superficialis muscle during indirect stimulation of the medianus nerve.

METHODS

The right forearm of the sitting subject was positioned horizontally with an angle of 90° to the upper arm, which was in line with the trunk; the arm and the palm were in a plane parallel to the sagittal plane of the body. Just proximal above the wrist the forearm was fixed by a metal clamp upholstered with foam rubber. The elbow rested in a pan-shaped support. During the experiment, the hand was held by a flat metal cuff around the metacarpal joints; the cuff formed part of the strain-gauge dynamometer arrangement for measuring the torque at the wrist. The distance between the axis of the wrist joint for flexion/extension and the cuff at the palm had to be adjusted to the anatomical properties of each subject. Torque calibration was achieved by using weights. The resonance frequency of the measuring apparatus itself was about 100 Hz.

For activating the flexor superficialis muscle by indirect stimulation the surface electrodes (8 mm ϕ each) were attached to the skin above the medianus nerve about 5 cm proximal to the elbow joint. The position of the electrodes was changed until the flexor digitorum superficialis muscle responded selectively with maximum force exertion. During direct stimulation of the ext. carpi ulnaris the electrodes were placed above the belly of the muscle on the site of best response. The stimulation electrodes were connected to a battery-operated stimulator (Ahrend von Gogh, ES — 1 D). Square pulses of constant current were applied; the duration was 0.5 msec, and the amplitude varied. In addition to single shock stimulation, paired identical shocks with variable interstimulus intervals were also applied using a second stimulator coupled with the first. Between each stimulus event a 1–2-sec break prevented muscle fatigue.

In the experiments on indirect stimulation the muscle compound action potential was picked up by a pair of EMG surface electrodes (8 mm^2 each) attached to the skin above the belly of the muscle. The position of the electrodes was changed until a stable and reproducible EMG signal was seen on the oscilloscope.

During the experiments, the stimulus and the muscle compound potential, as well as the exerted torque, were simultaneously recorded on an analog recorder (Ampex FR 1300). After the experiment the signals were played back, digitized, and stored in the core memory of a four-channel averager (Nicolet Instruments, NIC 7074). The force signal was then integrated ($\int F(t)\,dt$) and stored in the fourth memory division of the averager. Thereafter, the signals were printed out on an X–Y recorder one after the other.

RESULTS

Indirect Stimulation

A typical experimental record of indirect stimulation at nearly maximum intensity is shown in Figure 1*A* for various interstimulus intervals ΔT.

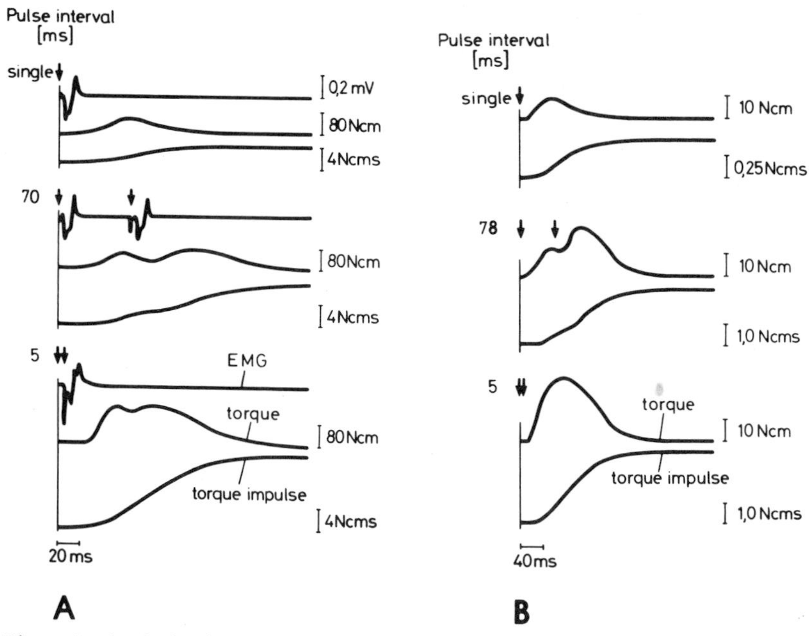

Figure 1. *A*, Activation of flexor digitorum superficialis muscle during indirect stimulation of the medianus nerve for different intervals between paired stimulus pulses of equal amplitude and duration. In each record the traces of EMG, torque and torque impulse are shown. *B*, Activation of the extensor carpi ulnaris muscle during direct stimulation, demonstrating torque and torque impulse. Stimuli are indicated by arrows. Records retouched.

The single stimulus response showed a well-defined EMG potential and a small torque amplitude as well as a small torque impulse. At an interval $\Delta T = 70$ msec the two EMG potentials were still clearly separated while the two torque responses showed a beginning overlap with two twitch maxima. Decreasing ΔT to 5 msec the maximum torque and especially the torque impulse were increased very much. The time course of the torque then formed a single mechanical event similar to a single twitch.

In all experiments the maximum torque amplitude was obtained at an interval of about 2.5–6 msec. This is illustrated in Figure 2, which also shows that the maximum torque and the torque impulse are a function of the interstimulus interval. Both values are related to the corresponding values obtained in single pulse stimulation.

The maximum peak of the torque was about three times, the torque impulse about four to six times the value of corresponding single shock responses. The effectiveness of double-pulse stimulation becomes evident when the mechanical responses to single stimulus pulses are compared

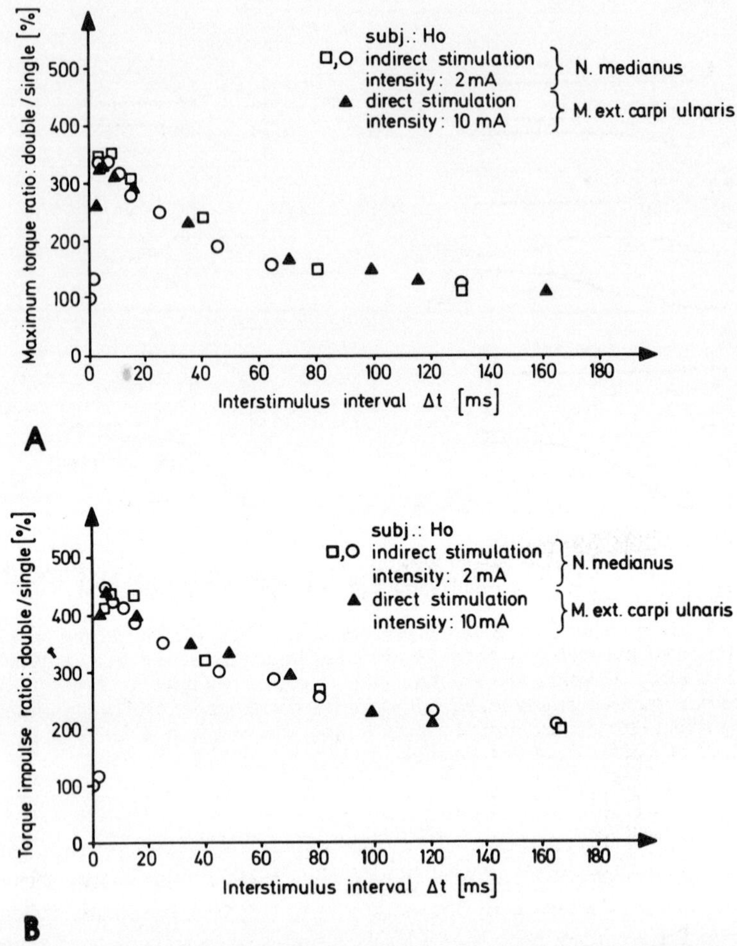

Figure 2. Facilitation of mechanical responses in double pulse stimulation as a function of interstimulus interval. Responses are shown as ratios of double pulse stimulation response to single pulse stimulation response expressed in percent of the single pulse stimulation response. *A*, maximum torque; *B*, torque impulse.

to those elicited by double pulses at various stimulus intensities. In Figure 3*A*, it can be seen that the maximum torque obtained with single pulse stimulation is about 80 Ncm at 9 mA stimulus intensity. But with double pulse stimulation the torque range is extended to 180 Ncm at 8 mA stimulus intensity, while still remaining below pain threshold. Torques above 80 Ncm could not have been elicited by applying single pulse stimulation in the above example.

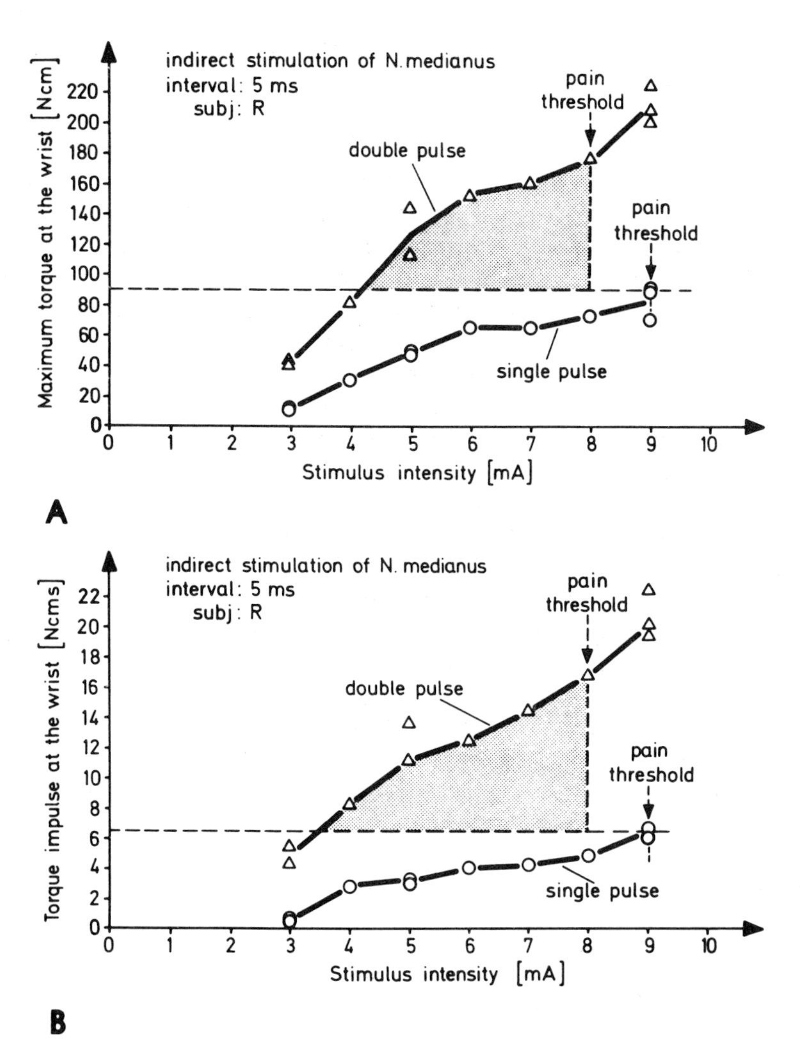

Figure 3. Mechanical response to single and double pulse stimulation as a function of stimulus intensity. *A*, maximum torque; *B*, torque impulse. Shaded areas indicate the extended range of response.

A similar range extension is also true for the torque impulse values as seen in Figure 3*B*.

Direct Stimulation

Records of the signals obtained during direct stimulation are shown in Figure 1*B* for comparison of the effects in indirect stimulation. Because of the heavy stimulation artifacts, the EMG signal is omitted here. The time courses of the torque and the torque impulse were very similar to the corresponding curves during indirect stimulation.

The maximum amplitudes of torque and the torque impulses at double pulse stimulation are plotted against the interstimulus interval in Figure 2 expressed in percent of the amplitude of the corresponding single stimulus response. Here again, the maximum mechanical response was to be observed at an interval of 2–6 msec. The torque exerted by paired stimulus pulse was then increased to about 300 percent, the torque impulse to more than 400 percent of a single twitch. In some cases the torque impulses exceeded 600 percent.

The absolute value of the stimulus intensity had to be chosen individually for each subject, e.g., according to skin impedance. In the subject used in Figure 2, e.g., the amplitude of the single stimulus pulse was increased from 5 mA (response threshold) to 25 mA (pain threshold), causing amplitudes of the maximum torque from 5 to 25 Ncm and of the torque impulse from 0.1 Ncms (newton-centimeter-seconds) to 0.7 Ncms. In double pulse stimulation at an interstimulus interval of 5 msec the maximum torque increased from 15 Ncm (5 mA) to 53 Ncm (15 mA) and the torque impulse from about 0.3 Ncms (5 mA) to 1.3 Ncms.

The results for indirect and direct stimulation as described were essentially similar and were observed for different forearm muscles and five different subjects. Only the absolute values of the stimulus intensity and the mechanical responses differed.

DISCUSSION

The mechanical responses elicited by a pair of electrical stimulus pulses do not sum up linearly because the maximum torque and the torque impulse exceeded 200 percent of a single shock response as was seen in Figures 2 and 3. The results are in good agreement with those recently reported by Lennerstrand (1974), who studied the isometric tension at the retractor bulbi muscle of the cat (whole muscle and single motor units) as a response to single and paired stimulation.

The facilitation of the response to the second stimulus because of the first one might have occurred for different reasons. (1) Facilitation could occur at the neuromuscular junction causing an increase in the number of active muscle fibers in the successive stimuli at the nerve as indicated by Norris and Gasteiger (1955). In this case, an increase in the EMG potential should also be observable as has been reported for the frog's skeletal muscle (Bobbert, 1969; Rau, 1971). The facilitation of transmitter release at the frog neuromuscular junction has been shown directly, e.g., by Magleby (1973) and Mallart and Martin (1968). In human skeletal muscles, an increase in EMG compound potential of the second response was found in some cases for isotonic contractions (Rau, 1971). But since the amplitudes of both EMG compound potentials in the indirect stimulation experiments of this study turned out to be constant, as can be seen in Figure 1, the marked increase in the second mechanical response could not be caused by the recruitment of additional muscle fibers. (2) Therefore, it can be assumed that the nonlinear summation of the mechanical effects takes place in the muscle tissues itself. This is supported by the results shown in Figure 2, where during direct muscle stimulation a percentage of facilitation was observable similar to that in indirect stimulation.

In practical applications, the mechanical response to double pulse stimulation can be increased to an amount that could never be obtained in single pulse stimulation because of the pain threshold. Furthermore, the response was still short and twitchlike. As indicated by Figure 3, a combination of the intensity variation and the possibility of single or paired stimulation increased the gradation in mechanical response markedly.

REFERENCES

Bobbert, A. C. 1969. The transfer function of frog myoneural junctions. Acta Physiol. Pharmacol. Neerl. 15: 289–328.

Lennerstrand, G. 1974. Mechanical studies on the retractor bulbi muscle and its motor units in cat. J. Physiol. 236: 43–55.

Magleby, K. L. 1973. The effect of repetitive stimulation on facilitation of transmitter release at the frog neuromuscular junctions. J. Physiol. 234: 327–352.

Mallart, A., and A. R. Martin. 1968. The relation between quantum content and facilitation at the neuromuscular junction of the frog. J. Physiol. 196: 593–604.

Norris, H. F., and E. L. Gasteiger. 1955. Action potentials of single motor units in normal muscle. Electroenceph. Clin. Neurophysiol. 7: 115–126.

Rau, G. 1971. A new gradation of skeletal neuromuscular transmission. IPO-Ann. Progr. Rep. 6: 85–87 (Inst. for Percept. Res., Eindhoven).

Modification for measurement of conduction velocities in motor nerves

V. K. Häkkinen
Central Hospital of Tampere, Tampere

P. Jaakola
Research Institute for Bioengineering, Tampere

After the studies by Hodes, Larrabee, and German (1948), conduction velocities in motor nerves have mostly been measured by stimulating the nerve at two different points (Figure 1) and by determining the latencies from the stimulus moment to the beginning of the corresponding muscle response (M-wave). The conduction velocity is calculated by dividing the interstimuli distance by the latency difference.

The method gives the conduction speed of the fastest conducting nerve fibers. Because of the complexity of the compound muscle response (due especially to different nerve fibers within a nerve bundle) such a method gives the most reliable result. Some problems, however, remain. For instance, there is the difficulty in identification of the motor unit innervated by the fastest nerve and the determination of its latency. Especially with surface recording, the origin of the M-wave is more or less curved, and it is often difficult to determine exactly when the potential begins. It has been estimated that even in proper conditions an error of 0.5 msec in each latency measurement is well within the unavoidable error of the method, and a latency difference of 1.0 msec might commonly occur (Simpson, 1964). Under clinical conditions the recording error could be even greater. The measurement of latency difference alone may cause an error of more than 10 percent in measurements of conduction velocities in motor nerves.

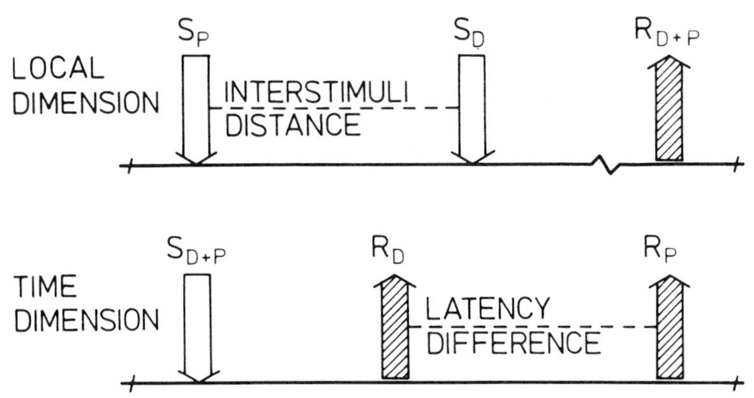

Figure 1. Local and time relations of stimuli and responses during measurement of conduction velocity in a segment of nerve. S_P, proximal stimulus; S_D, distal stimulus; S_{D+P}, proximal or distal stimulus, R_P, muscle response evoked by proximal stimulus; R_D, muscle response evoked by distal stimulus; R_{D+P}, muscle response evoked by proximal or distal stimulus.

A digital delay line system (Jaakola and Häkkinen, 1975), the delay of which can be chosen continuously, makes a modification for measurement of conduction velocities in motor nerves possible. This modification could be useful in measurements of the latency difference, and it is usable in quite simple clinical conditions.

APPARATUS

The main principle of the delay line system is shown in Figure 2. In order to make the system compatible with different kinds of apparatuses the delay line has been equipped with a lowpass input amplifier, the cutoff frequency of which is 15 kHz. The timing unit (B-E in Figure 2) has been constructed so that even very long delays can be produced easily and the delays can be chosen continuously. Like in the input amplifier

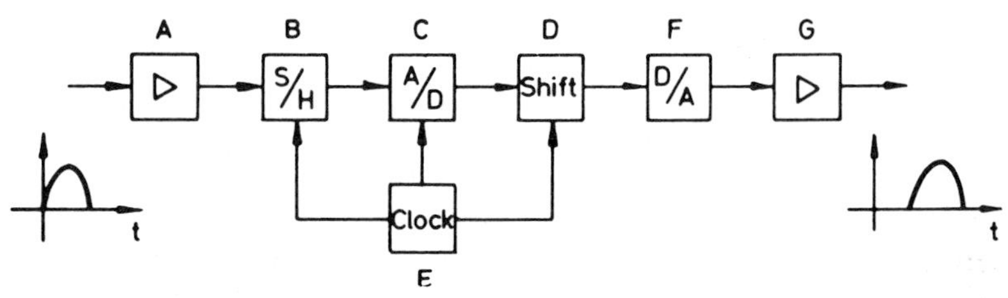

Figure 2. Diagram to show the main principles of the digital delay line system. The timing unit consists of the blocks *B–E*.

the device uses lowpass filtering in the output amplifier, the output impedance of which is very small.

For handling EMG-signals the delay line unit has the following characteristics: (a) sample frequency 50 kHz, (b) resolution 10 bits, (c) frequency range 0–10 kHz, (d) voltage swing ± 10 V, (e) delays can be chosen continuously from 2.0 to 20.0 msec, (f) the value of the delay can be read from the numerical display of the unit with an error of 0.05 msec.

In the present studies we have amplified the EMG signals on the Disa 2-channel Electromyograph (type 14 A 21) and used also a Tektronix 5103 N storage oscilloscope for both monitoring and recording.

PROCEDURE

By delaying the trigger pulse of the monitoring scope it is possible to move the "proximal" M-wave (R_p) towards the "distal" one (R_D) in time scale (see Figure 1) until the beginning parts of the two waves are superimposed. The latency difference can then be read directly from the numerical display of the unit.

Figure 3 shows how we have performed this procedure in practice. In the lower trace there is an M-wave evoked by distal stimulation of a

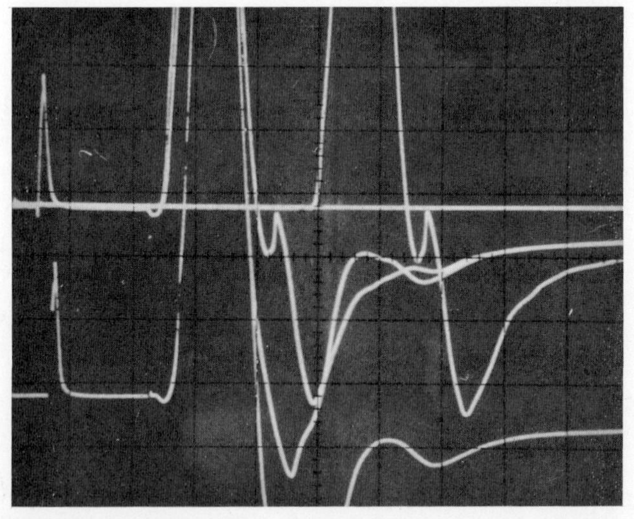

Figure 3. Surface recording from a thenar muscle following stimulation of the median nerve at the wrist (lower trace) and at the elbow (upper trace to the right). In the upper trace to the left the beginning parts of the two waves are superimposed. Calibration 1.0 mV and 2 msec per unit square.

nerve (in order to distinguish it afterwards from the "proximal" one). In the upper trace to the right there is an M-wave evoked by proximal stimulation of the same nerve without delay. The latency difference is first estimated on the scope screen. This estimation is used when the trigger pulse is first delayed. If the delay is not correct with the first sweep, it must be corrected and tested with one or more single stimuli and corresponding sweeps. In the upper trace to the left the beginning parts of the two waves are superimposed.

Especially when high amplifications are used, there may be a difference in the baselines of the two M-waves. This difference must be corrected before superpositioning. If the M-waves are very polyphasic, it seems sometimes useful to record the "proximal" muscle response with inverted polarity.

DISCUSSION

In principle this modification does not differ much from the original method. The modification also gives conduction velocity of the fastest nerve fibers. The procedures of stimulating the nerve, of recording the response potential, and of measurement of the distance between the stimulus points are of the same importance. Furthermore, there is reason to determine separately the distal latency, because of its inherent importance.

The modification does, however, have some advantages. For example, the latency difference can be determined by one instead of by two measurements. Thus the error should be smaller (if not systematic). Furthermore, the two response potentials can be very closely compared to each other. This helps to find the corresponding motor unit potentials.

More stimuli are needed for the modification than for the original method according to our experience. This is true especially if high amplifications are used and there is need for correcting baselines first. Disadvantages like this could be avoided if the potentials would first be stored in a memory and then be superimposed in a two-dimensional way.

We consider that a modification like this should help to diminish the variation of normal values of conduction velocities in motor nerves, but it could be especially helpful for teaching purposes.

REFERENCES

Hodes, R., M. G. Larrabee, and W. German. 1948. The human electromyogram in response to nerve stimulation and the conduction velocity of motor axons. Arch. Neurol. Psychiat. (Chic.) 60: 340–365.

Jaakola, P., and V. Häkkinen. 1975. A digital delay line system for studies of clinical neurophysiology. Proceedings of the III Nordic meeting on medical and biological engineering. Hämeen kirjapaino OY. 11.1–11.3.

Simpson, J. A. 1964. Fact and fallacy in measurement of conduction velocity in motor nerves. J. Neurol. Neurosurg. Psychiat. 27: 381–385.

Cinematographical approach to eye movement analysis during perceptual skill tasks

J. Terauds
University of Alberta, Edmonton

During perceptual-skill tasks in sports and other activities the movement and fixation point of the eyes may be a determining factor between success and failure. There have been numerous eye movement monitoring studies using methods ranging from observation to noncontacting photoelectric techniques in which records of eye movements along the horizontal and vertical are translated into shape. Unfortunately these methods have not been satisfactory for eye movement and fixation measurement in sports. The purpose of this study was to consider new cinematographical and video methods for determining eye movement during several selected perceptual-skill activities. Sports skills, where eye movement and proper fixation seem to be critical to success, were of special interest. It is possible that in some activities a certain eye movement pattern is closely related to executing the movement in a mechanically sound manner. These activities may include tennis, handball, soccer, basketball, volleyball, baseball, bowling, and many other sports, where the eyes play a dominant part in finding a target.

METHODS AND PROCEDURES

This study was conducted at the Research Center for Leisure and Sports at UTPB. The instruments used in the study included a NAC Eye Movement Recorder, Model 4, A Photo-Sonic 16 mm 1P camera, a Panasonic portable WV3085 video camera, and a Bolex H-16 5BM 16 mm camera.

By using the Eye Movement Recorder in combination with a motion picture camera, or a video camera, it was possible to record the general field of view as well as the discrete point of interest at which the eyes were trying to focus. The recording of a discrete point of interest is achieved by reflecting an illuminated spot off the eyeball cornea which is superimposed on the general field of view (Figure 1). The general field of view is picked up by the view photo lens and delivered to the motion picture camera or video camera through relay optics and fiber optics. The reticle from the spot lamp reflects off the cornea, then reflects off a see-through mirror into a spot lamp focus lens, and from there is relayed into the optics leading to the recording instrument. The operator of the eye movement recording apparatus can monitor the field of view as well as the fixation point by looking through a monitor finder. The Eye Mark Recorder procedure limits the field of view to approximately 43.5 degrees vertical and 60.0 degrees horizontal. The distance at which the subject has the object in focus ranges from 25 cm to infinity, without adjustments. Once the Eye Mark Recorder has been placed on the subject the parallax and spot focus can be adjusted along the X and Y axis. The tracking field varies considerably between individuals, depending on the shape of the eye cornea. The accuracy of the recorded fixation point is a function of the off axis angle and the shape of the eye cornea. Without excessive adjustments or calculations an accuracy of 1.0–5.0 degrees can be expected.

Three male and two female subjects participated in the experiments involving a forehand drive in tennis, a set shot in basketball, catching a basketball, batting a baseball, reading, and pursuit rotor tracking. In each case, the Eye Mark Recorder mask was adjusted to suit the shape of the subject's head and adjusted so that the light spot (eye mark) intended for marking the fixation point was on the object. In the case of objects moving across the field of view a grid was used. The grid, larger than the field of view in both horizontal and vertical directions, was essential in measuring the error between the eye mark within the field of view and the object of fixation.

The movie cameras were operated at 24 frames per second, while the video recording was conducted in real time. In all cases every effort was made to make the subject as comfortable as possible, and each time the subject was asked to try and ignore the (600 gram) mask strapped on the head.

The set shot in basketball and the basketball catch were recorded on film and video tape. The set shot was taken at a distance of 4.5–5.5 m from the basket by the subject who was instructed to shoot in a natural manner. The position of the eye mark was checked before and after each shot. Whenever the eye mark was out of place, the mask was adjusted to bring the eye mark back on the object. For the basketball catch the

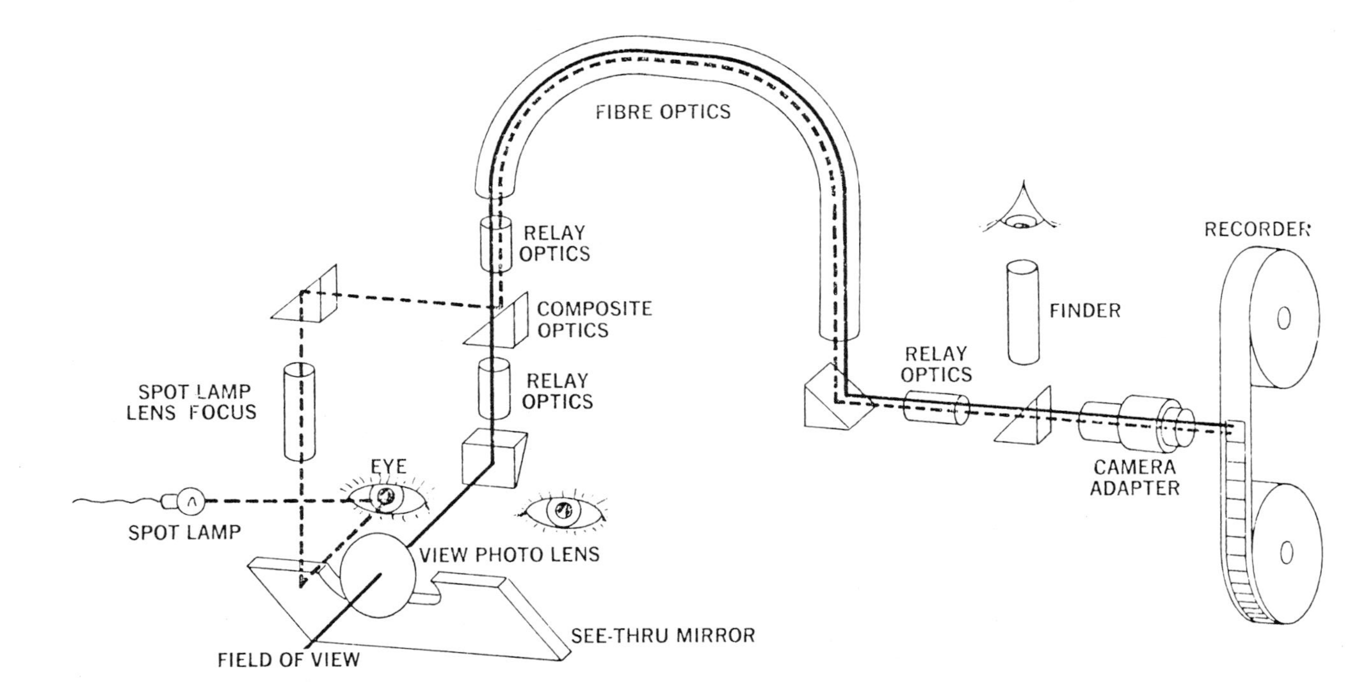

Figure 1. Eye movement recorder with cinematography or video method.

assistant throwing the ball was positioned 8.0 m from the subject. As before, the eye mark position was checked for accuracy before and after each trial.

For the tennis forehand the movie camera was mounted on the subject's back in the least disturbing position possible. A tennis ball was driven to the subject's forehand from a distance of approximately 13.0 m. The camera was started 5–7 sec before the event and stopped several seconds after the event. Again, the accuracy of the eye mark within the field of view was checked and necessary adjustments were made.

For batting a baseball the subject was placed 18.4 m from the pitcher. The batter was told to carry out all activity as if to bat for real, but to stop the bat without a follow through (to prevent damage to the camera on a tripod behind him). Again the accuracy of the eye mark was checked before and after each trial.

The relatively passive activity of working with a pursuit rotor and reading was recorded on both film and video tape. The accuracy of the eye mark was checked for each trial.

The 16 mm film and the video tape were analyzed on a Vanguard Motion Analyzer and TV monitor.

RESULTS AND DISCUSSION

On the whole the use of the Eye Mark Recorder, in conjunction with a 16 mm movie camera or a video camera, produced encouraging results. During the basketball set shot and the basketball pass, the eye mark was visible within the field of view throughout each trial. However, on several trials the mask had moved, resulting in an erroneous eye mark position for part of the trial. During the shot some discomfort to the subject during execution was caused by the protruding parts of the eye mask. This was noted for subjects accustomed to shooting close to the head. The protruding parts include the *XY* adjusters, the parallex adjustment knob, and to a lesser extent the optical fiber bundle. The basketball catch presented no equipment problems.

During the tennis forehand drive the Bolex camera on the back of the player seemed to produce some discomfort. Approximately one-half of the trials were valid. On the other trials the mask had moved. The eye mark mask did not seem to be a problem as long as the movements with the head were not quick. During quick movements the mask had a tendency to move out of position. For the reading and pursuit rotor trials the Eye Mark Recorder seemed to be the ideal instrument for tracking the subject's general field of view as well as the fixation point.

On the basis of these experiments the following recommendations can be made for the Eye Movement Recorder with cinematographical or video procedures: (1) this combination of instruments has great potential in the analysis of eye movement during activities that do not require peripheral vision, (2) this is an excellent method for recording eye focus during reading and similar activities, (3) during activities with sharp head movement some accuracy is destroyed due to face mask movements, which change the position of the eye mark, (4) the accuracy of the eye mark is to some extent determined by the shape of the individual's eyeball cornea from which the light is reflected into the relay optics, (5) the error of the eye mark on most subjects with slow head movement is in the range of 0.0 to ± 5.0 degrees, while in other cases it increases to ± 15.0 degrees; however, most of the error can be removed by using a grid and calculating the error out of the measurement, (6) a bite plate secured to the mask may be a convenient way to prevent the mask from moving out of position during quick head movements, (7) the Eye Mark Recorder is an excellent instrument for tracking the general field of view and with a movie camera attachment could be used for observations in a large number of activities, (8) with the presence of sunlight there is some difficulty in adjusting the eye mark within the field of view, (9) this method of observing and measuring eye movement would be excellent to make comparisons between the expert and novice in a large number of perceptual skill tasks.

High-accuracy analysis of movements in the spine with the aid of a roentgen stereophotogrammetric method

G. Selvik
University of Lund, Lund

T. H. Olsson
University Hospital, Lund

S. Willner
General Hospital, Malmo

Both for purely scientific purposes and to ascertain the effect of orthopedic treatment of diseases of the spine, a method that renders it possible to study in exact terms the movements between vertebrae is highly desirable. For these and other studies of movements in the skeletal system of living subjects, a roentgen stereophotogrammetric method has been developed by Selvik (1974). The method is easily performed in a clinical setting when appropriate equipment is available, and the results obtained are both easy to interpret and highly accurate.

METHODS

The roentgen stereophotogrammetric method requires that radiopaque indicators (markers) be inserted in the skeletal segments to be studied. As indicators we utilize tantalum balls 0.8 mm in diameter. They are inserted either during operation on the spine, in cases where a study of the postoperative course is planned, or else pericutaneously by the aid of fluoroscopy. The special instrument (Figure 1) and technique described

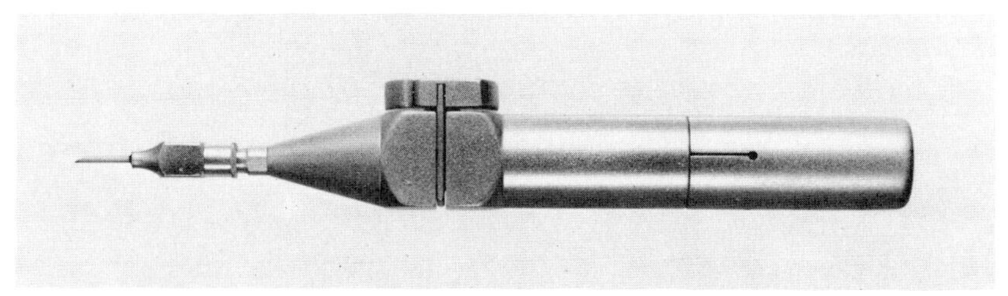

Figure 1. The instrument for implantation of tantalum indicators.

by Aronson, Holst, and Selvik (1974) are utilized for these implantations. Each skeletal segment to be studied (i.e., in spinal studies of vertebra, or, when a lysis has occurred to the elements of the vertebra) has to be supplied with three or more indicators. The indicators of a vertebra must not lie in a straight line, since the rotation about such a line cannot be investigated.

The roentgen stereophotogrammetric recording is accounted for in Figure 2. The stereophotogrammetric method obviously requires that the subject's spine markers be exposed from two roentgen tubes, but it is arbitrary whether the images are recorded on one film (as shown) or on two films. The important requirement is that reference indicators in a calibration cage (of glass) are double-exposed on the same film(s) as the spine markers. The mutual positions of these reference indicators are determined when calibrating the cage (Selvik, 1974, Chapter 3). These cage markers define a laboratory coordinate system, in relation to which the positions of the film(s) and roentgen foci can be calculated. However, we do not directly measure the position of the film(s) or the foci in this laboratory coordinate system, but ascertain it indirectly by data processing, after the films have been measured.

The measurement of the films is performed on a drawing table attached to an instrument generally used in cartography, an Autograph A8 (Wild, Heerbrugg, Switzerland). The image transmission, however, is modified utilizing a closed television circuit. Measured coordinates of spine and reference marker images are automatically recorded on a paper tape, which constitutes the input for the computation process. The computations are performed in a high-speed computer, the Univac 1108. The primary result is that the three-dimensional coordinates of the subject's indicators are obtained in the coordinate system determined by the calibration cage, as described by Selvik (1974, Chapters 2 and 4). This information is then transformed by another computer program to data for the rotations and translations of one vertebra in relation to another.

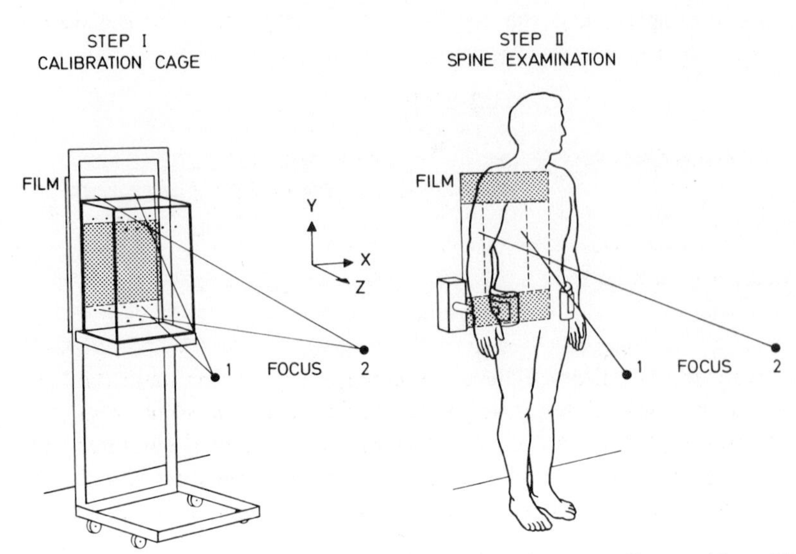

Figure 2. The investigation procedure of a patient in a standing position. The calibration cage with reference indicators, placed on a mobile frame, is exposed (step I) prior to the patient (step II) without moving the roentgen tubes or the film. The shaded areas of the calibration cage (step I) and the film (step II) are sheltered by lead sheets during the two steps of the investigation. The hips of the patient are supported by two adjustable arms. The directions of the coordinate axes of the laboratory coordinate system are also shown.

It should be stressed that identical orientation at two examinations of one vertebra chosen as a reference is not needed, because the computer program automatically rotates and translates the reference vertebra back to its position at the first examination. The relative motions between vertebrae can thus be fully studied.

The information presented by the last-mentioned computer program is a detailed account of the rigid-body movements of one vertebra in relation to another. Thus the finite rotations about the main axes of the body, the translations along these axes, the rotation angle, and the direction and position of the screw (rotation) axis, and also information on the validity of the rigid-body model, are determined. Generally when studying a subject we determine these parameters for lateral bendings to the left and right, flexion and extension, and axial rotations to both sides, as compared to the erect standing position. The time for measuring the seven roentgenograms for these various positions amounts to about 2 hr and the computation process costs about $15. The estimated absorbed gonad dose is 0.3 rad for one double exposure when examining the lumbosacral spine, and the corresponding mean bone marrow dose is 0.01 rad when examining the thoracic spine.

MATERIAL AND RESULTS

The clinical material in the present study, carried out in 1973–1974, consisted of ten patients. Two patients were investigated 1.5–2 yr after operation with anterior lumbosacral fusion because of clinical suspicion of pseudoarthrosis. In both cases a pseudoarthrosis was verified, and the movements between the vertebrae involved were analyzed, as well as the direction and position of the axes of motion. Three patients operated upon with posterolateral fusion in the lumbosacral spine were prospectively studied after operation. Among seven operated vertebral pairs, significant motion was demonstrated for all pairs but one. Unrestricted residual mobility was found on three operated levels. The three diseased levels were comparatively rigid about 1 yr after the operation, although this was not the case 3 months postoperatively. Two patients in the last-mentioned group, together with two other investigated cases that were not subject to fusion, showed lysis of L5 or L4. In these cases an unstable pattern of motion was found for movements such as those occurring between the supine and the erect standing positions, or other movements having no dominant pattern. Thus when the vertebra of lysis rotated forwards, the vertebra craniad to it rotated backwards, and significantly different rotations were also found about the longitudinal and sagittal axes. Three teen-age girls with scoliosis operated with fusion between an upper thoracic vertebra and L2 using Harrington's instrumentation were examined on three occasions during the postoperative year. An increase of the magnitude of 5° in their scoliotic deformities was found, with axial rotation and changes in both the sagittal and frontal planes. In addition, residual mobility in the fused area was detected 1 yr postoperatively in all three cases.

In addition to the clinical investigations, animal experiments have been performed to study the development of experimentally provoked scoliosis and the healing process of posterior spinal fusions in pigs. In the first case it was found, among other things, that axial rotations moved the vertebral bodies toward the convexity of the scoliotic curves. In the second case it was possible to correlate postmortem macroscopical examinations with the degree of mobility studied by roentgen stereophotogrammetrics. Complete agreement between the bone fusion findings and the degree of mobility or rigidity roentgenologically ascertained was shown. For a detailed discussion of the investigations mentioned, the reader is referred to a thesis by Olsson (1975).

The precision of the method in the clinical cases was ascertained by reevaluating one complete examination (seven films) from each of five patients. For 14 investigated segment combinations, the least satisfactory standard errors found were 0.30°, 0.25°, and 0.07° for the rotations and 184, 202, and 204 μm for the translations about and along the *x*-, *y*-, and *z*-axes, respectively. For a more optimal positioning of the segment indicators, the corresponding standard errors were 0.12°, 0.13°, and 0.07° for rotations and 26, 23, and 58 μm for translations (14 degrees of freedom).

As an example of the method used a reexamination of the first patient investigated is presented. This patient was a woman 54 yr old who had been operated upon with anterior fusion between L5 and S1 using autologous bone transplant 2½ yr earlier. She has suffered from pain since a disk herniation at age 50, and the pain has persisted in spite of extirpation of the herniated disk and the attempted fusion. At a first roentgen stereophotogrammetric investigation, a pseudoarthrosis between the operated segments was found (Olsson, 1975, p. 108). In the second roentgen investigation the direction of the axis of rotation is not constant. The significant motions ($p < 0.01$) found are a positive rotation of 1.5° about the *z*-axis (a sagittal axis) at lateral bending to the right, a positive rotation of 2.8° about the *x*-axis (a transversal axis) at flexion, fairly free rotations about all axes at extension, and finally a negative rotation of 0.4° about the *y*-axis (a longitudinal axis) at axial rotation to the right. The significant values are found, using Student's t-test, from a reevaluation of seven films of this patient, which showed the standard errors 0.26°, 0.07°, and 0.18° for the rotations about the *x*-, *y*-, and *z*-axes, respectively. The directions of the motions are those intended for all motions, but extension, the last motion, is not easily performed by the patient. Thus it can be concluded that rigidity in the operated area has not developed. The rotations measured, as well as those for a control vertebral pair (L4-L5), are shown in Figure 3.

CONCLUSION

It is concluded that the present roentgen stereophotogrammetric method is the first exact clinical method for the study of movements of the spine. It is a method for functional analysis and not primarily for morphological studies, as the indicators in the latter case must be placed in very well-defined anatomical sites in the vertebrae.

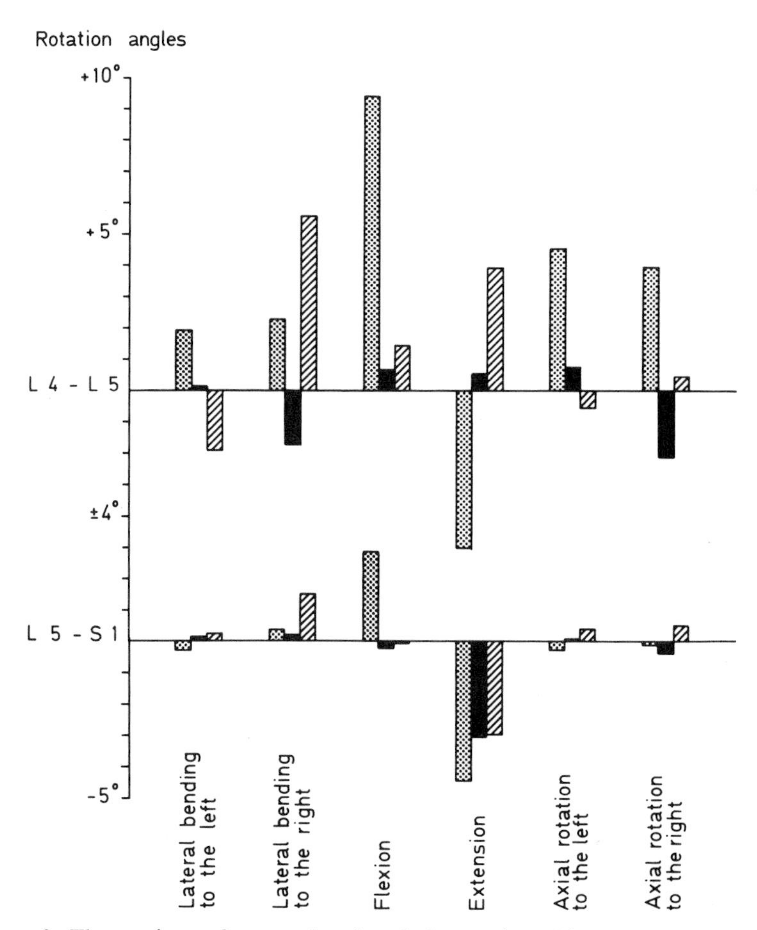

Figure 3. The motions of a vertebra in relation to its subjacent vertebra. Each staple group is a graphic representation of the rotations about the x-axis (a transversal axis), the y-axis (a longitudinal axis), and the z-axis (a sagittal axis). Stippled staples stand for the rotations about the x-axis for the motions given in relation to the erect standing position. Filled staples stand for corresponding rotations about the y-axis, and striped staples for those about the z-axis. The operated level is L5-S1, and L4-L5 is a control vertebral pair.

REFERENCES

Aronson, A. S., L. Holst, and G. Selvik. 1974. An instrument for insertion of radiopaque bone markers. Radiology 113: 733–734.

Olsson, T. H. 1975. A roentgen stereophotogrammetric study of the kinematics of the spine. AV-centralen, Lund.

Selvik, G. 1974. A roentgen stereophotogrammetric method for the study of the kinematics of the skeletal system. AV-centralen, Lund.

The Fifth International Congress of Biomechanics

Acknowledgments

The Fifth International Congress of Biomechanics would not have been a successful professional meeting without the competency of the members of the Organizing Committee and persons involved in the various tasks during the Congress preparation and conduct. Appreciation is therefore extended to Esko Karvinen, Matti Pulli, and Erkki Rutanen, whose work on the organizing committee was outstanding. Antti Arstila and Bengt Jonsson are acknowledged for their extensive help in the Scientific Program Committee. Mrs. Helena Karvinen, Mrs. Maija-Liisa Hirvi, and Mrs. Raija Komi are thanked for their excellent organization of the ladies' program. Valuable assistance was provided by the personnel of the Kinesiology Laboratory and by various other persons assisting the Organizing Committee or serving as secretaries to the scientific session chairmen.

Special thanks and appreciation are extended to Mrs. Irja Nurminen, who performed her task as a Congress secretary with exceptional efficiency and devotedness. Her cheerful appearance during the Congress will certainly be remembered for a long time.

The organizers were fortunate to receive support and encouragement from the Faculty of Physical and Health Education through Dean Kalevi Heinilä and from the Main Administration of the University of Jyväskylä, through Rector Ilppo Simo Louhivaara. The following other organizations and offices are thanked for their promotion of the Congress: University Association, City of Jyväskylä, The Finnish Society for Research in Sport and Physical Education, Ministry of Education and its two sections, Department of Sport and Youth Affairs, and Department of International Affairs. It was greatly appreciated that the Congress could receive financial support through these two departments of the Ministry of Education. Various commercial firms also served as co-sponsors and assisted the Congress through their contributions.

The organizational problems of the Fifth Congress were substantially eased by a thorough review of the conduct of the Fourth Seminar held at Pennsylvania State University in 1973. It was fortunate that Richard C. Nelson, Chairman of that meeting, so openly gave his advice for the V Congress. Appreciation is also extended to Jurg Wartenweiler, President of the International Society of Biomechanics, for his guidance during all stages of the Congress preparation.

Finally, the contribution of all the Congress registrants is greatly appreciated. Their excellent contribution certainly made the Fifth International Congress of Biomechanics both professionally and socially a rewarding experience.

Paavo V. Komi, Ph.D.
Congress Chairman

Organization

SITE

Department of Biology of Physical Activity
University of Jyväskylä
SF 40 100 Jyväskylä 10, Finland

MAIN THEME

Progress in fundamental and applied research in Biomechanics.

ORGANIZING COMMITTEE

Paavo V. Komi, Chairman
Esko Karvinen
Matti Pulli
Erkki Rutanen
Irja Nurminen, Secretary

SCIENTIFIC PROGRAM COMMITTEE

Paavo V. Komi, Chairman
Antti Arstila
Bengt Jonsson (Umeå)
Esko Karvinen

SPECIAL CONSULTANTS

R. C. Nelson (University Park, Pa.)
J. Wartenweiler (Zürich)

Sponsors, contributors, and commercial exhibitors

The organizing committee wishes to thank the following organizations for their support during the Congress. Appreciation is also expressed to the various companies which exhibited their publications during the Congress.

LOCAL SPONSORS

Ministry of Education (Finland)
Department of Biology of Physical Activity, University of Jyväskylä
City of Jyväskylä
University Association, Jyväskylä

CO-SPONSORS AND CONTRIBUTORS

Finnish Society for Research in Sport and Physical Education
Huhtamäki Oy-Polarpak, Hämeenlinna
Jyväs-Hyvä, Jyväskylä
Jyväsleipä Jalanne & Co, Jyväskylä
Karjakunta, Jyväskylä
Karjaportti, Jyväskylä
K-Kaupat, Jyväskylä
KOP, Jyväskylä
Merck Sharp & Dohme (MSD), Helsinki
Mäki-Matti Osuuskauppa, Jyväskylä
Orion Lääketehdas, Helsinki
OTK, Jyväskylä
Oy Gustav Paulig Ab, Jyväskylä

Oy Scan-Auto Ab, Jyväskylä
Oy Sokle Ab, Jyväskylä
Ruthin Konditoria, Jyväskylä
Saastamoinen Yhtymä Oy, Jyväskylä
SOK, Jyväskylä
STAR, Tampere
Suomen Sokeri Oy, Helsinki
Tukkukauppojen Oy, Helsinki
Valio Meijerien Keskusosuuliike, Jyväskylä

EXHIBITORS

Finn Metric Oy
Ahertajantie 6 D
02100 Tapiola, Finland

Foto Oy Ab
Mikonkatu 1
00100 Helsinki 10, Finland

Kistler Instruments AG
Eulachstrasse 22
Postfach 61
8 408 Winterthur, Switzerland

Selcom Ab
Box 30
S-43 121 Mölndal 1, Sweden

Ulmaelectro Oy
Pälkäneentie 20
00510 Helsinki 51, Finland

NORTHERN MICHIGAN UNIVERSITY LIBRARY
3 1854 000 226 715